Scientific Sources and Teaching Contexts Throughout History: Problems and Perspectives

BOSTON STUDIES IN THE PHILOSOPHY AND HISTORY OF SCIENCE

VOLUME 301

For further volumes:
http://www.springer.com/series/5710

Alain Bernard • Christine Proust
Editors

Scientific Sources and Teaching Contexts Throughout History: Problems and Perspectives

Springer

Editors
Alain Bernard
EHESS-CNRS and UPEC
Centre A. Koyré UMR 8560 and ESPE
Paris, France

Christine Proust
Université Paris 7 – CNRS
Laboratoire SPHERE UMR 7219
Paris, France

ISSN 0068-0346
ISBN 978-94-007-5121-7 ISBN 978-94-007-5122-4 (eBook)
DOI 10.1007/978-94-007-5122-4
Springer Dordrecht Heidelberg New York London

Printed on acid-free paper

Springer is part of Springer Science+Business Media (www.springer.com)

Acknowledgements

This volume is the outcome of a research project on the study of ancient scientific sources within teaching contexts in various historical periods. Began around 2005, the annual workshops were first hosted and financed by two French research institutes, REHSEIS (now SPHERE at Paris Diderot University and CNRS) and the Centre Alexandre Koyré (EHESS, MNHN and CNRS), within the framework of the interdisciplinary 'mathématiques et histoire' project (EHESS, principal investigator Giovanna Cifoletti). In addition to the colleagues who finally contributed to the present volume, many others actively took part in our workshops and contributed to the precise definition of our key issues. We specially thank Fabio Acerbi (CNRS), Bruno Belhoste (Paris I University), François Charette (Chester Beatty Library), Samuel Gessner (University of Lisbon), Catherine Jami (CNRS), Agathe Keller (CNRS), Sabine Romevaux (CNRS), Maryvonne Spiesser (University of Toulouse), Renaud d'Enfert (INRP, now IFE), Bernard Vitrac (CNRS), and Niek Veldhuis (University of California, Berkeley) for their insightful remarks and help.

As for the edition of the volume itself during the period 2008–2012, the translation and preparatory workshop was financially supported by the above research institutes (REHSEIS-SPHERE and Centre Alexandre Koyré), the 'mathématique et histoire' project (EHESS), the University Paris Diderot, the interdisciplinary laboratory HASTEC (Paris) and by the CAPHES (CNRS and ENS Paris). We particularly thank Richard Kennedy for his careful translation or polishing of the various contributions.

Contents

Contributors

Alain Bernard UPEC (ESPE) and Centre A. Koyré (EHESS-MNHN-CNRS UMR 8560), Paris, France

Andrea Bréard Université Lille I, Laboratoire Paul Painlevé, Paris, France

Florence Bretelle-Establet Laboratoire SPHERE, CNRS & Université Paris Diderot, Paris, France

Eva Caianiello Ecole des Hautes Etudes en Sciences Sociales (EHESS), Paris, France

Anne-Marie Chartier Department of the History of Education (SHE), IFE-ENS-Lyon: Formerly INRP (National Institute for Educational Research), Lyon, France

Karine Chemla Laboratoire SPHERE, CNRS & Université Paris Diderot, Paris, France

Giovanna C. Cifoletti Ecole des Hautes Etudes en Sciences Sociales (EHESS), Paris, France

Philippe Clancier UMR 7041 Arscan, Université Paris 1 Panthéon-Sorbonne, Paris, France

Stéphane Lamassé Université Paris 1 Lamop, Paris, France

Christine Proust Laboratoire SPHERE, CNRS & Université Paris Diderot, Paris, France

Alexei Volkov National Tsing Hua University, Hsinchu, Taiwan

Chapter 1
General Introduction

Alain Bernard and Christine Proust

For the past few decades, developments in History of Science have strongly emphasized the importance (if not necessity) of studying the cultural and intellectual contexts in which the sources were produced and used. Following this general trend, the core of the problematic addressed in this volume stems from two basic facts recognized by the contributors. The first is that, too often, sources of scientific knowledge have been studied without taking into account the various 'contexts' of transmission, specifically the "teaching context" in which this knowledge was elaborated, used and transmitted. The second is that other sources have been considered–sometimes dismissively and sometimes mistakenly – as relating to teaching and learning activities with little attempt to offer precision on, and demonstration of, the existence and nature of the underlying 'teaching context'. Accordingly, the various contributions to this volume offer a more nuanced approach to teaching and learning activities as well as a reflection on what these terms meant for the actors themselves. Indeed, the complexity and variety of such activities appear when various periods and contexts are compared and confronted with each other. Thus, the present volume involves several disciplines and brings together a team of researchers in the history of ancient, medieval and modern science and in the history of education, all of whom share the above epistemological and methodological concerns despite the diversity of their subject areas. They have agreed to contribute to finding a better grounded approach to their specific subject areas, which extend from mathematics and astronomy to medicine, divination and religious lore.

A. Bernard (✉)
UPEC (ESPE), Centre A. Koyré (EHESS-MNHN-CNRS UMR 8560), 27 rue Damesme, 75013 Paris, France
e-mail: alainguy.bernard@gmail.com

C. Proust
Laboratoire SPHERE UMR 7219 (case 7093), CNRS & Université Paris Diderot, rue Thomas Mann 5, 75205 Paris, France
e-mail: christine.proust@univ-paris-diderot.fr

A. Bernard and C. Proust (eds.), *Scientific Sources and Teaching Contexts Throughout History: Problems and Perspectives*, Boston Studies in the Philosophy and History of Science 301, DOI 10.1007/978-94-007-5122-4_1,

We will here present the content of the volume by distinguishing four main groups of contributions, according to the four different interpretative strategies employed and that might prove useful for similar inquiries (Sect. 1.3). But, as a necessary preliminary to this presentation, we give an idea of the vast scope as well as the complexity of the *questions* that the contributors have addressed. Indeed, this complexity alone explains why each contributor had to adopt a definite *strategy,* that is, a specific order to the questions raised that generally correlates with the nature of the corpus of sources examined. For this purpose, we will first outline the course of inquiry that the team of contributors followed during the preparation of this volume and that led them to distinguish three different meanings of the notion of historical 'context', as applied to the notions of "teaching and learning" (Sect. 1.1); we will then use this analytical framework to delineate the various issues addressed in the contributions (Sect. 1.2). This survey might also be seen as a first presentation of the contents, before we come to the variety of approaches followed by the contributors (Sect. 1.3).

1.1 Various Teaching and Learning 'Contexts'

The basic question that we (editors) initially addressed to the colleagues who later became contributors to the present volume[1] was the following: could historians of ancient science have missed something important by neglecting the teaching context when studying their sources? The very first reactions almost immediately led us to more fundamental questions: is the project of studying ancient sources in a teaching context legitimate and why? We soon realized that the answers differed, simply because the problems experienced in the course of each research project were not the same. This led us to the dual problems mentioned above: either ignoring the teaching context when it should be taken into account, or taking it for granted when it should be justified and studied more carefully. Thus, our initial questions had to be specified: 'teaching' and 'learning', in fact, mean many different things and types of activities, as we shall see later. Similarly, the notion of 'context' has now become very broad and is so commonly used by historians that it has become polysemic and actually refers to very different levels of analysis.

One of the most basic pitfalls of using vague terms, when addressing the kind of questions that interested us from the outset, is simply anachronism –a particularly perverse kind of anachronism. Just as Henri-Irénée Marrou, in his simple and enlightening language, remarked that many notions and aspects of ancient (Greek and Roman) education have become invisible or absurd to us, *because we are no longer trained in the same way as the ancients*, it is all too easy to take the pedagogical dimension of such or such a source as obvious, *solely on the basis of*

[1]For a brief and historical outline of the research project that lead to the preparation of the present volume, see the acknowledgements. More colleagues than the present contributors were involved.

our own experience of learning, although this experience is itself the product of particular settings, methods and worldviews. The risk of anachronism is no less important in the case of sources that can be proved to have been produced for pedagogical purposes or used at some point in a teaching context: precisely because they are, in a sense, addressed to learners (or might be interpreted to do so), it is often very difficult to dissociate one's position as a historian reading a source, and one's position as an intelligent human being, simply learning from the sources.

Our initial questions, therefore, even when specified according to the corpus under consideration and the prevalent historiography, obliged us to adopt a more nuanced and better qualified approach to teaching and learning activities. We also needed, obviously, to cultivate a consistent and comparative reflection on what these terms meant for the actors themselves. We therefore defined a series of new questions that served as the initial basis for the volume, namely the following: To which textual, social, cultural, practical and institutional environments does the expression 'teaching and learning contexts' refer precisely? Is the reference to teaching or learning helpful or even legitimate? How does studying teaching or learning practices enable us to reach a better understanding of ancient or more modern sources? Conversely, what kinds of evidence on teaching or learning practices do our sources provide?

This new point of departure entailed new discussions, in which the contribution of historians of education like Anne-Marie Chartier was decisive. Following her lead, we will paraphrase and generalize here the useful distinction she introduced in her own contribution on the meanings of the word 'context'. Indeed, the original, restrictive meaning of this word, which is still in use, refers to the 'textual' surroundings of a given word or passage and that contributes to defining or changing its meaning. But it can also be understood in a much wider and overused sense, so that it actually refers to different levels of analysis that should, as far as possible, be differentiated here.

Context as *environment*. The word 'context' can (and often does) first refer to a general set of cultural and traditional values, received norms and habits that define a general environment and a system of shared objectives for the activities and sources under scrutiny. If, for example, one had to study a certain number of traditional omens used in studying to become an 'exorcist' in ancient Mesopotamia, as Philippe Clancier's contribution makes clear, this general objective of 'professional training' of experts and priests is part of the context in the sense that it is a crucial element in the cultural environment.

Context as *situation*. The word 'context', however, might also refer to a set of concrete gestures, procedures and codified practices that are attested in varying degrees by the written sources. Such activities and the underlying settings are either 'directly' attested by simple narratives (like the autobiographic narrative of Hamadou Ampate Bâ's first education presented in the first contribution); or indirectly, through the traces of these activities left in the sources. This type of 'context' might be called a *situation*, following Chartier's suggestion.

The context as *meaningful textual system*: Finally, when it is understood in a very strict sense that is close to its etymological meaning (*contextus*), the word "context" might be understood as the *whole system of textual sources, as well as intellectual activities* that makes a set of signs, meaningless in itself, a coherent and meaningful entity. This coherence might be attained by the fact that texts or pieces of text are used, composed or read in parallel to other texts that together build meaning and direction, like a traditional text accompanied by its marginalia, or the tablets that had to be used in a concomitant way in the elementary learning of mathematics in the Old Babylonian period. This system that makes a text a meaningful entity can also stem from the concomitant activity of reading associated with the "mentalization" of the contents, as in the procedures described in Chartier's contribution.

This basic analytical framework appears to be useful for the broad classification of the main issues addressed in this volume, as we shall see below (Sect. 1.2). For the time being, let us emphasize the importance and usefulness of this preliminary distinction. What is indeed crucial about these three *different* meanings of 'context' is that the third, 'restrictive' meaning is not sufficient to build a convincing interpretation of the sources. In other words, their mere textual content, even when the environment is taken into account, does not provide enough information on their actual meaning or use for the historical actors. Therefore, the omission or occultation of one or more of these levels generally leads to distorted or even false interpretation. This is particularly true for the first two levels (environment and situation) but is also valid for the third (meaningful textual system), since a source might be misleadingly considered in isolation from the complex 'whole' that provides it its meaning. Moreover, the second level (situation) is of particular interest for us as it is the most neglected in the historiography and the most accurate for analyzing the motivations of actors. All in all, a historically convincing interpretation might be defined as one that takes into account, in one way or another, the complex system made by these levels. One important point is that it should do so, even when one of the levels is unattainable for lack of relevant sources: in this case, the absence has to be taken into account.[2] Each of the present contributions, as we shall see, provides examples of how to deal with the required complexity of interpretation.

The above distinction between the possible meanings of 'context' might apply in general to various objects of historical study. But they appear to be particularly useful and relevant in the case of teaching and learning contexts, because they help to distinguish between very different realities and activities that are usually confounded under such very generic terms as 'teaching', 'learning', 'school', 'education', and other such terms.

[2]For mathematically oriented readers, this conceptual situation might be compared to the familiar situations of statistic and probabilistic model-making: the very fact that you do *not* have enough information about the causes, environment or other deterministic elements of a given event has to be taken into account as possibilities that have to be tested against the available information.

More precisely, 'learning' and 'teaching' might be understood primarily as *activities* implying the simultaneous presence of experts and novices (including, but not exclusively, masters and students) and involving specific acts and procedures whose oral and transient character are often difficult to reconstruct. But these words might also be primarily understood with reference to intellectual *projects*. Such projects are typically incorporated into coherent writings (treatises, handbooks, textbooks, manuals, etc . . .) where the very object of the science involved cannot be disassociated from 'pedagogical' purpose and design. As such, these projects almost inevitably refer, by their own definition, to the cultural environment and the set of values attached to it.

An important part of such projects might eventually be represented by the design of certain treatises, which might be related to teaching and/or learning situations *or not*. It is indeed perfectly conceivable that didactic projects, as has been proved in certain cases, were adopted by an author in spite of the fact that they did not correspond to any 'real' teaching activity. The didactic 'outlook', in such a case, was basically just a way of presenting an intellectual project and ideas in a form that was acceptable in a given environment, or because it corresponded to a particular philosophical and/or political project. A very good example of this is found during the period of the French revolution with Condorcet's elementary treatise on arithmetic.[3] Even in cases where the authors were recognized teachers, a situation for which famous examples abound, such as Monge with his *Géométrie descriptive* in the nineteenth century,[4] the treatise did not represent a direct reflection of the author's approach to teaching, but a new project partly related to it, and partly related to a much wider agenda.

Just as a 'didactic treatise' might have been influenced by real teaching activity, or not, it might have served as a real tool for teaching and learning – including, in certain cases, self-learning. In even more complex cases, it might have served this purpose after slight, subtle but decisive transformation of the text itself, as in the case of the addition to canonized texts (like biblical psalms) of *ornamented marks* making it possible or easier to memorize the subject matter. In other words, our point here is not to deny that, in some cases, treatises or sources that have a 'didactic outlook' were not indeed created for the sake of serving didactic situations; but the above distinctions and examples make it *possible* to understand that the *relationship* between explicitly didactic claims and teaching activities *is not straightforward*. Therefore also, the inference from one to the other is only valid when supplementary historical evidence is found. Conversely, it was not because a text is not explicitly designed to serve a didactic purpose that it was not *used*, at some point and in certain situations, for this. To take up the striking and provocative formula used by François Charette in one of the workshops that prepared for this volume, *any text is pedagogical* – this meant, of course, that any text can, potentially, be turned into a pedagogical device, even if the content has no 'evident' relationship to the underlying teaching situation.

---

[3]Condorcet ([1799] 1988).

[4]Belhost and Taton (1992, 269–317).

These examples show the usefulness and advantage that the notions proposed above have for the historian, when he has to identify and analyze complex case studies in which different levels of analysis are involved. We will now go slightly deeper into the questions that might be raised at each of these levels of historical analysis, by following the lead of the actual questions that are explicitly raised in the contributions to this volume.

1.2 A Tentative Survey of the Issues Raised Within This Volume

As suggested above, the questions that we initially addressed to the contributors are of such nature that they do not only lead to plausible answers, but to *many more* questions, each calling to specific issues or methods in historical research. We do not mean to develop in detail here what these research issues and their related domains or references are: such information can be found, or alluded to, in the detailed bibliographies of the contributions. More modestly, we propose only a tentative survey of the wide scope of various issues that are raised within this volume. This survey will serve as an index for the contributions in which those questions are addressed or raised.[5] We will broadly classify these questions according to the threefold scheme of 'context, situation and environment' presented above (Sect. 1.1) and inspired by Chartier's contribution.

1.2.1 Questions on the Context as "Environment"

Beginning with the simplest (but essential) questions concerning the ***general environment*** that must be taken into account for the interpretation of sources, there is one central question that is addressed in all the present studies and that must indeed be addressed in every case: namely, what did teaching and learning mean for the historical actors themselves? In other words, what is their 'cultural' purpose and nature, not in a broad sense, but *as it is defined and understood by contemporary actors*? As is shown by many of the contributions, answering these questions often calls for slight but legitimate anachronisms that are justified by the relative permanence of cultural values and ideals. Thus, it is possible to use the knowledge available on the teaching of arithmetic to Italian merchants in the beginning of fourteenth century to account for later works (see Stéphane Lamassé's contribution) or earlier texts (see Eva Caianiello's contribution): in terms of 'environment', they all indeed belong to the same 'cultural' horizon. Similarly, something can be learned from our knowledge of the ancient Chinese systems of examinations for the elite

[5]This list was established on the basis of our editorial exchanges with the contributors.

in the definition of the Vietnamese context analyzed by Alexis Volkov, and for an appreciation of the situation in the provinces in the far South analyzed by Florence Bretelle-Establet.

Two sub-questions appear in some studies. The first is whether the sources examined reflect, in one way or another, the underlying cultural meaning of teaching and learning? This question is central in Giovanna Cifoletti's contribution, as it deals with sources that have a heavily informed 'philosophical' component, each author defining their own view of what teaching should be. The same phenomenon is also found in some of the manuscripts discussed in Lamassé and in studies discussing highly developed prefaces (Caianiello, Bernard, Bretelle-Establet). The 'cultural environment' and its underlying values, in other words, can have a 'local' component that might explain deep differences from other contemporary sources.

The second sub-question concerns the professional context: are we dealing with sources that reflect the values of a very specific professional training? This question is of course addressed here in the studies in which such an environment is obvious, as it is for the exorcists in Clancier, the merchants in Lamassé and Caianiello and physicians in Bretelle-Establet. In other cases one might suspect such a background but it remains unclear, like for the audience for Theon's teaching in Alain Bernard's contribution (astrologers? Or students in liberal studies?) or for the readership of Renaissance treatises in Cifoletti's (Noblemen? University or college students? Craftsmen?) – but this ambiguity might, in turn, become a possible element for interpretation (as in Cifoletti's chapter).

Finally, questions on the cultural environment specifically include, in some cases (here in Clancier, Bretelle-Establet, Cifoletti, Bernard and Lamassé), an inquiry into the social status and self-representation of teachers. While teachers might be regarded as practitioners, effectively and actually involved in teaching activities, they might also be viewed as *persona*, as figures with their own image and place in society, which might be reflected by their self-representation. What is, then, this representation and what are its main components and features?

1.2.2 Questions on Teaching and Learning "Situations"

Such issues are generally raised, of course, only when we have good reasons to believe that we can identify and define such situations. A preliminary question on this aspect is, of course, the following: which sources can provide some kind of knowledge on these situations? This question is specifically addressed in Chartier and Clancier, and also in Christine Proust's contribution to some extent. In most of the other studies, it is addressed through reference to relevant secondary studies.

There are two specific questions touching on teaching and learning situations that challenge the written character of our sources. The first is whether the sources help us in defining an *oral* (non-written) activity specifically attached to the type of source under investigation? This issue is central to Chartier's approach to the techniques formerly used in reading apprenticeship; the vocalization and "oralization" of knowledge here appears to be a necessary prerequisite and is explicitly expressed in the sources (mainly testimonies) that describe the process. But the relationship to orality is also discussed at length by Clancier, Bretelle-Establet and Andréa Bréard, as well as by Proust, Volkov and Lamassé to some extent.

The second question, closely related to the first, is on memorization: does teaching and learning imply a strong component of memorization, whether or not this process is related to rote learning or to some kind of understanding? This crucial question, addressed in Mary Carruthers's classical study on memory in Latin Middle Ages,[6] is again central in Chartier's approach to teaching and learning of reading skills, but it is also discussed in all the contributions dealing with Chinese or Vietnamese material (Volkov, Bréard, Bretelle-Establet) and in Proust and Lamassé for some aspects of the corpus they investigate.

Formal teaching and learning naturally involve both students and teachers: there are questions raised about how much and what we know on these actors: as far as students are concerned, the question is addressed here in Chartier, Clancier, Proust, Bernard and Cifoletti. For the teachers, the question is addressed in Chartier, Clancier, Bretelle-Establet and Caianiello.

Finally, there is an interesting question about the possible use of *modern* anthropological and educational research studies: to what extent can modern anthropological testimonies improve our understanding of past situations underlying a historical source? The problem hidden in this question is due to the chronological discrepancy –and therefore anachronism- that such an approach inevitably entails, as most of the available anthropological studies generally concern the contemporary period. In spite of this unavoidable flaw, some of the present studies show that such an approach, when used with care, can be useful and enlightening: such is the case of Chartier's approach, which relies on the testimony of Hamadou Ampate Bâ, and of Bretelle-Establet's contribution, which calls on E. Hsu's studies on the teaching of medicine in modern China, and of Volkov's contribution, which relies on the tools of modern research on mathematics education.

1.2.3 Questions on the "Meaningful Textual System of Sources"

This is, in a sense, the most complex and detailed set of questions, as the interpretation of written sources is, in general, the main object of study of the contributions to this volume. The first and most important question, which is addressed by all the chapters in the volume, concerns the cross-referencing of information contained in the sources so as to highlight any overall system, that they appear to be part of:

[6] Carruthers (1990).

thus, do we have sources other than those principally investigated, that enable us to define a larger picture and environment, and possibly a system of coherent texts? This question, of course, refers directly to the two main questions related to the environment and situations above.

A second, more specific question concerns the crucial difference between the *material nature* of the sources, and the *text* they contain. Does the source under investigation, as a *material object* (and not as the indifferent substratum of a significant set of signs, that is a text), have a paratext, that is, traces and marks that might be significant for the definition of its actual use, in particular its potential use in a pedagogical situation? Not surprisingly, this question appears central in the studies on archaeological sources contained in this volume, namely Clancier's and Proust's studies on Mesopotamian tablets. In the latter, this distinction between the textuality and materiality of sources is a central issue and the main key for the interpretation developed. The question is also addressed at length in Chartier's, Bretelle-Establet's and Lamassé's studies.

The next questions, also very important, concern the variety of erudite or, in general, intellectual activities that might be associated with a given source. In a nutshell, can one plausibly imagine that the sources were meaningful for activities that were not (strictly) pedagogical? This question is addressed in several of the studies: Clancier and Proust (copying), Bretelle-Establet (checking information through encyclopedias), Cifoletti (preparing a printed edition), Bernard (commenting), Lamassé (building new series of problems on the basis of a traditional corpus). Indeed, to speculate that the source(s) investigated is/are meaningful within a teaching and learning context does implicitly mean, that other possible contexts of activity are disregarded or considered irrelevant; in some (in fact, in many) cases, several such 'intellectual' contexts might appear relevant and non-exclusive.

There are several sub-questions that help to develop this general inquiry. The first is on the *internal organization* of the texts contained in the sources. Indeed, almost all the contributions address the following and apparently straightforward question: what does the text contain, how is it organized and what can be interpreted from this in terms of pedagogy? While this question is apparently simple, the answers it yields give no indication of how the text was used and in which situation. In fact, they give an analytical picture of the organization of the text and little else. This organization, considered independently of any question of 'actual teaching', might of course be labeled 'pedagogical' in its own right. This point of view is developed at length in Lamassé's contribution through his original concept of a 'writing project' (*projet d'écriture*), which identifies a consciously designed organization in certain texts (a series of problems being a typical example) that might *potentially* and in *a differed manner* be related to pedagogical practice (or not). The same question is addressed in Cifoletti. But, in general (and in most of the other contributions in this volume), these organizational features might be interpreted with reference to other intellectual activities and situations than just pedagogical ones: checking the organization and content then appears as a criterion to 'test' the various interpretations and activities that one might speculate are associated with the source.

The second sub-question concerns the *history of the texts* found within the sources. The circumstances in which a text was produced might be very different from those in which it was later used, especially when it came to be used in a pedagogical context. The question, then, is simply to check, when possible, the history of the texts contained in the sources examined. This issue is addressed in Clancier, Chartier, Proust, Cifoletti, Lamassé and Bréard.

The last sub-question concerns the possibility of comparing the contents of the sources with similar content but that might refer to different intellectual activities or circumstances. This kind of comparison is crucial to the contributions by Lamassé, Bréard and Cifoletti and is also found in Proust: we will come back to this issue below (Sect. 1.3), when dealing with the 'comparative' interpretative strategy they have in common.

1.3 The Various Interpretative Strategies Developed in the Volume

Given this rough survey of the wealth of questions addressed in the various contributions, we can now characterize the various ways, in which these questions have been addressed in each case: in which order, and for what basic purpose.

Indeed, this book is naturally not meant to cover all the fields, historical periods, cultural areas, sources or authors potentially concerned by these questions: such a titanesque purpose would have been meaningless on such a broad topic. On the contrary, we have only looked for the maximum of diversity and have asked the contributors to keep overspecialized remarks to a minimum, so as to make each argument accessible even to readers who are not specialists of a given corpus, period, or domain. Our true purpose, then, has been to build a heuristic tool by looking for *suggestiveness* and by making the *principal questions* and *interpretative strategies* adopted in various situations as clear as possible. As has been shown above on a theoretical level (Sect. 1.2) and is amply demonstrated in the contributions, there is no "simple case" for interpretation: the range of parameters, testimonies, circumstances and issues that have to be checked is so vast that to make sense of the whole generally precludes a more simplistic approach. To borrow a current anthropological concept for modern societies, sources that are interpreted in learning and teaching context define a "technologically dense environment". Although modern, this concept can also be applied to ancient periods. Moreover, *historical* sources specifically raise the problem that most of the time, by their very nature, we lack the knowledge of all the circumstantial elements that would help one to reconstitute such "dense environments". Quite on the contrary, then, when faced with similar problems of interpretation, the contributors had to develop their own interpretative strategy based on a variegated series of questions, most of them listed above (Sect. 1.2) and taking into account even those data that they cannot verify for lack of relevant sources.

As a result, we have discovered that these strategies broadly fall into four relatively distinct categories: (a) the 'holistic' approach aimed at the restitution of a coherent system; (b) the critical approach of commonly accepted assumptions; (c) the 'comparative' approach; (d) the 'zooming' approach.

(a) The 'holistic' restitution of a coherent system is the method followed by Chartier and Clancier and is probably the closest to traditional studies in History of Education. In these two cases, there is hardly any doubt about the fact that the sources and activities discussed related to teaching and learning. The key problem is to understand how and to restitute as precisely as possible the nature and consistency of the underlying didactical system. Thus, Chartier takes her point of departure from the bizarre misunderstanding sometimes expressed by modern minds when faced with the ancient material and methods used for teaching to read. She explains this modern misunderstanding on the basis of a 'holistic' examination of all the various levels of context that must be taken into account in order to make sense both of the "ancient" methods for learning to read and of the subsequent changes that made these old methods into *bizarreries*. Similarly, Clancier begins with a rich documentation, namely the scholarly libraries of Seleucid Uruk. This documentation has an identified provenance and is complex, diverse and interrelated enough to make a reconstitution of exorcism learning and teaching techniques possible, as far as the plausible reconstitution of oral activities and the genealogies of teachers, at least. In her study of the material found in Old Babylonian scribal schools of Nippur, Proust comes close to a holistic approach.[7]

In general, then, the holistic approach might be recommended in all cases, in which the documentation is sufficiently rich to make such reconstitutions possible, or where modern anthropological studies can legitimately be used to complement the lack of information.

(b) The critical approach, as followed by both of us.[8] The critical approach takes its point of departure from commonly accepted assumptions on certain sources: Proust took three types of old Babylonian tablets with mathematical content that are all reputed to be related to learning or teaching activities in some sense, and showed that in *each* case the common wisdom must be nuanced at least, when it is not found to be entire wrong in certain cases. Bernard took a well-known commentary on Ptolemy's *Almagest* that is traditionally reputed to be associated with teaching the *Almagest*, not least because the preface to the aforementioned commentary explicitly says so: it is then shown that while this notion is possible, doubts may be raised on the crude inference from the commentator's declaration to the reality of his teaching. There are indeed many other ways of understanding his initial statements, some of them having little to do with any actual teaching. We strongly

---

[7] See Part I of her contribution, as well as her other articles on the subject. Parts II and III of Proust's contribution fall in the next category.

[8] This might be due to the fact that we were both co-editors of the volume and, at the same time, contributors to it, so the general problems we were thinking about may have influenced our approach to the sources.

recommend the critical or 'cathartic' approach, then, to any reader who has been or will be confronted, like we were, with sources for which their interpretation in a teaching or learning context is taken to be self evident and has therefore has not been sufficiently questioned. This critical approach, in other words, is very suitable for those kind of sources in which there is a clearly prevalent interpretation derived from the traditional historiography: in the two cases presented here, the traditional categories discussed mainly come from Neugebauer's pioneering work for the old Babylonian material, and Rome's critical edition and studies for the Greek.

Lamassé's contribution, although we decided to categorize it as being "comparative", also comes close to this approach. Indeed, his point of departure is a corpus for which a historiographical tradition already exists, with strong presuppositions that deserve to be re-visited.

As with any other historical study, the critical approach does not (and should not) amount to hypercriticism: the basic problem and purpose, is not to criticize traditional assumptions for the sake of criticizing them, but really to keep the best of traditional discussion and refine the picture with a more accurate and well-grounded approach.

(c) The comparative approach takes an interesting point of departure: instead of questioning *one* source, like a given text or a set of more of less homogeneous sources, it focuses on a collection of sources having apparently homogeneous content that can be broadly interpreted in didactic terms, but that actually presents internal diversity. Such an approach is mainly represented, within this volume, by the contributions of Stéphane Lamassé, Andréa Bréard and Giovanna Cifoletti. Lamassé focuses on a specific corpus of arithmetical works for merchants produced in the Franco-Provencal area around the fifteenth century. What makes this corpus of ten or so works especially interesting is, on one hand, its obvious relationship to the arithmetical treatises produced and used in the Italian context from the late thirteenth century and known to have been used in the professional training of merchants; and, on the other hand, the chronological gap between this reference period and the fifteenth century, a period during which significant changes in the world of merchants and commerce occurred. This corpus, then, requires more careful examination, following Carlo Ginzburg's method of checking various indices, to account for its internal diversity and check the precise reasons for which some (not all) of its aspects might indeed be related to educational purposes. Bréard, similarly, focuses on various examples of versified mathematics which, by its very nature, might all be related to pedagogical concerns connected to the traditional exercise of memorization, so central in Chinese classical education. But a closer examination of various examples of such versified material shows that the underlying educational purposes and techniques might be quite different, from one type to another. Cifoletti, finally, takes as her object a series of mathematical treatises in arithmetic and algebra published during the Renaissance period in Europe. While they all reflect the predominant Renaissance ideal for a new humanistic approach to the arts and sciences and therefore strongly connect to pedagogical ideals and practices, her close investigation reveals significant differences in the underlying intellectual projects and in the target readership – she shows, in particular, that the choice of language, the organization of the treatise and the conscious use of innovations in the printing industry, enable one to perceive the evolution and development of this literature and to identify the various types of reader.

Bretelle-Establet's conclusion comes close to a comparative approach, since she finally alludes to the possibility of comparing the object of her study to another treatise of the same period but of a different design.

The comparative approach, then, appears useful when the interpreter is faced with a corpus that presents, either a clearly unifying characteristic that is broadly related to a pedagogical dimension or an internal diversity that calls for further questions. It might be viewed as a form of critical approach, but here the emphasis is less on a methodological question that calls for the choice of a 'test' corpus, than on a specific corpus that calls for a refinement of methodology.

(d) Finally, the 'zooming' approach is ultimately focused on one specific source for which little 'directly contextual' data are available, but for which we have general and plausible reasons to interpret in a pedagogical context. This approach is represented, within this volume, by the contributions of Eva Caianiello, Alexei Volkov and Florence Bretelle-Establet. Thus, Caianiello concentrates on Fibonacci's intriguing and influential treatise, the *Liber Abaci.*[9] While we have very good reason to believe that, both as a student and as a teacher, Fibonacci followed a complex and interesting curriculum, which is partly characteristic of the contemporary milieu of Pisan merchants and which is reflected within his *Liber Abaci*, there is too little documentation to prove it. The problem therefore, is to make the best of the many 'circumstantial' elements that can be studied on Fibonacci's curriculum, and then check how (and to what extent) such an inquiry might help us to better understand his text (in the study in question, Caianiello only focuses on Fibonacci's long and detailed preface). Similarly, Volkov discusses a Vietnamese mathematical treatise compiled at the beginning of the nineteenth century by a Vietnamese political figure and for which we have no date and little evidence on the direct circumstances of its production or use in actual teaching. On the other hand, we have, here also, very good reasons to believe that its general structure reflects a general context, heavily influenced by the ancient Chinese examination system, in which the critical variation on parameters in a series of mathematical problems played a central role and was the focus of pedagogical efforts. Volkov then shows that this special attention to parameters might be retrieved through a detailed analysis of the treatise itself. Finally, Bretelle-Establet focuses on "the *Essentials of Medicine*" a medical 'textbook' written around 1800 in South China. Here, again, there is good reason to believe that this text was produced and used in a pedagogical context and that its structure is influenced to some extent by the standard curriculum stemming from the traditional, centralized teaching of medicine in China; but there

[9]We have chosen here and throughout the book to write the title of Fibonacci's treatise as *Liber Abaci* and not *Liber Abbaci*, following W. Van Egmond's suggestion (Van Egmond 1980, 5). The reasons for adopting this uniform spelling are explained in Caianiello's contribution.

is, at the same time, little 'direct circumstantial' evidence that would enable one to check the internal analysis of the text – and this discrepancy is a concern with this approach.

Bernard's contribution on Theon comes close to this 'zooming' approach, in Parts III and IV of his study, when he gives some hints of more general and cultural data that might help us to better understand the content and purpose of Theon's approach.

In general, then, the zooming approach might be recommended in any case where we have good reasons to interpret a given source in pedagogical terms, but in which there is little hope to do this through the 'holistic' approach described above. Quite on the contrary, in such cases, one is obliged to speculate on a possible but ill-informed context, the gist of the interpretative game being to control, as far as possible, one's reconstruction on the background of more general data. This is probably the approach that appeals most to refined and "internalist" approaches to the sources, 'external' data nevertheless playing a role in checking the potential validity (or weakness) of such analyses.

To conclude, neither the questions (Sect. 1.2) nor the types of interpretative strategies (Sect. 1.3) that we have listed are meant to be exhaustive: more studies of the same kind might certainly have revealed more questions, strategies, and methods. More modestly, the present lists are only meant to serve as a guide for the reading of this book. Karine Chemla's postface, written from a wider point of view, will usefully complete our introduction. She proposes indeed to comment on the various contributions by highlighting their signification in the framework of the most recent debates and evolutions in the field of history of science.

What use then – at least in the best of our hopes – could be made of this book? As mentioned above, what we really imagined was to create a heuristic tool: the range of these contributions is meant to provide the reader with a suggestive spectrum of possible interpretations of scientific sources in a teaching context as well as a sense of the complex questions that should be asked when confronted with such difficult interpretations. We hope, then, that it might provide our reader the elements for similar investigations.

Acknowledgements The present volume is the outcome of a research project on the study of ancient scientific sources within teaching contexts in various historical periods. Began around 2005, the annual workshops were first hosted and financed by two French research institutes, REHSEIS (now SPHERE at Paris Diderot University and CNRS) and the Centre Alexandre Koyré (EHESS, MNHN and CNRS), within the framework of the interdisciplinary 'mathématiques et histoire' project (EHESS, principal investigator Giovanna Cifoletti). In addition to the colleagues who finally contributed to the present volume, many others actively took part in our workshops and contributed to the precise definition of our key issues. We specially thank Fabio Acerbi (CNRS), Bruno Belhoste (Paris I University), François Charette (Chester Beatty Library), Samuel Gessner (University of Lisbon), Catherine Jami (CNRS), Agathe Keller (CNRS), Sabine Romevaux (CNRS), Maryvonne Spiesser (University of Toulouse), Renaud d'Enfert (INRP, now IFE), Bernard Vitrac (CNRS), and Niek Veldhuis (University of California, Berkeley) for their insightful remarks and help.

As for the edition of the volume itself during the period 2008–2012, the translation and preparatory workshop was financially supported by the above research institutes (REHSEIS-

SPHERE and Centre Alexandre Koyré), the 'mathématique et histoire' project (EHESS), the University Paris Diderot, the interdisciplinary laboratory HASTEC (Paris) and by the CAPHES (CNRS and ENS Paris). We particularly thank Richard Kennedy for his careful translation or polishing of the various contributions.

References

Belhoste, B., and R. Taton. 1992. L'invention d'une langue des figures. In *L'Ecole normale de l'an III, Leçons de mathématiques*, ed. J. Dhombres. Paris: Dunod.

Carruthers, M. 1990. *The book of memory, a study of memory in medieval culture*. Cambridge: Cambridge University Press.

Condorcet, Jean-Antoine-Nicolas de Caritat. [1799] 1988. *Moyens d'apprendre à compter sûrement et avec facilité*. Critical apparatus, studies, notes and commentary by C. Coutel, N. Picard, and G. Schubring. Paris: Art, Culture, Lecture.

Van Egmond, W. 1980. *Practical mathematics in the Italian renaissance. A catalogue of Italian abbacus manuscripts and printed books to 1600*. Firenze: Instituto e Museo di Storia della Scienza.

Part I
Holistic Approach

Chapter 2
The Teaching Context and Reading from the 16th to the 19th Centuries: The Role of the Memorization of Texts in Learning

Anne-Marie Chartier

It is particularly difficult to interpret historical texts used in teaching when little is known about the way in which these materials were exploited: careful reading of the content of the texts gives no indication of how they were used by the teachers and by the pupils. The following contribution shows the importance of the "teaching context", in the particular case of the religious texts used to teach reading to beginners from the sixteenth to the nineteenth centuries.[1] Prayers, psalms,

[1]This contribution is part of research conducted since 1996 into the history of school subjects by the History of Education Department of the INRP (National Institute for Educational Research). Elementary learning (reading, writing and arithmetic) poses an interesting methodological and epistemological problem as it does not have a 'reference discipline' (there are no university chairs of 'reading, writing and arithmetic'). They are considered as technical skills having no particular purpose in themselves, relevant in the past for passing on empiric information and today for theoretical modelling in the fields of psychology, linguistics and cognitive science. Regarding reading skills, historians treated this as a two-stage indicator of literacy (being literate or illiterate), before discovering an intermediate stage, that no longer exists today (being able to read but not to write), which denied literacy tuition in former times some of the motivations and the effects which are expected today (Graff 1981). Hence further investigations into reading practices over the centuries. Some studies refer to the act of reading in its broad cultural context (the social status of the written word in political, legal and religious spheres, the conflict between oral and written culture and between high and popular culture); others record the material constraints imposed by the writing itself (texts written on scrolls, codices, coded phonetically or with ideographs, continuous or segmented, handwritten or printed, reproduced orally or in silence, etc.) For a perspective from Antiquity to the twentieth century, see (Carvallo and Chartier 1999). The third line of research which I am engaged in is the history of learning and teaching methods, within and outside the school setting. This article summarizes, with new critical scrutiny, previously published

A.-M. Chartier (✉)
Department of the History of Education (SHE), IFE-ENS-Lyon: Formerly INRP (National Institute for Educational Research), Lyon, France
e-mail: anne-marie.chartier0054@orange.fr

A. Bernard and C. Proust (eds.), *Scientific Sources and Teaching Contexts Throughout History: Problems and Perspectives*, Boston Studies in the Philosophy and History of Science 301, DOI 10.1007/978-94-007-5122-4_2,

catechism lessons and Bible extracts were all used for purposes of both religious instruction and teaching literacy. How was this possible? How could texts of religious significance, spiritual meaning and liturgical function be used as supports for teaching literacy to children? Today's researchers were taught to read in a very different way when they were children. As it is difficult not to project evidence from these memories and from generally accepted beliefs about reading onto historical situations, it is necessary to take some methodological precautions. To identify any anachronisms that can produce these retrospective projections, we will start by comparing contemporary and historical sources on this subject.

When future historians carry out research into how children learned to read in the year 2000 (an archaic time when they taught without software or computer screens), they will find an abundance of three types of data. Firstly, theoretical texts on learning how to write (scientific studies, didactic instructions). Then, empirical data on the procedures used and their results (descriptions of reading lessons, debates on the methods and the results, statistics evaluating performance). And finally a wide variety of teaching materials, booklets and reading guides [*primers* and *readers*], exercise books, teaching aids and teacher handbooks.

Historians have no equivalent sources when they carry out the same work for teaching literacy in Antiquity, the Middle Ages or the Renaissance, as few examples of writing by learners have been found.[2] From the seventeenth century,[3] literacy for beginners was addressed in school rules[4] and in some pedagogical treatises.[5] Educational curricula[6] detail the content and the exercises planned for advanced pupils (Grammar, Latin, Rhetoric, Philosophy and Mathematics), but the documentation is more laconic on the practices required for the crucial step of the introduction to the written word. It was as if learning to read, so essential to schooling, was such a rudimentary step that there was very little to say about it. It was presented as a mere exercise in memorization, in which the student repeats a text over and over until he can read it without making a mistake, before moving on to another text.

---

data on the history of reading methods, official requirements, textbooks and student workbooks (Chartier and Hébrard 2001; Chartier 2004a, b, 2007, 2009).

[2]Transitory writing is nearly always destroyed by the authors. The texts written by schoolchildren (Mesopotamian tablets, Hellenistic papyri, monastic copies, Renaissance students' exercise books, eighteenth century Latin schoolboy texts) that have been found represent only a very small fraction of the work undertaken while learning, but they do provide invaluable information to limit anachronisms when attempting to reconstruct the 'teaching context'.

[3]For England, see (Cressy 1980); for the USA, (Monaghan 2005); for France, (Chartier et al. 1976; Chervel 2006); for Italy, (Grendler 1989); for Spain, (Viñao Frago 1999); for Portugal, (Magalhães 1994); for Sweden, (Lindmark 2004). For a comparative perspective (Graff 1981).

[4]Jacques de Batencourt, *L'école paroissiale*, 1654, Jean-Baptiste de La Salle, *Conduite des écoles chrétiennes*, 1720 in (Chartier et al. 1976).

[5]John Locke, *Some Thoughts concerning education*, 1693; Jean-Jacques Rousseau, *Emile ou de l'Education*, 1761.

[6]See for example, *Ratio studiorum* [1599] in (Demoustier and Julia 1997).

According to some authors, no qualifications were necessary; it was enough to be able to read yourself to patiently supervise these repetitions. A famous English textbook advised craftsmen to supplement their incomes in this way: "tailors, weavers, shopkeepers, seamsters [...] mayst sit at thy shop board, at thy looms, or at thy needle, and never hinder thy work to hear the scholars, alter thou hast once made this little book familiar unto thee."[7]

Everything written today on the teaching of reading contradicts such assertions. Knowledge of linguistics, psychology and teaching are obviously necessary to guide beginners. But this is not always sufficient, as we see in the noticeable failure rates found even in countries with university-educated teachers. How was it possible for schoolchildren in the past? The very simple process by which anyone could teach seems repetitive, tedious and strange today. While all the textbooks in the year 2000 present the various difficulties gradually, placing the letters/sounds to be studied in short, easy to decipher and understand sentences, the ancient texts are far removed from the children's interests and their capacity to understand. They presented learners with religious texts (prayers, psalms, and liturgical passages) or moral instruction ("Civilité"), often in Latin, and the difficulty of these texts did not seem to pose a problem to anyone.

Today's specialists: cognitive scientists, psycholinguists and educationalists find it difficult to understand how children managed to learn to read using a method which seems to hinder access to fluent, autonomous reading rather than accelerate it. By the nineteenth century, the first studies into the history of teaching literacy,[8] while denouncing the weak network of schools, high school fees and the negligence of the Church, blamed the illiteracy of the time on the 'archaism' of the teachers, questioning school organisation and the methods used to teach reading.

Firstly, school organization: Teachers practiced an 'individual method' of teaching. Each student was called to the teacher's desk in turn, so the amount of time spent with each pupil was inevitably very short. Then the pupil was sent back to his seat to do his exercises alone. However, as Jean-Jacques Rapet, Inspector General of Public Instruction, wrote in 1860:

> What can we have a poor child do when he can neither read nor write? There is no other way of keeping him occupied: because to pretend that one will succeed by putting a primer in his hands and telling him to study is a delusion that is not suffered by teachers enlightened by experience. He will be able to hold his primer in his hand, twist it around, gnaw at it, but he will not be able to learn from it because that, for him, is impossible. In order to learn to read, as long as you aren't nearly able to read properly, it is absolutely necessary to have the help of a teacher.[9]

---

[7] Edmund Coote, *The English Schoole-Master*, 1624, [48 publications from 1596 to 1696], quoted by D. Cressy (1980, 38).

[8] One of the first studies carried out in France was by James Guillaume, author of the very well informed article 'Lecture' in the *Dictionnaire de Pédagogie* edited by Ferdinand Buisson, published in 1887. For England and the USA, see (Mathews 1966; Banton Smith 1965 [1934]).

[9] Rapet, "De l'organisation de l'enseignement dans les écoles primaires", *Journal des instituteurs* 34, 19 August 1860, 118, 119. (Chartier 2007).

Next, the didactic choice: the learner had to spell out each word, syllable by syllable before reading the entire word. This segmentation prevents you from catching the sense of the words and of the sentences. The pupil had to read and re-read each text by himself until he knew it by heart. He learned to recite rather than to read; to learn but not to understand.

We have hypothesized that while these methods seem absurd compared to our current methods, it is because we no longer understand them. Could we find their internal coherence by reconstructing the cultural, scholastic and educational context in which they were used? For that, we will have to temporarily suspend our current thinking on teaching literacy, on the relationship between written and spoken production and on the link between memory and understanding. It is this progression we will present in three stages.

Firstly, we will present a number of teaching aids used from the sixteenth to the eighteenth centuries, in order to re-establish the 'discursive context', the didactic instructions and discuss the criticism condemning the old methods. Then we will present the accounts of two learning experiences, which demonstrate the effectiveness of these methods in their 'working context' (in the family in one case and in school in the other). Finally, we will evoke the 'social operating context', which gave them meaning, made them effective and that eventually led to their abandonment in the nineteenth century. It will then be possible to understand how the circumstances that led to this change[10] came about, and why, once passed from living memory, these methods that were so meaningful to our ancestors, have become so puzzling and seemingly absurd to the eyes of subsequent generations.

2.1 Teaching Materials for Beginners Between Sixteenth Century and Eighteenth Century in Western Europe

2.1.1 Description of Primers and Readers

The early teaching materials used to teach reading have almost all disappeared because they were very fragile and had no commercial value. Three principal tools

[10]A discordant or even a crisis situation is created when reading acquires a new function that forces a change in learning objectives without the teachers being able to formulate new teaching methods. This was the case when the old methods, suitable for reading 'timeless' and 'classic' texts (e.g. religious, moral, literary), continued to be used for reading ever-changing, ephemeral types of text, e.g. the press. The search for new methods stimulates the pedagogical imagination, and some of the solutions become imperative and lead to the establishment of new standards. New technical innovations (the hand press, then the steam printing press, low-cost pulp-based paper, metal pen nibs, typewriters, photocopiers and computers linked to printers) can, in certain cases, be adopted quickly when they do not disrupt teaching practices, or conversely, be slow to enter the school system because they disrupt the couple between the (unchanged) ends and the long established means employed to arrive there.

could be found throughout Europe: horn-books, ABCs and readers. The *horn-book* or *battledore* (*carta* or *tavola* in Italy, *charte, carte, palette* or *tablette* in France) is a small wooden board with a handle where the alphabet is written once, in small and again in capital letters, below the design of a Cross. A list of vowels came below the alphabet, sometimes, but not always, followed by a list of consonants and a few syllables set in columns, for instance A, E, I ,O, U combined with B, C, and D. Finally, there was the prayer formula for the sign of the Cross (*In nomine Patri et Filio et Spiritus Sancti, Amen*), a psalm verse, and/or a longer text, like the *Pater Noster* in Latin or in the vernacular, according to the country and the religious denomination.

The *ABC* (or *psalter, primer, crisscross/Christ cross row*) was a very cheap booklet made up of only a few pages, sometimes illustrated with woodblock prints. Whereas the horn-book often belonged to the family, an *ABC* was often given to an individual child as their first and sometimes only schoolbook. It was called a *psalter* (*salterio* in Italian and Spanish, *psautier* in French) more often than an *ABC* or *primer*, because of the texts it contained: the PaterNoster, the Ave Maria and the Credo in Latin in catholic countries and the Lord's Prayer, the Creed and the Ten Commandments in the vernacular elsewhere. Other prayers included the Confiteor, the seven Penitential Psalms, and various religious hymns depending on the different religious denominations. Unlike the psalters printed for religious use alone, the psalters used as primers included, on the first page, alphabets in upper-case and lower-case letters, a list of syllables presenting the 'consonant-vowel' and 'vowel-consonant' combinations ('ba', 'ca', 'da', and 'ab', 'ac', 'ad', etc.). The Pater Noster, printed in larger type than the following prayers, was often syllabicated, that is, divided into syllables (Pa-ter Nos-ter).

The third type of book, readers (prayer books, books of hours, catechisms), introduced longer texts. Comparing primers and readers, still in use today, shows the difference between the two stages. Readers were made up of well-known texts, which had not been written especially for children: Veni Creator and the Vespers of the Virgin in catholic countries, translations of Gospels and Psalms in protestant ones. There was a listing of the sacred truths that had to be remembered. A new literary genre appeared in the sixteenth century, called *Civilité*.[11] Uniquely these books combined religious with secular subjects as they also dealt with manners and civility. They were often printed in 'civilité type',[12] imitating cursive writing, which helped children to become accustomed to reading manuscripts.

[11]Erasmus wrote *De Civilitate morum puerum*, in 1530. It was translated or imitated in many vernacular versions, which had a great success in schools from the seventeenth to nineteenth centuries.

[12]Jimenes (2011).

2.1.2 Curricula for Beginners

Thus, according to the above descriptions,[13] beginners learned successively:

- To recite the alphabet and to name all the letters by heart: "*The usual way to begin with a child when he is first brought to school is to teach him to know his letters in the horn book, where he is made to run over all the letters in the alphabet or Christ cross row, both forwards and backwards, until he can tell any one of them which is pointed at.*"[14]
- To associate the vowels and consonants on a table of syllables (BA, Ba, etc.): "*The scholar shall learn perfectly, namely, to know the letters by their figures, to sound them aright by their proper names, and to join them together, the vowels with the vowels in diphthongs and the consonants with vowels in others syllables.*"[15]
- To spell out prayers by following the procedure ('P A Pa, T E R Ter, PA-TER'): "*When you are certain they have mastered the alphabet, teach them the syllables, using the Lord's Prayer as your text. All this time pay close attention to their pronunciation, and do not allow the boys to slur or drawl their vowels and consonants in the manner of their natural speech, but make them separate and distinguish the sounds clearly from one another, as is done in Latin diction.*"[16]

  In Lutheran countries at that time, prayers were still in Latin: "*What is the first letter in 'Pater'? It is a 'p'. Show me the 'p' in the alphabet. What comes after the 'p'? an 'a'. Show me the 'a'. And so on.*"
- To read words by whole syllables (Pater / noster / qui / es / in / coelis ... Our/ Father/ who/ art/ in/ heaven ...): "*After they have got some knowledge of their letters and a smattering in some syllables and words in the horn book, to turn them into the ABC or primer, and therein to make them name the letters and spell the words till by often use they can pronounce at least the shortest words at the first sight.*"[17]
- To read by groups of words (Pater noster/ qui es in coelis / sanctificetur nomen tuum), longer texts, in Latin or vernacular, like Psalms or other texts from the Bible: "*When he can read any whit readily let him begin the Bible and read over*

[13]As oral exercises leave no trace, these curricula can only be described from teaching instructions and testimonies (from teachers or former pupils) when they exist. However, once the beginner knows how to read, the teaching curricula can be seen in the list of texts given to be studied, as shown in the *Ratio Studiorum* for studying Latin. Cf. (Compère 1985, 76–77).

[14]Charles Hoole, *A new discovery of the old art of teaching schoole*, 1660, quoted by D. Cressy (1980, 41). The spelling in the English text has been modernised.

[15]William Kempe, *The education of children in learning*, 1588, quoted by D. Cressy (1980, 22).

[16]Here is an example of a teaching practice described in "*the school ordinance*" in Lutheran Saxony, in 1580, quoted by (Strauss 1981, 102).

[17]Charles Hoole, *op. cit.*, in (Cressy 1980, 21).

the book of Genesis and other remarkable histories in other places of scripture which are most likely to delight him, by a chapter in time."[18]

- To read books in the vernacular (Catechism, Christian Civility).
- To read manuscripts[19] in order to become accustomed to cursive letters.

Thus, scholars seem today to have good reason to doubt the effectiveness of these tools. Using them, students probably became familiar with the religious culture of their community; they became able to read or to recite prayers or the catechism without making mistakes, but it seems impossible, according to our present knowledge, they became 'true' readers.

2.1.3 Discussion of the Criticisms of Old Teaching Methods for Young Readers

Information from several sources makes it necessary to re-examine criticisms of long-condemned practices formerly employed for teaching reading.[20] Firstly, we know that a large number of competent readers were produced using methods that today seem inefficient, but satisfied the teachers and parents of the time. The 'spelling method' was used by teachers in schools for the lower classes as well as by the personal tutors who taught the children of the elite. Its pertinence was not doubted; it was the method used since Antiquity; nobody questioned it.

In effect, the texts used to instruct how to introduce reading to beginners remained the same from Antiquity to the eighteenth century. They followed the progression 'letters – syllables – words – texts' presented in *De Institutione Oratoria* in the first century of the Common Era. Quintilian recommends presenting each letter to children, making the link between its sound, name and shape ("*just as they do with men*") without reciting the alphabet, which has little value

> At any rate I do not like the procedure (which I see very common) by which children learn the names and sequence of the letters before their shapes. This is an obstacle to the recognition of the letters, since they do not when the time comes pay attention to the actual outlines, because they follow the promptings of their memory, which rush ahead of their observation. [. . .] It will be best therefore for them to be taught the appearance and the name side by side: it is like recognizing people. But what is an obstacle in learning letters will do no harm when we come to syllables. [. . .] With syllables, there is no short cut. They must all be learned; there is no point in the common practice of postponing the most difficult questions relating to them, to be discovered only when we come to write words. We must beware also of trusting the first memory too readily: it is better to have repeated syllable-drill over a long period and not be in a hurry to achieve continuity or speed in reading either, unless the sequences of letters are produced without hesitation or doubt, and

[18]Charles Hoole, *op.cit.*, in (Cressy 1980, 21).

[19]Stages 4, 5, 6, 7 are described in J.-B. de La Salle, *Conduite des écoles chrétiennes*, 1720 (Chartier et al. 1976).

[20]Chartier (2000).

> anyway without the child having to stop and think. Only then let him begin to construct words with the syllables themselves and form connected sentences with the words.[21]

This order continued through the Middle Ages. "The instruction of young children normally includes first the study of letters, then of syllables and then gradually leads to the knowledge of words and phrases",[22] so wrote Remi d'Auxerre in the tenth century. The Distichs, a compendium attributed to Cato the Elder[23] but dating from Marcus Aurelius, offers a collection of 'short readings' (couplets) for beginners, its use persisting even up to the nineteenth century in Spanish colonies.[24] However, for young monks the first book was the collection of the 150 psalms, "read the psalter" meant "learn to read". This is why the use of the word "psalter" continued for the ABCs printed in large number after the invention of the printing press, while the psalms were replaced by other prayers.

The reading texts are therefore very familiar; they are part of our common religious heritage, but this does not allow us to understand the sequence of exercises they supported. What did the pupil 'learn' by reading and re-reading a text that he already knew by heart like the Lord's Prayer? Could he learn to read using texts written in Latin, a language he did not know and would remain unknown by all those who were not destined to do long years of study? Why was it necessary to spell out each syllable before pronouncing it? To understand these practices we will study documents that shed light on the old methods. As we do not have direct testimony from the time, we will use two twentieth century experiences of learning carried out using the old methods.

2.2 Accounts of Learning

Autobiographies have the advantage of showing learning as a singular experience, told from the learner's perspective. By recounting the activities or the 'events' that characterised his schooling, the former student recreates the feelings he had for the exercises presented to him and whether he found them important or trivial, easy

---

[21]Quintilian, *The Orator's Education*, ed. and transl. by Donald A. Russell, Loeb Classical library, 2001, 77 and 79. "My aim is the education of the perfect orator," said Quintilian at the beginning of his famous *De Institutione Oratoria.* Learning how to read is dealt with in four paragraphs in the first of the twelve chapters making up book 1. Most of book 1 deals with the care that should be given to very young children, the choice of nannies, tutors and the language to be spoken and written. He thinks it preferable to start by teaching a child Greek, before moving on to Latin, which is less difficult. This means that little space is given to this question in the twelve books that make up the complete work.

[22]Rémi d'Auxerre (circ 850–910), in *Patrologie latine*, 131, 845, quoted in (Riché 1979, 223).

[23]Dioniso Caton, *Libri Minoris et primum Catonis disticha Moralia, com Antonij Nebriissensis annotattionibus. Apud Inclytam Apuli*, s.i., 1545.

[24]*El Sabio Catón, Avisos y ejemplos del Sabio Catón Censorino Romano*, reprint in Puebla de los Angeles, México, 1815, quoted by C. Castañeda et al. (2004, 35–66).

or difficult, interesting or boring. The risk is, of course, that all reconstructions are made *a posteriori*. Even when sincere, they are often filled with false memories as the author projects his current knowledge, judgements and feelings onto a far distant past. In the cases chosen, this risk is minimal. The first testimony (that of Jean-Paul Sartre) is based on an account set very early in the family memory, the second testimony (that of Hampaté Bâ) is corroborated by other observations: it concerns teaching methods still practiced today in Koranic schools.

2.2.1 How Jean-Paul Sartre Learned to Read

In *Les Mots*, Sartre tells of how he developed the desire to read. At that time,[25] there were picture books and illustrated alphabet primers,[26] but not storybooks for very young children. Books for young people were long stories, intended for older children who were already good readers.

It was one of these books that Anne-Marie, Sartre's mother, opened one day instead of making-up a story as she usually did, inventing the scenario as she went along. Much to his surprise, the young Jean-Paul saw her stock-still like a statue and speaking in a voice he was not accustomed to. He understood that "*it was the book speaking*". It was, he said, like a revelation:

> I came to prefer the prefabricated tales to improvised ones; I became sensitive to the unchanging sequence of words: they would return each time you read them, always the same ones and always in the same order, and I would wait for them. In Anne-Marie's stories, the characters lived haphazardly, as she herself did: they acquired fates. I was at Mass: I was present at the endless reiteration of names and events.
>
> Then I became jealous of my mother and decided to usurp her role. I seized upon a work called 'Tribulations d'un Chinois en Chine' and took it away to a box-room; there, perched on a folding bedstead, I pretended to read: my eyes followed the black lines without skipping a single one and I told myself a story out loud, taking care to pronounce every syllable. I was discovered – or I let myself be discovered; there were cries of admiration and it was decided that it was time I was taught the alphabet. I was zealous as a catechumen; I even gave myself private lessons: I climbed onto my folding bedstead with Hector Malot's 'Sans famille',[27] which I knew by heart, and, half-reciting, half-deciphering it, I went through every page, one after another: when the last was turned, I knew how to read.[28]

[25]Jean-Paul Sartre was born in 1905. After the premature death of his father, he and his mother Anne-Marie went to live with his maternal grandfather, Charles Schweitzer, who was a German scholar and author of several books on language teaching. The event took place in around 1908–1909 when he was between 3 or 4 years old, otherwise the family would not have spoken about his exceptional precocity.

[26]Le Men (1984).

[27]*Sans Famille* by Hector Malot was a successful novel (1878) and soon became a classic for young people.

[28]Sartre (1964, 16–17). Translated by I. Clephane (1964, 32).

This testimony from the 1900s gives some key elements to understanding how the methods used under the Ancien Régime could develop early reading skills. The first step: the child discovers how written stories differ from those transmitted orally, be they made-up or recounted from memory. The latter are subject to paraphrasing and random variations,[29] while the former remain the same with every repetition. The young listener is fascinated by this immutability and by the ritual particular to oral diction, "*taking care to pronounce all the syllables*." He spontaneously attributes a sacredness to the book ("*I was at Mass*") which is not given to oral stories. The second step: seeing the child copy his mother and pretending to read, he is "*taught the alphabet*", meaning the letters and the first elements of syllabication, by his family circle. The third step: the little boy can then teach himself to read out loud by himself, "*zealous as a catechumen*", thanks to a book that he has heard read so often that he knows it by heart. How does he do it?

Jean-Paul Sartre employed the method evoked but not described in the books quoted above: by reading/reciting, he is not trying to understand a story he does not know, he is deciphering one he already knows 'literally'.

He establishes the link between the spoken phrase held in his memory and its word by word written transcription; he breaks down each word, syllable by syllable, the same as we decipher, note by note, the musical score of an already-memorized melody. Training in this way allowed him to commit to memory a sort of 'mental dictionary' of the combinations of syllables and words,[30] which he would use as a tool for reading other texts. It is as if, at a time when the 'spelling method' was disappearing,[31] that Sartre rediscovered the role of the literary memory of a text in learning that comes to us from Antiquity. It allows the pupil to 'learn by himself', not by deciphering an unknown text in order to understand its meaning, but by discovering, word by word how to transcribe a text known by heart.

His family extolled young Jean-Paul's success as a glorious achievement, and half a century later he would recount the story that he had heard told so many times during his childhood and conjure up the image of his industrious, painstaking, and then triumphant re-reading of his favorite book.

---

[29]On the difference between oral and written accounts and the dichotomy between societies with and without writing, see (Goody 1987). The thesis that Goody defends is that literal repetition becomes the rule in societies with writing and that ethnologists often assume the consistency of oral accounts without having proof. He examined two versions that accompanied the same Bagré ritual held many years apart in Ghana and noted differences so great as to bring into question the reliability of the textual analyses carried out on corpora of myths collected from a single source.

[30]We find the same principle in the 'Global method' invented at the end of the nineteenth century, based on fairy stories the children 'knew by heart' (The Little Red Hen, The Gingerbread Man) and widely used by families. In class, the teacher either introduced an explicit procedure for analysing the words (the word method) or he waited for the children to find it intuitively and transfer it from known words to words they did not know (the whole word method) cf. (Banton Smith 1965 [1934]).

[31]Use of the spelling method declined from 1850. It is probable that Charles Schweitzer, Jean-Paul Sartre's grandfather who he lived with, learned to read in this way.

Once this initial step had been taken, he could throw himself without fear into the immense family library. Moreover, delighted by such precocity, his grandfather wanted the child prodigy to start in a lower grade at high school, but his spelling test was a disaster and this ambition had to be postponed: the reading virtuoso barely knew how to write.

A posteriori, the writer philosopher could see a premonitory clue in this self-learning: for him, "*words*" were bought to life through reading (hence the title of his autobiography, *Les Mots*). Did Sartre suspect that, just like him, thousands of little "*catechumen*" had learned to read by deciphering their catechisms? They had heard the catechism recited so often by the older pupils in class that they new it by heart. Like him they learned to read without knowing how to write.

2.2.2 *How Hampâté Bâ Learned to Read*

The second testimony, exactly contemporaneous to the first, sheds light on some other aspects of this traditional way of learning, in a very different cultural setting: not that of a Parisian family from the academic world, but a rural family in an African Muslim society. Amadou[32] received his first lesson in Mali, in 1908, when he was between 7 and 8 years old. After the welcoming ritual, the teacher made him say the *shahada* ("there is no God but Allah and Mohammed is his prophet").

> Then he took the board and wrote out seven letters of the Koranic alphabet [...]. He made me hold the board so that the top rested on my left forearm and the bottom rested on my right thigh. I was ritually prepared to receive the teaching of the sacred book. With religious respect, Tierno Kounta traced each of the seven Koranic letters with his own right index finger, each time saying out loud the name the Peuls had given it. These seven letters were those that make up the Koranic formula Bismillah (in the name of God) that we find at the top of each sura in the Koran and that Muslims say before each important act in their life.
>
> Tierno Kounta repeated the lesson for me seven times, and seven times I repeated it after him, after which he sent my parents away. I had to stay in a corner of the courtyard and repeat the lesson four hundred eight times, following the letters with my finger. This took me about two hours [...]. Once back home, full of pride in knowing my first lesson, I deafened everyone shouting it out over and over again. It took nothing less than my father's intervention to silence me.[33]

[32]After attending a Koranic school, then a French school, Amadou Hampâté Bâ (1900–1991) occupied several posts in colonial administration. In 1942, he was appointed to IFAN, (French Institute of Black Africa) in Dakar by Theodore Monod and worked in the 'Ethnology' section. He embarked on a collection of oral traditions in Senegal, Guinea, Niger, Upper Volta, Sudan, Mauritania and northern Cote d'Ivoire. The documentation of these surveys makes up the Amadou Hampâté Bâ archive in Dakar. From 1962 to 1970, he was a member of UNESCO's executive council.

[33]Hampaté Bâ (1991, 196–197).

This introduction to reading also illustrates the uses invoked by the old *primers:* ritual making the writing sacred, memorising the formula to be read, spelling out each letter and repeating the text until it is known perfectly. The first difference to Sartre's account; this comes from training in memorization, imposed as a constraint, with which the pupil had to comply, whatever the cost (408 repetitions!). The second difference; the first lesson to be learned was not a story but a formula the child had heard said a thousand times by those close to him. This *Bismillah* ('in the name of God') is the equivalent of the formula presented in all the Christian ABCs ('In nomine Patri et Filio . . . ') where each lesson started with the ritual Latin formula that accompanies the sign of the cross.[34] The pupils at Christian schools, like those at Koranic schools, noted that to read is to pray: All this is poles apart from Jean-Paul Sartre's cultural world.

This account also allows us to understand the convenience of individual teaching even in a school having numerous students. Amadou was taken to school the moment his parents felt he was ready for the rite of passage ritual to 'bury his early childhood': there was not a collective entry to the school year; you could join the group at any time. During each lesson, the child had to follow the teacher's finger on the board and, at the same time, pronounce each letter and repeat its name (in the local language and not in Arabic) before reading the word. The pupil only advanced onto the next sura when the teacher was satisfied: it was impossible for the pupil to be inattentive, impossible for the teacher to be unaware of what the pupil knew. As the pupils were all then busy working by themselves, for lengths of time that varied according to the length of the text and the rapidity of their memory, the teacher could, without collective progression, guide each pupil from sura to sura at each pupil's own rhythm.[35] This is what Hampâté Bâ shows by describing the continuation of his education that took place later with another teacher, Tierno Bokar:

> Niélé woke me before sunrise. I quickly washed myself, said my morning prayer, and then ran to school where my board was waiting for me, still bearing the Koranic text I had written the day before. I sat myself down in a corner and recited the text out loud to learn it by heart. Every student roared out his text at the top of his voice, without a care for anyone else and making an indescribable din, which strangely, didn't bother anybody. At around

[34]The cross drawn at the start of each alphabet explains the popular name *croisette* (little cross) and *Croix-de-par-Dieu* (God's cross) in French, and *crisscross* (Christ's cross) *row* in English.

[35]No matter where children work with their parents, be it in town or in the country, it is very difficult to impose punctuality and attentiveness, as J-B de La Salle testified in the seventeenth century, and as appears in UNESCO reports on developing countries in the twenty-first century who are all supporters of 'modern' collective and simultaneous education. What is disturbing to the contemporary paradigm is that teaching may be collective (with the teacher in charge of a group of children), while at the same time remaining individual (each pupil working through the same book at their own pace). In the 'simultaneous mode', the teacher has a group of pupils working simultaneously on the same page of the book and insists that everyone does the same exercise at the same time.

> seven o'clock, if I knew my text well I took my board and ventured up to Tierno. He usually stayed in the vestibule of his home, more rarely in his bedroom. "Moodi [Master]! I said to him, I have learned my lesson". I squatted down next to him and recited my text. If he was satisfied I could go and wash my board and write new verses on it, from an example he gave me. Otherwise I kept my lesson from the day before and revised it until the next day [. . .] After copying the new text, I presented it to Tierno. He corrected the text, then read it out loud, while I followed with the tip of my index finger. Once back in my corner, I repeated the text ten or fifteen times, which took me until eight o'clock in the morning. Tierno then gave me permission to return home.[36]

While the school day was short (Amadou was free after 8 o'clock in the morning), the duration of the schooling depended on the time the pupil took to finish working through the Koran in its entirety. This event was marked with a solemn ceremony during which the son demonstrated his ability to 'recite the Koran' to his father. What does this mean? Hampâté Bâ teaches us that 'reciting your text' means following, with the tip of the index finger, the text of a sura copied on the board and reading out loud without taking your eyes off the words as you read them. We find the sequence of actions mentioned again; reading/recitation out loud of a memorised text aided by a written support, copying a new text, 'model reading' by the teacher and re-reading by the pupil 'following with his finger', the re-reading/repetition of the text without the teacher's help until it is committed perfectly to memory.

Knowing the lesson well means being able to co-ordinate the eyes, the finger and the voice, with neither error nor hesitation. Each recitation consolidates the association of the letter with its sound and emphasizes, if it is needed, that it is God who speaks in the Book and not the reader. Obviously, while working one-to-one, the teacher may give brief explanations, translate the sura and comment on it, the principal objective is to ensure that it is known literally. It is in this way that the same sura can be found afterwards in the sacred book, that it can be recited at the mosque and can be quoted from memory when necessary.

However, this testimony does not enable us to know exactly when the pupil becomes able to understand a sura as he reads it for the first time. Moreover, there is nothing to indicate that Tierno Bokar encouraged this autonomous reading. It is only at the second or third reading of the Koran (the stage Hampâté Bâ's elder brother found himself) that the pupil becomes an 'autonomous reader' (this does not mean freedom of interpretation). For the Koranic teacher, a good student certainly did not devour books in the way the young Sartre did.

We can, from these two situations, try to reconstruct the contexts which gave the texts meaning and made them effective for beginners of the sixteenth century, before looking for the 'contextual changes' that lead to their being abandoned in the nineteenth century.

---

[36] *Ibid,* 230–231.

2.3 Reading and Memory: From Literal Memory to Semantic Memory

2.3.1 The Different Levels of Contextualization

The above testimonies both show the complex interaction of three 'contextual' levels, found in different analytical levels: the didactic context of the exercise, the learning situation: in the family or in school, and the cultural environment in the social space.

1. The didactic context allows us to understand how the different learning activities work together with the text. 'Context' is used in the strict sense of the word; it is impossible to make the inference from the analysis of the content of the text alone. In both cases the children learn the principle of syllabication and this basic deciphering technique gradually consolidates and widens as each child advances through his book. The order of the letters/syllables to be deciphered depends therefore on the text, while in modern school textbooks, the opposite is the case.[37]
2. The learning situation allows us to understand how the 'individual method' of teaching is compatible with having responsibility for a group of children (with a single child, the method can only be 'individual'). The pattern of alternating between (short) periods for lessons and (long) periods for study is very convenient for the adult who plays his role as guide and supervisor 'on demand'. It is convenient for parents who follow academic progress closely and always know where there child 'is up to' in their book. It is also very convenient for the schoolchild: he is not disturbed by others who dawdle or are absent, he can advance very quickly or very slowly without the pressure of his class-group, he is sure his progress will be monitored face-to face by his teacher.
3. The cultural environment helps us to understand the importance given (or not) to the literal memory of texts, not only in school, but more generally in the society. In the case of Jean-Paul Sartre, the text known by heart served only as a temporary way of learning the mechanisms for deciphering, before reading a lot, quickly and alone: semantic memory makes literal memory redundant. For Hampaté Bâ, 'being able to read', was 'knowing how to recite', it was being able to remember a text literally and there was no question of 'reading without the teacher'.

These different contextual levels are not independent of each other, but form a 'system': changing one can unbalance the whole. If the teacher taught collective lessons, the same pace would be imposed on everyone and the slower pupils would fall behind. If *primers* offered 'easy texts for beginners', the pupils would be held

[37]Modern textbooks are characterized by a progressive study of sounds and how they are written, using specially written texts (/a/: "the cat and the rat ran; the cat ran at the rat", *McGuffey's Eclectic Primer*, New York, American Book Co, 1909, p. 8). Banton Smith (1965 [1934]).

back in a childish school culture, instead of entering into the written culture of adults. If the teacher did not demand written reproduction of the text, but was content just to check that the child understood it (that is to say, the child has 'processed' the information' well), all the work of self-learning by repetition would be called into question.

The central question is, therefore, to understand what this memorization of texts corresponds to in a society that uses writing. In effect, much as it appears necessary to cultivate oral memory in a society without writing, it appears pointless to do this when writing is available as 'artificial memory' and textual supports are abundant, as has been the case since the invention of printing. Why learn a text by heart when we have it in writing in front of us? Why hasn't the presence of books put an end to the memorization of texts? It is this point we must investigate as a matter of priority by (temporarily) leaving these two testimonies behind.

2.3.2 *The Cultural Context: Literary Memory and Intensive Reading*

The two accounts highlight two different concepts of the act of reading. Sartre's family wanted to awaken the child's curiosity and his appetite for stories, so that he could read any book, be it instructive or amusing, easily and by himself: it is the idea of 'extensive' or 'generalized reading'[38] which prevails in our schools and in our society today. On the contrary, Tierno Bokar's role was to deepen the understanding of just one book (the Koran): the 'restricted' but 'intensive' reading, for which he, the teacher, is the authorised guide, is at one and the same time academic study and spiritual instruction.[39]

This certainty prevailed in Europe from the sixteenth until the eighteenth century. Reading did not consist in processing information, but in learning and memorising word by word. Learning by rote was not a tool to be used only by children or the uneducated. It was employed in the training of the clergy at all stages of the curriculum, and many 'art of memory'[40] techniques were undertaken to facilitate the accumulation and use of knowledge.

When the Protestant Reformation tried to bridge the gap between the great number of *illiterati* and the small number of clerical *literati*, this mode of reading became

[38]The opposition "restricted vs. generalized reading" comes from Jack Goody (1977). For (Engelsing 1974), the distinction between 'intensive' and 'extensive' reading was the basis of a reading revolution in the eighteenth century.

[39]Hampaté Bâ (1980). Hampâté Bâ considered him his spiritual master and visited him regularly until his death. One could, of course, link this way of teaching, continually analysing and commenting on the same memorised text, to the teaching by Talmudic teachers and medieval school teachers (Carruthers 1990).

[40]Yates (1966).

an objective for all Christians. As Luther wrote,[41] "nothing should prevent believers from having direct access to the Scriptures, and laymen should be able to read the truths of faith by themselves, in common language". The Catholic Church followed in his footsteps after the Council of Trent,[42] and every catholic had to become a reader in order to recite prayers without making mistakes. Literal repetition avoided errors in dogma, hence the importance of the words used in prayers and catechisms in this time of schism, religious persecution and theological conflict.

How could the pupils understand what they read? In schools, religious texts were introduced with brief explanations during catechism lessons. In his *Small Catechism* (1529), Luther explained how the head of the family should, in a simple way, teach his household the Ten Commandments, the Creed and the Lord's Prayer. So, when children used a prayer to spell out letters and to separate words into syllables, the text had already been explained. For instance, for the Lord's Prayer:

Give us this day our daily bread.

- What does this mean?
- Answer: God gives daily bread, even without our prayer, to all wicked men; but we pray in this petition that He would lead us to know it, and to receive our daily bread with thanksgiving.
- What is meant by daily bread?
- Answer: Everything that belongs to the support and wants of the body, such as meat, drink, clothing, shoes, house, homestead, field, cattle, money, goods [etc].

What seemed 'natural' in 1550, seemed 'impossible' in 1850. Since the early nineteenth century, educators have made the distinction between learning by rote and reading, and between memorization and literacy: "*Because oral cultures must obviously depend on memory and hence value memory highly, such valorisation has come to be seen as a hallmark of orality, as opposed to literacy. This has led to a further assumption that literacy and memory are 'per se' incompatible*",[43] wrote Mary Carruthers, whose research belies this view. As a result of this prejudice, 'modern' educators have argued against 'mechanical recitation' that a literal knowledge of texts would produce, as being inconsistent with 'intelligent reading' that would allow understanding.

2.3.3 From Intensive to Extensive Reading

This instruction in reading was in line with a cultural transmission based on the Scripture. What happened when secular culture became pervasive in

[41] Martin Luther, *M.L.O.*, IX, ed. Labor and Fides, Genève, p. 111. After the Peasants' War (1525–26), Luther did not encourage adults or children to read alone, but only under the supervision of a pastor (Gilmont 2003).

[42] The Council of Trent (1545–63) condemned protestant heresies, affirmed the major points of truth in catholic faith and supported the publication of catechisms (Julia 2003).

[43] Carruthers (1990, 10).

eighteenth-century publishing? Besides traditional reading, which was collective, solemn, slow, repetitive and restricted to religious texts,[44] a new way of reading emerged, which extended to secular texts and was individual, fast, pleasurable, and favoured novelty. This new way of reading[45] did not imply the disappearance of earlier ones, as many more people had been taught at elementary school and could read the Bible or their Psalter. But reading for pleasure was in contradiction to the "intensive reading" which educators used to teach. The emergence of "extensive reading" correlates with the growth in the production of gazettes and novels.[46]

Anybody, man, woman or child, servant, soldier or artisan could thus become a competent reader, without becoming a "professional user" of written material. While religious literacy was at one time a controlled, collective practice, dominated by the clergy, novels and newspapers now allowed reading to be practiced on an individual level, in which books were chosen from one's own social network or through recommendations published in the press.

This change exerted an influence on schools. The way educators understood reading, its patterns, aims and impact was transformed. To prepare their pupils for 'extensive reading', they had to first teach them all the syllabic combinations. Only when they had succeeded in mastering all of these, were children allowed to test their skills on a sentence, or a text they had never heard before. An unknown text seemed the 'natural way' to discover the meaning of the sentences through the syllabication of the words, i.e. to read (according to what 'reading' is to-day). But the spelling process, effective for learning the written code, was not as effective for understanding the meaning of unknown texts. When the old 'spelling method' was used to read a new text, its inadequacy was evident: it became a barrier that caused a growing number of children to fail in learning to read. Teachers became more and more critical, as the children's failure seemed to stem from the 'bad' old method and not from its use in a "new context". As pupils could no longer work alone to decipher a text they knew by heart it was essential there was a teacher to help them (as Rapet wrote).

2.3.4 The Learning Situation and Didactic Context: Learning to Read with and Without Learning to Write

Another important difference between the two testimonies is the role played by writing: This time Jean-Paul Sartre is found on the side of archaism, Hampâté Bâ with modernity. In effect, in order to read the suras he had to learn, the young Amadou copied them out onto a wooden board that was washable like a slate. When he started his instruction, he had the required dexterity and his tools (reeds, artisanal

[44]Hall (1996). The Pietists gave a good example of "intensive reading" of the Bible.
[45]Wittman (2003).
[46]Lyons (1987).

ink and a wooden board) were unsophisticated, easy to use and very inexpensive. Eventually, he learned to read and write in the same script, Arabic italics, while Jean-Paul Sartre only knew how to read print: this precocious reader failed a writing test for beginners. This was also the case for seventeenth century pupils in (Christian) Europe, who learned to write after having learned to read, while under Koranic teaching methods, the two were taught together, as happens in all schools today. What are the causes and consequences of these differences?

The combination of several factors explains this separation between learning to read and learning to write. Social reasons: weak external pressure. For most parents, being able to write was not a priority (many took their children out of school as soon as they could read) and the Church's only concern, as we have seen, was to facilitate the reading of religious texts. Only the children of the educated classes and the children of shopkeepers and craftsmen 'needed' to know how to write.

The learning situation: the physical teaching conditions (buildings, writing supports and, instruments) presented major obstacles. 'Writers' needed tables, and when the classrooms only had benches, the students rested their books on their knees. Paper was expensive and had to be used sparingly. Goose quills were difficult to use and so difficult to sharpen that "the writers" called on the services of an adult, a impossible task for a teacher already in charge of keeping discipline and teaching reading. The learning of writing was therefore reserved, for the end of their studies, for those who had the capacity or the need. Consider the case of Henry Facy, a good student who had first held a quill pen at 14 years old, a few months before his death in 1667. As his schoolmaster testified, "Henry did learn what was taught him very well, as well as could be expected from a boy of his age, about 13 or 14 years old", and "he was able to write a good hand after a copy and could read in the Bible". He had been learning to write for 5 or 6 months at the time of his death and "had only recently achieved full literacy."[47]

Didactic context: The alphabets and lists of syllables that started in primers were printed in block capitals and lower case characters. Learning to write was to learn a different script, very different to the first. Because the up-strokes, down-strokes and ligatures (continuous strokes) of cursive script could be confusing to the untrained eye, a person might be able to read the 'block letters' (cast in an iron mould, then printed) without being able to read manuscript. Hence, the role of the 'Characters of Civility' whose rounded italics resembled handwriting; giving a smooth transition from one form of writing to another. The written techniques for basic arithmetic (calculations, keeping accounts) also required the use of a quill. In urban schools, the teachers usually used manuscripts (contracts, leases, wills) to teach the children of craftsmen and merchants how to decipher secular documents, whose form and content were very different to those of religious texts. But only reading (and not writing) was necessary for everyone: to learn the catechism and receive communion (catholic) or confirmation (protestant) and join adult society.[48]

---

[47]Cressy (1980, 29).

[48]Chartier (2007, ch. 1 and 2).

Spelling out loud, letter by letter, is a direct consequence of this de-coupling of reading and writing. In effect, in copying out a sura to read, the pupil reproduced it letter by letter. Producing these copies everyday practiced the act of reproducing letters on the page, fixed the sequential order of the letters in the memory and automated the association of letters and sounds. How do you give someone who learns to read without writing as clear an appreciation of the constituent elements of each word? The solution that came naturally was to spell the letters out loud, as witnessed by this school dialogue recounted by William Coote[49]:

John: How do you write 'people'?
Robert: I cannot write
John: I mean not so, but when I say' write', I mean 'spell', for in my meaning they are both the same.
Robert: Then I answer you, 'p-e-o-p-l-e'.

The 'spelling' method also played a role by orally replacing the act of writing. We have circumstantial evidence: the 'spelling' method declined nearly everywhere throughout Europe from the 1850s, when new tools disrupted the teaching of writing.[50] Holding slates on their laps and using chalks allowed young writers to practice tracing the sticks and the rounds (that is, the 'i's and the 'o's), metal nib pens that didn't rust and were practically indestructible replaced goose quills. As these could be used from 6 years old and the low price of pulp-based paper made exercise books affordable to all, the introduction to writing could take place at the same time as the introduction to reading. Cursive writing exercises rediscovered the progression invoked by Quintilian[51] in the first century. The simultaneous reading-writing method, re-invented at the end of the nineteenth century, did away with the spelling out of "B-A BA": which passed into posterity under the name of the 'syllabic method' as each syllable is pronounced as a single feature.

To the question, "Do you know how to read and write" posed ritually during conscription, fewer and fewer soldiers answered, "Only to read"; school attendance was compulsory under the IIIrd Republic and everyone learned reading and writing at the same time. Consequently, middle-class children who, in the 1900s were 'put to letters' by their mothers at the age of four, were the last to learn to read without learning to write at the same time. They had not yet acquired the manual dexterity necessary to write the letters and they were too young to have to write out the calligraphic lines of letters and syllables. They were then, like Jean-Paul Sartre, the last witnesses to a vanishing teaching method.

---

[49]Edmund Coote, *The English Schoole-Master*, 1624, quoted in (Cressy 1980, 21).

[50]On teaching materials and "reading methods", Anne-Marie Chartier 2007, *op. cit.*, chapters 5 et 6.

[51]The young Romans evoked by Quintilian learned reading and writing in cursive script at the same time and used a bone or metal stylus on a wax tablet, the equivalent of our slates: the wax was heated to erase the copied text.

2.4 Conclusion

We have seen, therefore the teaching methods (repetition *vs.* explanation, constraint *vs.* motivation, memorization *vs.* understanding) that historians of education have often said ensue from educational choice (authoritative *vs.* liberal pedagogy) only make sense in reference to the learning goals: what type of reader do we want to produce? These goals do not need to be enunciated today because they are imposed 'naturally' as a result of the role and status of writing in the world around us. It is through the teacher's intervention that this imposition is carried out, through the writings definitively associated with the beginner's first acts of reading. How then to qualify these complex interdependencies, linking the schoolchild with his textbook open in front of him in his micro-world, with the goals of his teacher, representing the 'written world' and guarantor of its value?

The philosopher Pierre Hadot, rejecting the "*acronyc*" readings of ancient philosophy, declared, "*Any work should be placed in the praxis from which it emanates.*"[52] In using the word "*praxis*", he aims for more than just 'the context', that is, the discursive corpus in which the work lies. The *praxis* evokes "*the specific conditions when they were written; on one hand the spiritual conditions, meaning the philosophical tradition, rhetoric or poetic, and on the other the physical conditions, meaning the scholastic and social settings, constraints coming from the physical writing medium, historical circumstance.*"[53]

The contemporary reading *praxis* is far removed from these uses and even today's expert readers could not imagine that they ever existed: they view the casual act of reading as the processing of information, motivated by its utility or entertainment value. The "*Systematicity*" of a discourse seems to them to be an obvious sign of an aspiration to conceptual theorization, while Pierre Hadot detects (for Antiquity) a pedagogical training method.[54] In trying to reconstruct the "*praxis*"[55] surrounding the old methods of teaching reading, we have, ourselves, also rediscovered reading as a mnemonic device. The introduction to reading Christian texts was enlightened by the account of a literary initiation (Jean-Paul Sartre) and instruction by reciting

---

[52]Hadot (2001, 116).

[53]Pierre Hadot, *ibid.*, 149.

[54]We can give as an example for Christianity in Antiquity: the short treatise on dogmatic theology that is the Apostles' Creed or Credo. Its wording was modified over time by the Councils of Nicaea (325) and Chalcedon (451), but it cannot be understood outside its specific uses; catechistic (summaries memorised by each convert), sacramental (formulae spoken during baptisms), liturgical (profession of faith by the congregation at each act of worship).

[55]It could have been possible to invoke other models, if the object of the analysis had been learning to read as a process in which the identification of the medium is made impossible to ascertain from the message, by evoking the construction of a habitus (Bourdieu), inculcation devices (Foucault), the assimilation of schema (Piaget), or the internalisation of cultural tools (Vygotski), etc. The term 'praxis' employed by Pierre Hadot, instead, puts the focus on what was the object of this analysis: not the 'process of incorporating' the texts, but the 'physical and spiritual' environment which has made reading exist historically and pedagogically as a mnemonic device.

the Koran (Hampâté Bâ). Undoubtedly, we could have made use of an ancient pagan religion or another book. We have thus raised some enigmas surrounding the introduction to reading for beginners: from his first steps in writing, a still-illiterate child, had to learn how to practice a spiritual exercise like his learned teacher. That leaves the question, but it is beyond the scope of this contribution, what is 'in practice' a spiritual exercise?

References

Banton Smith, Nila. 1965 [1934]. *American reading instruction*. Newark: International Reading Association.

Carruthers, Mary. 1990. *The book of memory, a study of memory in medieval culture*. Cambridge: Cambridge University Press.

Carvallo, Gugliemo, and Roger Chartier (eds.). 1999. *A history of reading in the west*. Cambridge: Polity Press.

Castañeda, C., L.E. Galván, and L. Martínez-Moctezuma (eds.). 2004. *Lecturas y lectores en la historia de México*. México: CIESAS.

Chartier, Anne-Marie. 2000. Réussite, échec et ambivalence de l'innovation pédagogique : le cas de l'enseignement de la lecture. *Recherche et Formation, Innovations et réseaux sociaux* 34: 41–56.

Chartier, Anne-Marie. 2004a. Introduction: Teaching literacy: What practices, when and why? In *Handbook of children's literacy*, ed. Terezinha Nunes and Peter Bryant, 493–498. Dordrecht/London: Kluwer Academic Publishers.

Chartier, Anne-Marie. 2004b. Teaching reading, a historical approach. In *Handbook of children's literacy*, ed. Terezinha Nunes and Peter Bryant, 511–538. Dordrecht/London: Kluwer Academic Publishers.

Chartier, Anne-Marie. 2007. *L'école et la lecture obligatoire*. Paris: Retz.

Chartier, Anne-Marie. 2009. The teaching of literacy skills in Western Europe an historical perspective. In *The Cambridge handbook of literacy*, ed. David Olson and Nancy Torrance, 449–467. Cambridge: Cambridge University Press.

Chartier, Anne-Marie, and Jean Hébrard. 2001. Literacy and schooling from a cultural historian's point of view. In *Literacy and schooling from the cultural Historian's point of view*, ed. Thomas Popkewitz, Barry M. Franklin, and Miguel Pereyra, 263–288. New York: Routledge.

Chartier, Roger, Dominique Julia, and Marie-Madeleine Compère. 1976. *L'Éducation en France du XVIe au XVIIIe siècle*. Paris: SEDES.

Chervel, André. 2006. *Histoire de l'enseignement du français du XVIIe au XXe siècle*. Paris: Retz.

Compère, Marie-Madeleine. 1985. *Du collège au lycée 1500–1850*. Paris: Gallimard.

Cressy, David. 1980. *Literacy and the social order: Reading and writing in Tudor and Stuart England*. Cambridge: Cambridge University Press.

Demoustier, Adrien, and D. Julia (eds.). 1997. *Ratio studiorum* [1599]. Bilingual ed. In Latin and French, Adrien Demoustier and Dominique Julia. Paris: Belin.

Engelsing, Rolf. 1974. *Der Bürger als Leser. Lesergeschichte in Deutschland, 1500–1800.* Stuttgart: Metzler.

Gilmont, Jean-François. 2003. Protestant reformations and reading. In (Cavallo and Chartier 1999), 213–237.

Goody, Jack. 1977. *The domestication of the savage mind*. Cambridge: Cambridge University Press.

Goody, Jack. 1987. *The interface between written and the oral*. Cambridge: Cambridge University Press.

Graff, Harvey J. (ed.). 1981. *Literacy and social development in the west*. Cambridge: Cambridge University Press.
Grendler, Paul F. 1989. *Schooling in renaissance Italy. Literacy and learning 1300–1600*. Baltimore/London: Johns Hopkins University Press.
Hadot, Pierre. 2001. *La philosophie comme manière de vivre*. Paris: Albin Michel.
Hall, David D. 1996. *Cultures of print. Essays in the history of the book*. Amherst: University of Massachusetts Press.
Hampaté Bâ, Amadou. 1980. *Vie et enseignement de Tierno Bokar. Le Sage de Bandiagara*. Paris: Seuil.
Hampaté Bâ, Amadou. 1991. *Amkoullel, l'Enfant Peul*. Arles: Actes Sud.
Jimenes, Remi. 2011. *Les caractères de civilité. Typographie et calligraphie sous l'Ancien Régime*. Méolans-Rével: Atelier Perrousseaux.
Julia, Dominique. 2003. Reading and counter-reformation. In (Cavallo and Chartier 2003), 238–268.
Le Men, Ségolène. 1984. *Les Abécédaires français illustrés du XIXe siècle*. Paris: Promodis.
Lindmark, Daniel. 2004. *Reading, writing, and schooling: Swedish practices of education and literacy, 1650–1880*. Umeå: University of Umeå.
Lyons, Martin. 1987. *Le triomphe du livre. Une histoire sociologique de la lecture dans la France du XIXe siècle*. Paris: Promodis.
Magalhães, Justino. 1994. *Ler e Escrever no Mundo Rural do Antigo Regime. Um Contributo para a História da Alfabetização e da Escolarização em Portugal*. Braga: Universidade do Minho.
Mathews, Mitford. 1966. *Teaching to read, historically considered*. Chicago/London: University of Chicago Press.
Monaghan, E. Jennifer. 2005. *Learning to read and write in Colonial America*. Amherst: University of Massachusetts Press.
Riché, Pierre. 1979. *Écoles et enseignement dans le Haut Moyen Âge*. Paris: Aubier.
Sartre, Jean-Paul. 1964. *Les Mots*. Paris: Gallimard. Trans. Irene Clephane. London: Hamish Hamilton.
Strauss, Gerald. 1981. Techniques of indoctrination: The German reformation. In (Graff 1981), 96–104.
Viñao Frago, Antonio. 1999. *Leer y escribir, Historia de dos prácticas culturales*. Mexico: Fundación Educación, voces y vuelos, I.A.P.
Wittman, Reinhard. 2003. Was there a reading revolution at the end of the eighteenth? In (Cavallo and Chartier 1999), 284–312.
Yates, Frances A. 1966. *The art of memory*. Chicago: University of Chicago Press.

Chapter 3
Teaching and Learning Medicine and Exorcism at Uruk During the Hellenistic Period

Philippe Clancier

Abbreviations

BaM	Baghdader Mitteilungen
BM	British Museum
CAD	Chicago Assyrian Dictionary
CAMS	*Corpus of Ancient Mesopotamia Scholarship*. http://oracc.museum.upenn.edu/cams/gkab/corpus/
DN	Divine name
GKAB	*The Geography of Knowledge in Assyria and Babylonia: A Diachronic Analysis of Four Scholarly Libraries*. http://oracc.museum.upenn.edu/cams/gkab
KAR	Keilschrifttexte aus Assur religissen Inhalts
PN	Personal name
SAA	State Archive of Assyria
SpTU	Spätbabylonische Texte aus Uruk
UVB	Vorläufiger Bericht über die Ausgrabungen in Uruk-Warka

3.1 Introduction

Mesopotamian societies in the first millennium BCE left traces of learning and teaching activities and provide evidence of both a theoretical pedagogical program and practical exercises.

P. Clancier (✉)
UMR 7041 Arscan, Université Paris 1 Panthéon-Sorbonne, Paris, France
e-mail: philippe.clancier@mae.univ-paris1.fr

A. Bernard and C. Proust (eds.), *Scientific Sources and Teaching Contexts Throughout History: Problems and Perspectives*, Boston Studies in the Philosophy and History of Science 301, DOI 10.1007/978-94-007-5122-4_3,

The most revealing material known today comes from the circle of "scholars" working in the temples. The word "scholar" refers to the scribes who were masters in their field of knowledge. Those fields were exorcism/medicine, astronomy/astrology, divinations, rituals, mathematics and so on all disciplines being held by priests. The most able of them were recognized by the title of *ummânu* which, in that context, could be translated by "master of knowledge".[1]

These scholars wrote tablets showing their teaching practices. In the first millennium BCE, a master mainly taught students (the first being his sons)[2] in his own house (no schools are attested). Such private houses have sometimes been unearthed and, not surprisingly, the written material found in the buildings shows that the disciplines taught were scholarly and concerned divination, medicine, mathematics, astronomy, etc. and also legal matters (such as writing contracts and so on). It is on this specific social class and these learning practices that this article will focus, due to the availability of documentation.

The possibility of discovering school texts and exercises in well managed archaeological excavations, linked with indentified buildings and library owners, is particularly important in reconstructing a social, intellectual and, to a certain extent, the political context that allows modern historians to enter, at least a little, into specific cultural practices. Research in the Ancient Near East depends heavily on the high quality of the archaeological registration of tablet locations, the context of their discovery (private house, temple or palace) and their status (still in use when the library or the archive was abandoned or already rejected). Moreover, this archaeological data enables Assyriologists to work on real libraries and archives (meaning that the tablets were physically kept together).[3]

Tablets from the first millennium BCE, especially from the second half of that period, unearthed in such contexts, show both practical and theoretical teaching and learning practices for a particularly important profession of the time: the exorcist-doctor.[4] Data for their training is found in three forms: student's exercises, some master's "lecture" notes and a theoretical text, the so-called *Āšipu*'s handbook.

As the "handbook's" authors claimed[5] that their text was a pedagogical program, a comparison with the actual remains of a house where exorcist-doctors were trained could allow the confirmation, or not, of whether its orientation influenced the training of scholars in the late first millennium BCE. Such a comparison would allow us to interpret the different types of texts unearthed in the houses. So, our object will be to present both the theoretical and practical aspects of the training of the exorcists-doctors.

---

[1]The Assyriologists used to translate *ummânu* by scholar. It is this habit that we will follow here.

[2]There are no attestations of girls being trained in scholarly disciplines in the first millennium BCE.

[3]It is still possible to work on virtual archives or libraries using the names of the main owner of a group of texts but this method is not as good as the one consisting of studying a collection of texts which were found together and thus kept together.

[4]*āšipu*.

[5]Line 1 of the obverse and 7 of the reverse (Appendix 1).

For that purpose, we will, as a first step, take a closer look at the available documentation. Then we will see the learning practices from the handbook's theoretical approach through to the practical exercises carried out by the students. After this, we will see what teaching meant when compared to the "handbook", through the specific example of the "commentary" texts and their references to oral traditions.

3.2 The Urukean Documentation

Documentation found in the southern Mesopotamian city of Uruk is one of the best sources for studying learning and teaching practices in the first millennium BCE. The reasons for this are the very good archaeological context for the way texts used in such a perspective were kept and the abundance of material. Moreover, all this material can be used to see if the *āšipu*'s handbook really had an influence.

3.2.1 The Archaeological Data

Archaeologists in Uruk have unearthed a house where two scholarly families lived, but not at the same time, the first at the end of the Achaemenid Empire and the second during the Hellenistic era (that is from the end of the fifth century to the end of the third).[6] What they discovered is a locus where a house was built and then rebuilt where two different families lived at two different periods.[7] The first family were descendants of Šangu-Ninurta, who may have arrived from Babylon during the fifth century as the family is not recorded in Uruk before this time.[8] The descendants of Šangu-Ninurta may have lived in the house between 445 and 385 B.C.[9] Legal and scholarly tablets unearthed in their home attest the activities of three generations: Šamaš-iddin, his sons Anu-ikṣur (the most documented scholar in the collection by the tablets he wrote) and Rimut-Anu and, for the last generation, his grand-son Anu-ušallim.[10] The second family are descendents of Ekur-zakir and are attested in the city long before the Hellenistic period. They lived in the same

[6]Oelsner (1986, 77–97) and Pedersén (1998, 207).

[7]For the excavations, see (Schmidt 1972, 56; Hoh 1979, 28). The excavations took place between 1969 and 1972.

[8]Indeed, the only other Šangu-Ninurta family known is attested in Babylon from Nebuchadnezzar to Xerxes (Jursa 2005, 71–72). It is possible, even if it is only an hypothesis, that the family moved from Babylon to Uruk because of the trouble occurring to the old Babylonian notability during the reign of Xerxes. For the question of revolts at the beginning of the Xerxes' reign, see (Waerzeggers 2003–2004).

[9]Clancier (2009a, 59).

[10]See the complete genealogy in Appendix 2.

archaeological locus but after the Šangu-Ninurta family and so on a later occupation level. The Ekur-zakir family lived in the house from as early as the reign of the Hellenistic King Philip III Arrhidaeus (323–316) and possibly earlier (but in any case, after 350) until 229 B.C.,[11] and left legal and scholarly texts in the locus. Four generations have been identified from texts in this house: Ištar-šum-ereš, father of Iqiša (the principal owner of the collection), grand-father and homonym of Ištar-šum-ereš, the father of Ša-Anu-iššu, the last attested descendent of Ekur-zakir in the locus.[12] We will refer to the locus containing the two houses as the *āšipu*'s houses, an *āšipu* being an exorcist-doctor.

Two different collections of tablets have been found in the *āšipu*'s houses: one from the earlier Šangu-Ninurta family, the second from the later, Ekur-zakir family. The two collections were completely separate as the tablets of the first family remained undisturbed in the older house that had been abandoned for some time before the descendants of Ekur-zakir settled and built new buildings in the same locus. Both contain legal (mostly contracts) and scholarly texts (literary compositions, hymns, prayers, omens, exorcisms, medical recipes, mathematics, astrology and astronomy and so on). The texts were written on clay tablets. We only have this kind of writing media but it is well known, from colophons and some archaeological remains, that wax tablets were also extensively used for scholarly texts. The absence of wax tablets, destroyed by the passage of time, makes the reconstruction of the complete collections problematic, but it is to be noted that there were no differences between the types of scholarly text written on clay or on wax. We do not have texts on leather or papyrus and, in this case, it seems that the texts were very different from those on clay or on wax.[13] Indeed, they were, most likely, from the Aramaic or Greek traditions and so do not concern the present study.[14] Of course, not all the clay tablets which were once part of the two collections have been found but what we have is still a impressive corpus: of the 407 scholarly tablets (we are not interested here in legal texts), 131 can be attributed to the Šangu-Ninurta family; 157 to the Ekur-zakir family and 119 remain unattributed as later Parthian graves partly disturbed the original archaeological context by penetrating the earlier archaeological strata.

These collections are libraries formed by and for teaching and learning activities. Their owners did not intend to make collections of reference texts which already

---

[11]Tablet *SpTU* 4, 157.

[12]Clancier (2009a, 60–61).

[13]Indeed, while legal tablets regularly refer to legal texts and letters on leather (written in Aramaic or Greek script), this is not the case for scholarly tablets. This last category depended heavily on cuneiform writing and so, the clay or wax media was better suited to this than the flatter ones. However, two scholarly texts in cuneiform on *magallatu* (leather scrolls) are attested to: BM 41548 edited by (Leichty 1970, 200–201) and (Weidner 1966, 46). The colophons mention the fact that the end of both texts are on *magallatu*. This only concerns divination.

[14]For the question of writing media in the late Babylonian period, see (Geller 1997; Clancier 2005; Westenholz 2007).

existed in the city's temples.[15] The unearthed tablets show the school curriculum from the very first steps (like copying the very basic BE sign in learning how to write cuneiform script[16]) until the high scholarly level of the specific profession of exorcist-doctor. On their way, the students had to learn divination in its many forms (from haruspicate to astrological), incantations, hymns and prayers, mathematics, astronomy, historical and literary composition. But exorcism and medicine were the principal disciplines to be mastered by the *āšipu*, the exorcist-doctor. It is on this particular training that the present work will focus. The main purpose will be to identify what was taught and how, the learning processes and the extent and limitations of the teaching activities attested by the documentation especially by exploring the complex relationship between academic learning and professional training. Indeed, while the surviving documentation from the two families is extensive, it is often difficult to interpret, as a perfectly carried out exercise can easily be mistaken for a professional document. This aspect is central to any reflection on teaching and learning activities in this specific Urukean context. Indeed, what the tablets mostly show is the erudite axis of the knowledge to be acquired but the more practical aspects of the professional training of a doctor are a lot less clear. This last aspect may have been given orally for the most part, but even here the clues attest to scholastic teaching. This spirit can be found in an important document which will be commented on throughout these pages: the *Āšipu*'s handbook.

3.2.2 *The Āšipu's Handbook*

For that matter it seems to us that the "*Āšipu*'s handbook", that is the "handbook" of exorcist-doctors, is a good document to think about training in a well known archaeological context.[17] Modern scholars gave it its present name focusing on the objective announced in its first lines. Indeed, the "handbook" is a list of texts which are to be used "for learning, consulting and copying".[18] These three words sum up the different uses of a scholarly text:

1. *Learning*: the canonical series were used as reference works and a great deal of knowledge could be acquired from the tablets themselves. We will see later that input from a master was necessary to fully understand the texts. In contrast, a very personal task would have been for the student to learn them by heart.

---

[15] For the identification of the collections of the *Āšipu*'s houses as the remains of practices of training at home, see (Clancier 2009a, 81–101).

[16] This step is not well attested in the *āšipu*'s houses and will not be discussed here.

[17] An exemplar of the "handbook" was found in the library of the Šangu-Ninurta family (*SpTU* 5, 231) and is translated in the appendices.

[18] Line 7 of the reverse. The texts listed in the "handbook" are usually called "canonical series" by modern scholars and *iškāru* by the ancient scribes. The "handbook" is thus partly a kind of "bibliography".

2. *Consulting*: the canonical series also had a life after having been learnt. Indeed, the text is always used as the ultimate argument in a specific context. Its knowledge seems never to have been (officially) contested.[19]
3. *Copying*: students used to copy tablets to learn not only their content but also their specific signs and vocabulary. A very good copy could have been kept as a library exemplar to replace an older tablet in bad condition. Copying was the normal way to preserve the texts. Not only students but masters also copied texts which were often offered to the temples as a gift but also to always have readable copy of tablets. In short, canonical series were documents to be learnt and scrupulously copied as reference texts. The capability to understand, know and copy them was the basis of any scholarly training. They were the compendium of written knowledge.

So, the "handbook's" purpose was to guide the advanced exorcist-doctor students in their acquisition of *āšipūtu*[20] and all the knowledge necessary to be a true scholar, an *ummânu* in Akkadian. But the "handbook" is itself, at the same time, a canonical document as well, composed centuries before, learned and copied as a scholarly text with a status close to that of the main medical or divinatory series for example. So, such a text seems to serve a dual purpose: being created as a tool with a training program to follow, and a text to be learnt like any other canonical text in a scholarly curriculum.

3.3 Learning Medicine and Exorcism from Theory to Practice

In a first step, the purpose will be to work on the learning activities as programmed in the "*Āšipu*'s handbook". As noted above, the "handbook" was not only a guide for students but also a scholarly text, especially at the end of the first millennium BCE, and the training it gives needs to be verified. In short, the question is to know if it is possible to see the "handbook" as a true inspirational learning program.

To answer this question, we can compare the list of texts and competences given by the "handbook" for the training of students and what was found in the *āšipu*'s houses in Uruk. Indeed, the libraries unearthed here contained many texts closely connected to *āšipūtu* and it is possible to see if the handbook's spirit influence the later exorcists-doctors' curriculum or if it was simply a canonical text without any link to a training purpose.

[19] *SAA* 10, 58 for example, a letter from Balasî to the Assyrian king referring to bird omens taken from a series. Perfect examples of the real importance of the texts themselves are the incantations which were to be said without changing a single word.

[20] *āšipūtu* is the discipline of the *āšipu* – mainly medicine and exorcism.

3.3.1 Influence on Learning: The "Ašipu's Handbook"?

In Mesopotamia, from at least the end of the second millennium B.C., the different disciplines (exorcism, medicine, astrology, divination and so on) were partly organised and transmitted through canonical series.[21] These series were extremely important not only in the student's apprenticeship but also during their whole life as reference works. In that context, the *āšipu*'s handbook is a list of series[22] or competences[23] to be known by the future exorcist-doctors. Its general structure is not complex as it is divided in two different parts.[24]

The first part is on the two main aspects of *āšipūtu,* that is, exorcism and medicine.[25] Here, the handbook lists around 80 titles (*incipit* in fact) of *āšipūtu*. Some of them are problematic but the objective of the list is clearly announced: "learning and consulting".[26]

The second part is attributed to Esagil-kin-apli (reverse l.5) who is presented as the author of the program thereafter. He is also known to be the author of the *Sakikkû*,[27] the main text of *āšipūtu*. The *Sakikkû* is also cited through its incipit: "When an exorcist comes to a patient's house".[28] Esagil-kin-apli is said to be a descendant of Asalluhi-mansum, a scholar and sage to Hammu-rabi. This false genealogy makes a link between Esagil-kin-apli and the great Babylonian King Hammu-rabi (1792–1750) to improve his importance.[29] In fact, Esagil-kin-apli would have lived during the reign of Adad-apla-iddina (1069–1048).[30] The program in his "handbook" shows his advice to *āšipu* students through a list of fields of

---

[21]For a general introduction to the phenomenon, see (Lambert 1957).

[22]Such as line o.17. *Sakikkû*, the main series of exorcism-medicine.

[23]Competences differ from series as it is not texts that are said to be known but practices such as "Cures of epilepsy" (line r.12.) or even "the cures of the totality of illnesses" (line r.14.).

[24]See Jean (2006, 72–75) for a more complete commentary on the plan.

[25]There were two professions that could be designated as medical: *āšipūtu* and *asûtu*. The latter was that of the *asû*, mostly translated as "physician". The *asû* had disappeared by the end of the first millennium at least from the written documentation. For the relation between *asû* and *āšipu* cf. (Finkel 2000, 146; Abrahami 2003; Geller 2007).

[26]Obverse l. 1.

[27]The *Sakikkû* ("Ailment") is a compendium of around 3,000 omens on 40 tablets which form the canonical series *Enūma ana bīt marṣi āšipu illaku*. The series is divided into six sub-series: 1. Interpretation of observations made by the exorcist-doctor walking to the patient's house; 2. head-to-toe observation of the patient with symptoms and origins of the illness; 3. duration of the illness; 4–6. specific illnesses. The structure of the series shows the exorcist's view of a patient (Heeßel 2000: 19). For further details, see (Robson 2008, 458).

[28]*Enūma ana bīt marṣi āšipu illaku*. (Labat 1951; Heeßel 2000).

[29]Esagil-kin-apli, in an Urukean text from the Rēš temple, is said to be the scholar of a descendant of Hammu-rabi whose name is broken (*BaM* 2, 89, text dated from 165 BCE). This scholarly tradition of attributing scholars to famous Kings was highly developed in Uruk (see *infra*).

[30]For the question of the historicity of Esagil-kin-apli, see (Robson 2008, 458).

knowledge more than through a list of texts to know.[31] Concerning medicine, there is no clear allusion to specific series but directly to the illnesses to be cured.[32] So, at the end of his studies, the young exorcist had to know all the illnesses and their treatments.

Consequently, in line r.15 the first objective of the exorcist's training appears: "until you obtain knowledge of all *išippūtu* (...)".[33] So the *āšipu* had not only to know all the canonical series as listed in the first part of the handbook but also had to be a good practitioner in exorcism and medicine as is said by Esagil-kin-apli in the second part.

The second objective was to produce a scholar, something said in the lines r. 16–21. The *āšipu* had to understand the commentaries and glosses, the texts in Emesal (a Sumerian dialect), in Sumerian and, of course, in Akkadian and had to extend his competences further than incantations, rituals and medical texts. Indeed, the *āšipu* had to know everything concerning divination as it is meant by the *Enūma Anu Enlil* series (the main astrological text) or *Šumma ālu* (terrestrial divination).[34]

The end of the text indicates without any ambiguity that the professional training is not the only end and that the young exorcist had to join the small circle of *ummânu*, the "scholars", in which he could debate with erudition.

3.3.2 Learning Processes: Exercises in the Light of the "Ašipu's Handbook"

The numerous texts excavated from the *āšipu*'s houses in Hellenistic Uruk can be used to compare the "program" expressed by the *āšipu*'s handbook and the curriculum of Urukean students in a scholarly context. At the end of the first Millennium, learning seems to have been organised in the following schematic ways:

1. The first steps of the young student were in part made in the temples, where they learned to read, write and calculate. However some very basic exercises were also found in the *āšipu*'s houses.[35] That very first beginning of knowledge

[31] As in lines 8–9 of the reverse: "All existing against evil spell, *namburbi* (A kind of exorcism), signs from heaven and earth".

[32] Lines r.12-r.13: "Cures for epilepsy, (Against the demon) Master-of-the-roof, The hand of a god, the hand of Ištar and the hand of a ghost". Epilepsy appears in the *Sakikkû* (Stol 1993) but the series is not cited here. The hand of a god, a goddess or a demon is a designation for illnesses appearing in different series.

[33] *išippūtu* is here the word used for *āšipūtu*, "exorcism-medicine".

[34] Line 18 of the reverse.

[35] For example: *SpTU* 5, 277 written exercise of the signs "diš bad" (Van Buylaere 2009). For the first steps of the curriculum, cf. (Gesche 2001).

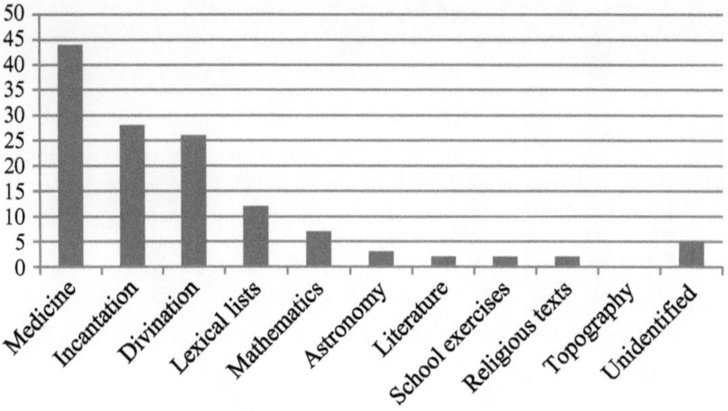

Fig. 3.1 The Šangu-Ninurta library

acquisition is not even mentioned in the "handbook" which focuses on what one could, anachronistically, call the "academic" level.

2. That "academic" level was taught by a master in his own home with the objective of producing a professional and a scholar.

The collections of texts in the houses attest this "academic" level. The fact that these libraries are the product of teaching and learning is deduced, for the learning aspects, from the huge number of lexical texts[36] and also by tablets written by students for their masters as exercises, as we will see. Following the statistics of the subject matter of the tablets, it is quite interesting to see that the "spirit" of the *āšipu*'s handbook is followed. Indeed, the majority of the tablets (with some differences between the two libraries) and the "handbook", are concerned mostly with medicine, incantation and divination and the other disciplines appear in very small proportions (Fig. 3.1):

In the first library, that of the Šangu-Ninurta family, there is a net predominance of medical texts with those on incantation and divination almost equal in size. These three disciplines, very close to each other, form the heart of *āšipūtu*. Moreover, within the general designation of "medical" texts the *Sakikkû* series which was at the crossroads of medicine, divination and incantation is found *en masse* (Fig. 3.2).

In the case of the second library, medicine is far less represented than incantation and divination. Indeed, the *Sakikkû*, so attested in the Šangu-Ninurta library, is much less present here. In the same time, astronomy and its divinatory counterpart, astrology, appear here in all their importance. In fact, the *āšipu* were also, with the incantation-priests (the *kalû*), the astronomer-astrologers working for the temples during the Hellenistic time, the period of the second collection. Such an evolution is clearly reflected here.

[36]Lexical texts are documents organising knowledge through lists and tables. Most of the time, Sumerian and Akkadian appear in the texts giving the translation of each term in both languages.

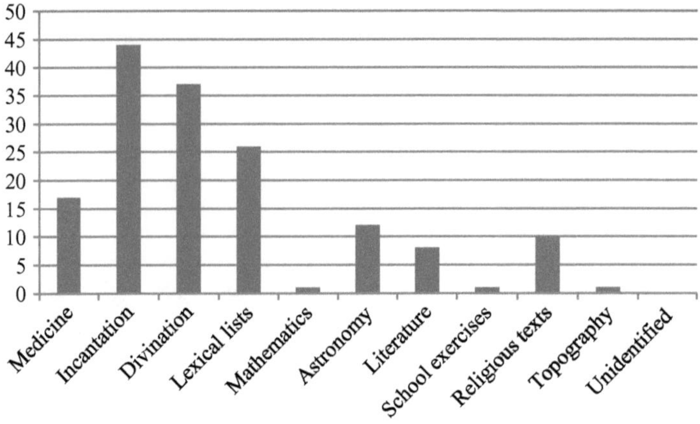

Fig. 3.2 The Ekur-zakir library

For the two libraries, we notice the huge number of lexical texts as well as dictionaries doubtlessly used by both masters and students to learn words and expressions.

If there is some similarity between the *āšipu*'s handbook and the general profile of the libraries unearthed in the *āšipu*'s houses, it is important to clarify its meaning. Indeed, one could say that if the series unearthed in the libraries are the same as those in the handbook, it is because the list given in the handbook shows that the exorcists had to know a great deal of knowledge beside their medical, incantational and divinatory specialties. However, the net dominance of medical, exorcismal and divinatory documents invites one to think that the "handbook's" program was *de facto* followed, and, even if it was written in the end of the second millennium or in the beginning of the first, its spirit still led the training of the *āšipu*. So the textual categories of the "handbook" express the current learning in the scholarly society of the first millennium BCE in all Mesopotamia, but adapted to the *āšipu*. The profiles of the collections unearthed in the *āšipu*'s houses are close to the stipulations of the "handbook". They can be compared to other libraries of the same period also found in Uruk. The best example is the lamentation-priests' (*kalû*) collection from the Rēš temple. It has a completely different profile[37] and so invites the thinking that the collections of the *āšipu*'s houses were not common but original and suited to the reality of a specific training.

[37]Clancier (2009a, 86–90). *Kalû*'s library at the Rēš temple is not the result of the training of the lamentation-priests but is the result of their professional activities inside the temple. Consequently, the large majority of the collection is religious texts with divinatory documents (especially astrology) in second position. There are no medical texts, just one incantation and only three lexical texts which are, in fact, more lists than true lexical documents.

3.3.3 Identifying Student's Tablets Through Colophons

It is difficult to identify a student's tablet when the young man was at the very end of his studies as the work was normally so good that the texts copied were without mistake. In such a situation, only the colophon may contain information about the status of the writer. In it, the young scribe wrote some details about the texts he just finished:

- Incipit of the series, sometimes incipit of the sub-series with the number of lines and so on. For example: "If a man's eye hurts him: hand of DN.[38] Second tablet of (the chapter) *When you approach a patient*; 4th tablet (of the series) *When an exorcist goes to a patient's house*. 136 lines recorded".[39]
- A formula referring to the original text used for the copy: *kīma labīrišu šaṭirma bari*, "written and checked as its original".
- The name of the tablet's owner: *ṭuppi* PN,[40] "tablet of PN".
- The name of the scribe if different of the owner: *qāt* PN, "hand of PN".
- The original writing medium (sometimes), for example: ^gis^*lē'u*, "writing board" that is a wax tablet. As noted above, no such medium was discovered in the *āšipu*'s houses.
- Less often, the place where the original tablet was kept. In Uruk, this is frequently in the main city temples.

To illustrate the information that could be found in a colophon, it is possible to present an example containing various topics[41]:

[*kīma*] *labīrišu šatir-ma bari u uppuš gabari lēʾi makkūr* [*Anu u Antu*]

[*tuppi*] *Anu-ah-ušabši mār ša Kidin-Anu liblibbi* [*Ekur-zakir*] *m*[*ašmašši Anu u Antu]*

[*šeš*]*galli ša Rēš Urukayu qāt* [*Anu*]*-balassu-*[*iqbi mārišu*]

[*ana*] *ahāzišu arāk ūmīšu balāṭ napišti*[*šu u kunnu išdêšu išṭurma*]

[*ina Uru*]*k u Rēš bīt bēlūtišu uk*[*īn* ...]

"Copied, collated and correctly established [according] to its original on writing board from the Properties of [Anu and Antu.[42]]

[Tablet of] Anu-ah-ušabši, son of Kidin-Anu, descendent of [Ekur-zakir], ex[orcist of Anu and Antu],

---

[38]"DN" is for Divine name. Numerous illnesses were designated by the name of a god, a goddess or a demon.

[39]Labat (1951, 44–45).

[40]PN: personal name. The name is given with the name of the father, sometimes the grandfather and the ancestor. For the exorcists of Uruk, the ancestor is most often Ekur-zakir during the Hellenistic period. In the first library we can see this was also the case for the Babylonian Šangu-Ninurta family.

[41]*SpTU* 1, 2.

[42]Anu and Antu were the main divinities of Uruk in the late Achaemenid period and during the Hellenistic time.

[*šeš*]*gallu*[43] of the Rēš temple,[44] the Urukean. Hand of [Anu]-balassu-[iqbi, his son].

He [wrote] (the tablet) [for] his understanding, lengthening of his days, his physical and moral well-being and the [stability of his position].

He placed (the tablet) [in] Uruk, inside the Rēš temple, temple of his (Anu) suzerainty".

The scribe, when different to the owner, is often thought to have been a student writing for his master. But there are exceptions. For example, in the astronomical tablet *ACT* 101, it was a father writing for his son. In this case, it is not an exercise at all. But generally, colophons help us enormously in reconstructing the history of people and relationships. For example, Anu-ikṣur, who established the largest part of the first library the main owner of the first library, can be followed from his apprenticeship when he wrote tablets as exercises for his father (his master), then as a young scholar working alone in text commentaries (cf. *infra*) and then as a master with new generations of students writing tablets for him. The fact that all Anu-ikṣur's tablets were unearthed together in such a good archaeological context also helped understand this progression.

Moreover, some indications exist concerning the relationship between the tablet owners and the writer when the scribe's status appears after his name in the colophon. The student recognizes his situation with the note *šamallû* that means "apprentice". The scribe can also be called, in the *āšipu*'s houses, *mašmaššu ṣehru*, "young/junior exorcist". Normally the masters have the title of *mašmaššu*[45] only, the *ṣehru* being an *āšipu* during his first few years of activity and not yet recognized as an expert, an *ummânu*. He frequently seems to still follow his master closely. But such detailed mentions are very rare. The normal formula is only "Tablet of PN. Hand of PN".

However, the texts of the *āšipu*'s houses offer opportunities for identifying students' work by other means, such as the differences between the copy of a complete series tablet, an "unfinished" copy and the abstract.

3.3.4 The Finished Copy, the "Unfinished" Copy and the Abstract

None of the copies of canonical texts found in the *āšipu's* libraries concern a complete series. There is no example of an extensively copied canonical text. Moreover, some texts were just copied as abstracts as in an example of the first tablet of the *Sakikkû*.[46] In any case, these copies seem to have the objective of showing

[43]The literal translation is "big brother". It is a very high function of priesthood.

[44]The Rēš, sanctuary of Anu and Antu, was the main temple of Uruk at that period.

[45]This title is widely discussed. It normally means "exorcist" but, during the Hellenistic period, it seems to be used as a specific designation of the exorcists working for a temple.

[46]*SpTU* 3, 87.

to the master that the student was able to understand a text and write it correctly. Indeed, if the objective was to set up a reference library, complete series should be attested in the houses. So, these copies and their form had some learning functions:

1. A *finished copy of one or several tablets of a canonical series*. We can compare two different copies of the 16th tablet of the *Sakikkû*, medical omens canonical series. The first, *SpTU* 1, 37, is badly damaged and the colophon is broken. Here is its first line: "If he is ill for 1 day and his head hurts him: *sunstroke*, (it is the) "Hand of his father's god", he will recover".

   The second example of the same text is tablet *SpTU* 2, 44. It is the same text and is probably a copy exercise. Indeed, the colophon is in good condition and records Anu-ab-uṣur, son of Anu-mukin-apli, descendant of Kuri as the scribe. The Kuri family was not specifically involved in the profession of exorcism, so, maybe we have here an example of complementary training. In any case, complementary or not, the owner of the tablet, Iqiša, son of Ištar-šum-ereš, descendant of Ekur-zakir, is very well known as the main tablet owner of the later library of the *āšipu*'s houses. He seems to have been the master of Anu-ab-uṣur.[47] If we take exactly the same first sentence the result is the following: "If he is ill for one day and his head hurts him: *sunstroke*, (it is the) "Hand of his father's god", he will recover, he will die". The expected text is only "he will recover".

   Here, the copy for the first sentence is unreliable. Indeed, how can we know the future of the patient in the case of one day's illness and a headache? The text was normally used to make a diagnosis (the name of the illness is "Hand of his father's god") and also a presage ("he will recover"). What happened here? As far as we can envisage, it is possible to present, at least, two different explanations for this example. The first is that the scribe made a mistake. The second is that the original tablet was partly broken at the end of the line and so, the scribe decided to show the two possibilities. This latter explanation seems impossible for reasons seen just below. The mistake is more likely and reveals two facts: while the scribe knew how to read the text, he was not familiar enough to see the problem he made in the text's internal logic and, secondly, he did not (yet?) know the series by heart.

   Another aspect of the copies found in these libraries suggests that the texts were not used for their content but more likely to show that the scribe was able to copy correctly. Indeed, a central notion in the work of the copyist was the absolute respect of the original's layout when copying it. This means that when breaks appeared on the original tablet, the scribe never tried to restitute the missing passage. In place he wrote *hipi* "break" with some degree of precision: *hipi labīru* for "old break" (already on the former original from which the present

[47] It is still possible that the table was copied for Iqiša by Anu-ab-uṣur without any student-master relationship. However, this explanation does not take care of the title given to Iqiša, *mašmaššu*, and the absence of the status of the scribe, Anu-ab-uṣur. This difference of level between the two men leads to thinking that the owner is already a master (or an *ummânu*, "an expert") and not the writer.

tablet was copied) or *hipi eššu* that is "new break" (if it was not previously noted on the original). The 4th tablet of the *Sakikkû* unearthed in the *āšipu*'s houses contains such mentions:

o.cii.5'.[48] "If his temples ^{break} are feverish at the same time: Hand of Sîn x x x Šamaš".[49]

Another example, even more explicit:

o.6'. [i]-⌈na⌉ ú-de-e 2-i šá kàš ta-na-rab$_4$ ki-i LU $^{\text{hi-pí}}$ UT x [. . .][50]

"In a second recipient, you dissolve (it); *when* LU ^{break} UT x [. . .]"

Around the break, the signs were sometimes difficult for the copyist to read and this is shown by the fact that the scribe added inaccurate signs like the LU and UT in the second example. This deep respect for a canonical text was a very real way of establishing the quality of a copy for the student as well as for the master. But it is important to see that the preceding examples where unearthed in an *āšipu*'s library and that having broken texts are not a help to the everyday practice of medicine and exorcism. Indeed, in the case of *SpTU* 3, 88 o.ii.5′, the first example, it is impossible to firmly establish the diagnosis as the complete description of the patient's symptoms are not available. In the second case, in *SpTU* 5, 253, f.6′, the consequences on the everyday practice of medicine is quite important. The text is a medical recipe and, here, the bad state of preservation of the original as copied by the scribe does not help at all in making the remedy.

2. *The "unfinished" copy* is another kind of document found in the corpus unearthed in the *āšipu*'s houses:

   "Tablet 2 of "If he is ill for 1 day and the nature of the affliction (is)", tablet 16 "When an *āšipu* goes to a patient's house" not finished.

   "If, at the beginning of his illness, he sweats, has boil(s) and this sweating (is) from the shin".

   Executed properly, written and checked from an old writing-board.

   Tablet of Iqiša, son of Ištar-šum-ereš, descendant of Ekur-zakir, the exorcist. Hand of Anu-ab-uṣur, son of Anu-mukin-apli, descendant of Kuri."[51]

The series were written on different tablets organizing the text into parts and chapters. Here the series is "When an *āšipu* goes to a patient's house",[52] the part is "If he is ill for 1 day and the nature of the affliction (is)" and the chapter "If, at the beginning of his illness, he sweats, has boil(s) and this sweating (is) from the shin". So, the tablet is the 16th of the series and the second of the part. It is the expected organization but the scribe notes that the tablet is "nu al.til" that means unfinished. The tablet is, moreover, copied from an already

[48]Obverse, column 2, line 5′.

[49]*SpTU* 3, 88.

[50]*SpTU* 5, 253.

[51]*SpTU* 2, 44, l. r.22′-top edge 3.

[52]That is also called *Sakikkû*.

very damaged original and there are a lot of mentions such as "break"[53] or "new break".[54] So, the original is not copied to its full extent and even the text itself is quite difficult to use in the everyday practice of medicine.[55] Its function was not to be a reference for medical proposes but for training objectives by validating (or not) the capability of the student to understand and copy a tablet from a specific series.

3. *The abstract* is the last type of document examined here. It is a tablet onto which only a part of the original text was copied. The function of such tablets is still discussed. The abstract does not summarize the text but shows an extract of the original document.[56] The series presented themselves as catalogues of cases and only some of them are chosen to be copied in the abstract. In the example of *SpTU* 3, 87, only 30 % of the cases were written onto a small tablet which was just big enough for that purpose. Why did scribes write this sort of tablet? It is quite difficult to answer this question but it is possible to propose some hypotheses. First hypothesis, the abstract could have been a copy exercise on only a part of the text as demanded by the master. Why, in this case, not copy the tablet from its beginning? Second hypothesis, the text attests a memory exercise and third, that the tablet was used as a reminder of a text known by heart. Other possibilities could be added but I would like to privilege here the question of orality in the *curriculum*. Indeed, many series had dozens of tablets and so if the text was the reference, it was not available outside the library. In that perspective, the solution, not at all surprisingly, was to know the complete text by heart. Maybe, the abstract had some role to play during the students training or even the professional life.

All these tablets show what the writing training of advanced students entailed. It was fortunate to find collections having so many students' exercises such as the perfect copy by a *šamallû* (an "apprentice"), a copy with mistakes and an unfinished copy not useful for professional practice. Orality, which was certainly a principal medium of the scholars' training, is not attested to but the abstracts perhaps play a key role here, as suggested above.

Until now we have focused on student's documents, trying to identify evidence of learning practices. It is not surprising, however, that the collections also preserved teaching documents produced, this time, by masters.

---

[53] *hipi*.

[54] *hipi eššu* as attested in lines 0.10, r.7′, r.12′, r.13′, r.19′ (when just "break" is mentioned like here, it possibly denotes an old break already noted on the original).

[55] The text can be seen in English translation by the author in the following address: http://oracc.museum.upenn.edu/cams/gkab/corpus/

[56] For an example, see, *SpTU* 3, 87, abstract of the *Sakikkû*, tablet 1 (CAMS). As a finished copy, the tablet is introduced by the incipit of the series or by the incipit of the part or of the chapter.

3.4 Teaching Processes and Objectives

Student's work was checked by their teacher whose presence could be seen in the colophons (as already noted above) and through a specific type of text: the commentary. Studying the *āšipu*'s libraries at Uruk helps us to understand, in that context, some of the teaching processes and objectives. As what we have are only texts, it is not easy to see the master's action in the training of the student. Indeed, the teacher's work must have been mostly oral. However, the written commentaries on the series allow us to see how the reference texts were taught and used. We will focus here on the scholarly training of the students, that is, the processes of making a scholar from an apprentice. The more practical aspects of the medical and exorcism apprenticeship will not be seen here as they need very precise examination of the vocabulary used in the commentaries, something not suitable here.

3.4.1 The Commentary

There are different types of commentaries called *ṣâtu*, *mukallimtu* and *ma'šaltu*. It is sometimes quite difficult to differentiate one from the other, especially in the late Babylonian period, but we can say that the commentaries can be divided into two main categories (sometimes mixed in one text). The first is the lexical commentary, explaining the meaning of specific words. The second is an erudite and esoteric commentary whose objective is to make the logic of the text and its "secret" knowledge understood. Both categories of commentaries can be illustrated by two examples.

The first kind can be seen in the text *SpTU* 1,27, a *Sakikkû* commentary:

> o.10'.o.11' [. . .] *uzabbalma* (means): "he will linger": *zubbulu* (is for): "to linger", meaning "he will become aggravating": $GA_2.GA_2$ (means): "to linger", (said) of an illness.

In this example, the Akkadian verb in the original text is conjugated as *uzabbalma*. Its infinitive is *zubbulu* and is explained and translated in the specific context of *SpTU* 1, 27 as "to linger". $GA_2.GA_2$ is the Sumerian ideogram used for this verb. So, here, the reader of the commentary explaining the *Sakikkû* series understands the contextual meaning of an Akkadian verb (*zubbulu*) and its Sumerian translation ($GA_2.GA_2$).

The second kind of commentary, the erudite and esoteric one, can be illustrated by *SpTU*1, 47[57] and concerns the diagnosis of an illness:

> o.2-o.5. The demon of the lavatory (is): Šulak. (A person) should not enter the lavatory (because): Šulak will beat him. Šulak of whom it is said (to be): ŠU (ideogram that means): "hand": lā (Akkadian that means): "not": KU_3 (ideogram that means): "pure". He enters the lavatory, his hand is impure, on account (of whom) it is said.[58]

[57]Commentary of the medical series called *If a person's skull suffers from fever* usually called UGU ("skull"), by Assyriologists.

[58]Reference to the "oral tradition" (see *infra* "The commentary, the master and the oral tradition").

Table 3.1 The commentaries in the Šangu-Ninurta library

Library of Šangu-Ninurta family			
Texts categories	Number of tablets	Commentaries	Proportion
Incantation	28	0	0
Medicine[a]	44	22	50 %
Astrology	9	2	22 %
Haruspicy	2	2	100 %
Divination *varia*	15	3	20 %
Astronomy	3	0	0
Lexical texts	12	0	0
School exercises	2	0	0
Prayers	2	0	0
Literature	2	0	0
Mathematics	7	0	0
Unidentified	5	2	40 %
Total	**131**	**31**	**24 %**

[a]Including *Sakikkû*

Here, the commentary is not about vocabulary but about the origin of an illness. The name of that illness is Šulak, a demon's name. To explain who Šulak is and its action the commentator gives a reconstructed etymology mixing Sumerian and Akkadian.

We noted above that the catalogue of the *āšipu's* libraries was similar to the one given in the *āšipu*'s handbook. However, the exorcists-doctors collections do not contains only exemplars of the canonical series (with many copies by students) but also commentaries on them. Moreover, the commentaries on the series are sometimes more numerous than the presence of the series themselves. As an example, for the whole library of the Šangu-Ninurta family, the oldest library of the two, it is possible to identify, at least, 31 commentaries on 131 tablets (Table 3.1):

As we can see in the table, the best example is from the *Sakikkû* series.[59] As the main reference in the discipline of exorcism during the first millennium BCE, its presence in the libraries is highly expected. On 18 *Sakikkû*[60] tablets that can be attributed to a member of the family, at least 13 are commentaries.[61] Almost $\frac{3}{4}$ of the main series of the exorcist-doctors are actually commentaries in the library of the Šangu-Ninurta family. No clue indicates that the series itself was, by extension, in the house. So we only can expect that the original text was copied in the temple library.

[59]Included in the entry "Medicine". (Labat 1951; Heeßel 2000).

[60]*SpTU* 1, 27; 28; 29; 30; 31; 32; 33; 34; 35; 36; 37; 38; 39; 40; 41; 42; *SpTU* 3, 100 and perhaps *SpTU* 5, 254.

[61]*SpTU* 1, 27; 28; 29; 30; 31; 32; 33; 36; 38; 39; 41; 42; *SpTU* 3, 100 and may be a 14th with *SpTU* 5, 254. These tablets are edited in English by the *Geography of Knowledge in Assyria and Babylonia* project (Clancier 2008–2009).

3.4.2 The Different Uses of the Commentary

A commentary is not a student exercise except for the "canonical" commentaries copied by students as for any other reference work.[62] Most of the time, in the *āšipu*'s houses, it is the main evidence of teaching activity, as it appears through the colophons. For example, the tablet *SpTU* 1, 50, commentary on a medical recipe, says:

> r.18-r.20. Commentary and oral tradition of "(If) *falling sickness* strikes a person." [from] the mouth of an expert, reading out of Anu-ikṣur, junior exorcist [. . .] descendant of Šangu-Ninurta.

What we can see is that the commentary (*ṣâtu*) is said to be from the "oral tradition" (*šut pī*) concerning a specific text. Next, the colophon makes reference to the origin of the commentary which is "the mouth of an expert" (*pī ummâni*). The use of the commentary is clearly illustrated by the expression "reading out of PN" (*malsût*). Here, the text seems to be from the scribe himself, Anu-ikṣur, already an *āšipu* and being trained to become an *ummânu*, a true scholar, by working on the commentaries.[63] The question here is the origin of the knowledge. Is it from a debate with the other *ummânu*-experts, as is said in the *āšipu*'s handbook (line 19 of the reverse), a lesson by a master or a commentary by Anu-ikṣur himself even as a junior?[64]

Other mentions in the texts themselves show the central importance of orality in commentaries. Indeed, the oral tradition appears through expressions like "it is said" or "they say". Back to commentary *SpTU* 1, 47, we have here an example:

"He enters the lavatory, his hand is impure, on account (of whom) it is said".[65]

Sometimes the verb *qabû* "to say" is in the plural for the impersonal form, as in tablet *SpTU* 1, 49,[66] making an indirect reference to the community of the *ummânu*-experts:

> r.5. "*illūru*-flower: (whose ideogram is) GURUN (that also means): "egg of *hahhūru*-raven", that is said".[67]

Here, the hearsay reference is highly valuable as it is the justification of the textual explanation. If the series is the reference for knowledge, the master is then the reference for the explanation of the series. His thoughts appear in specific

[62] See *infra*.

[63] Indeed, on line 15 of the reverse Esagil-kin-apli says that the student has to know all the disciplines of *išippūtu* that is *āšipūtu*, exorcism-medicine. Afterwards, from line 16, the junior exorcist will have to work on *The creatures of the steppe* (a way to name the commentaries upon the canonical series, cf. *CAD* Ṣ: 116) and other very difficult aspects of scholarship to be an *ummânu*.

[64] Anu-ikṣur wrote a lot of commentary tablets attesting that he was a specialist in that exercise both personally and as a teacher.

[65] *ana* UGU *qa-bi*.

[66] Medical recipe commentary by Anu-ikṣur.

[67] *ša qabû*.

notes inside the commentaries. Indeed, sometimes he specifies that, even if the oral tradition is clear on some points, it is not in accordance with his own experience. Tablet *SpTU*1, 32 is a good example of this kind of situation. Again, this text is a *Sakikkû* commentary from Anu-ikṣur:

> r.4. [...] (the ideogram) DU_{11} (means) [...] utterance: I did not see (the ideogram) INIM (meaning) *amāt* on (any) tablet.: ... [...].

Anu-ikṣur follows the normal explanation but adds that he has not seen it written anywhere on a tablet. In this case, it is quite interesting to see that the oral tradition is sometimes unrelated to the written one.

3.4.3 The Written Commentary

We have said that the commentaries come from the "oral tradition", but what we have are written documents. Indeed, part of the oral tradition was recorded on tablets as commentaries and we have a lot of them for all the first millennium BCE. However, it seems important not to give the same status to all of them.

The previous examples taken from Anu-ikṣur's own work could be thought of as the result of teaching. In a school context these tablets could have been pedagogic supports. Anu-ikṣur was a very clever "commentator" and as a master in the *āšipu*'s houses, it seems reasonable to think that the commentaries he wrote were used in his teaching activities.

Another type of commentary is one that the scholastic tradition recorded in a "canonical" form. The text was thus copied and transmitted like any series and was sometimes also copied by students in their curriculum.[68] The best example of the importance of such documents can be found in the later library, that of the Ekur-zakir family[69]:

> r.18'-r.21'. From a wooden writing board. Commentary to "When Anu and Enlil"[70] [...]. Written like the original (and) checked. Tablet of [...], man of his god Nusku, priest of his goddess mother, son of ... [...] Ur-Gula, ... lapis lazuli ... [...], descendant of Enlil-belšunu, *nêšakku*-priest of Enlil. Hand of [...]

This tablet is interesting for various reasons. Firstly, the commentary was copied from an older tablet in the same way as for a canonical series, i.e. with the aim of making a perfect manuscript. The formula "Written like the original (and) checked"

[68]See, for example, the tablet *SpTU* 1, 90, a commentary on the astrological series *Enūma Anu Enlil* ("When Anu (and) Enlil"), tablet 56. The tablet belongs to Iqiša, the owner of the later house, but was written by Anu-ab-uṣur, son of Anu-mukin-apli, descendant of Kuri (Edited by E. Robson on CAMS).

[69]*SpTU* 3, 101.

[70]Here again, it is the *Enūma Anu Enlil* series.

was the one then in use for the series copies. Secondly, this tablet is from Nippur,[71] a town in the centre of Mesopotamia well known for the erudition of its scholars', but was unearthed in the later *āšipu*'s library in Uruk. So the commentary belonged to the Ekur-zakir family and was kept as a reference work.

3.4.4 *The Commentaries in the Šangu-Ninurta Family Library: Teaching Scholarship*

The numerous commentaries found in the *āšipu*'s houses, especially in that of the Šangu-Ninurta family, attest the teaching of the scholarly aspects of *āšipūtu* but not to the more practical ones. For example, while there are a lot of medical recipes, some of them are simply unusable for day to day cures as the texts were copied in a scholastic way with all the breaks mentioned without any attempt at restitution. Here the main point is to know the text and its cuneiform writing but not the making of the recipe itself.

It does not mean that the professional practice of medicine and exorcism was not taught in the *āšipu*'s houses and in the temples of Uruk. Indeed, as one can deduce from the mentions of "oral tradition", most professional applications of the *āšipus*' knowledge, especially the actual treatment given to patients, would have been taught directly "from the mouth of an expert" to the student in front of the patient himself. No trace can really be expected from such a practice through the kind of documentation unearthed.

The tablets from the *āšipu*'s houses shed light on the importance of a master in this scholarly apprenticeship context, and this type of master had a special social status far beyond his scholarly capabilities. An important step in the construction of the scholar's image was made, in the second half of the second Millennium, when the series were first documented. At that moment, some scholar's names were recorded, transmitted and some of them became cited as ancestors by important urban families.[72] The memory of such masters is well illustrated by Esagil-kin-apli who was the (attributed) author of the second part of the *āšipu*'s handbook. He was also said in many disparate Mesopotamian traditions to be the compiler of the *Sakikkû* series. In a text published by I. Finkel, it is possible to see the importance of a scholar at the end of the second Millennium and the beginning of the first[73]:

---

[71] As shown by the prosopography of the tablet.

[72] For example, Sin-leqe-unninni, which was an ancestor's name in Uruk, was also, in the scholastic tradition, the author of the canonical version of the *Epoch of Gilgameš*. For the question of authorship, see (Lambert 1957).

[73] Finkel (1988, especially 148–150). The text in itself is a good example of a scholastic tradition transmitted in Assyria and Babylonia through the whole first Millennium BCE. Indeed, it is a catalogue of the incipits of the series *Sakikkû* and some of the series *Alamdimmû*, *Kataduggû* and

> o.18', r.19'-33'. Concerning that which from old times had not received an [authorized] edition, and according to 'twisted threads' for which no duplicates were available, in the reign of Adad-apla-iddina, King of Babylon, to work anew..., Esagil-kin-apli, son of Asalluhi-mansum, the sage of King Hammu-rabi, the descendant? of Sîn, Lisi and Nanaya, a prominent notable of Borsippa, exorcist of Ezida, *pašīšu*-priest of Nabû (...) the *išippu*-priest and *ramku*-priest of Nanaya (...), scholar of Sumer and Akkad, through the incisive intelligence that Ea and Asalluhi had bestowed on him, deliberated by himself, and produced the authorized edition of *Sakikkû*, from head to foot, and established it for knowledge. Take care! Pay [attention]! Do not neglect your knowledge! He who does not attain? knowledge must not speak the *Sakikkû* omens aloud, nor may he pronounce out loud *Alamdimmû*! *Sakikkû* (concerns) all diseases and all (forms of) distress; *Alamdimmû* (concerns) external form and appearance (and how they imply) the fate man which Ea and Asalluhi ordained in Heaven. The twin series; their arrangement is one. [Let the *āšipu*] who makes the decisions, and who watches over people's lives, who comprehensively knows *Sakikkû* and *Alamdimmû*, inspect (the patient) and check (the texts), [...], and let him put his diagnosis at the disposal of the King.

In this text Esagil-kin-apli appears as a master in his discipline and as a very good compiler. The text produced is explicitly written "for knowledge". The sage or the scholar is put alongside the King and had to protect the people and the monarch.

A new vision of the importance of the scholar in the transmission of Sumero-Akkadian knowledge and even in the stability of the whole country appeared with the end of local monarchies, especially in the Hellenistic period. Indeed, what is documented, in the *āšipu*'s houses as in the temples, is schematically a small society which recognized its members through the practice of "cuneiform culture".[74] This society is the one who, as the urban notability in Babylonia, welcomed Alexander and his troops in 331.[75] This particular position made it the local intermediary for the new Greco-Macedonian power. Until the second century BCE, it remained a powerful local ruler and as thus, reinterpreted its history and function.

In a text unearthed in Uruk, dating from the Hellenistic period and called by modern historians the *List of apkallu and ummânu*,[76] the scholars[77] were placed in a genealogy which begins with the antediluvian sages (the *apkallu*) sent by the gods to transmit all knowledge to humanity. In that perspective there is no surprise that the main teaching and learning traces we have for the late period are for scholarly knowledge.

associated texts (Finkel 1988, 143) and two copies are known: the first from the Assyrian city of Nimrud, the second from Babylon. The translation is inspired by that of (Finkel 1988, 149–150).

[74] This culture was not the daily one of Babylonia in the Hellenistic period even for those scholars who, as said M Geller, "may have spoken Aramaic to (their) children or Greek to (their) tax official" (Geller 1997, 45).

[75] Arrien, *Anabases*, III, 16, 3–9.

[76] *BagM. Beih.* 2, 89.

[77] Some known as authors of canonical series like Sin-leqe-unninni or Esagil-kin-apli.

3.5 Conclusion

Teaching in old Mesopotamian societies is to a large extent almost invisible to the eyes of modern historians. Exceptions occur when explicit traces are found and one of them is the *Āšipu*'s handbook on the apprenticeship of exorcists-doctors, the *āšipu*. This text, started before the end of the second millennium BCE and possibly completed at the beginning of the first, gives a bibliography of canonical series with an extensive list of precise competences to be mastered. The objective of this training, as is written in the "handbook", is to become a good exorcist-doctor and an *ummânu*, an expert.

While it was difficult to use the "handbook" as a pedagogical program, the text was, however, copied throughout the first millennium like any other canonical text. So, we compared it with the tablets unearthed in two scholarly libraries in late Achaemenid and Hellenistic Uruk. These two collections were created by and for teaching and learning activities by masters whose intention was to train *āšipu*. Consequently, a scholarly text, the "handbook", itself attested to in the older library, could be analyzed in the light of day-to-day practical documentation. The unearthed tablets of both students and masters revealed a close profile with the disciplines said, by the "handbook", to be learned. So the spirit of the "handbook's" program was followed.

The students were trained to have a command of the canonical series which were the collection of all ancient Mesopotamian knowledge. They had to show their accomplishment to the masters by copying parts of the series. Sometimes these exercises were so brilliant that only mentions in the colophons can identify them as student works. However, the apprentices also produced far 'poorer' texts with many mistakes. Other types of documents were discovered such as incomplete copies and abstracts, here again analyzed in a training context. Not only the students but also the masters' activities are attested by the remaining tablets of the two libraries. Their teaching work is visible through the commentaries. These scholarly works explain the vocabulary of the series and their esoteric or secret meanings as well. Some direct references to the oral aspects of the transmission of knowledge and wisdom in the society of scholars appear in them. All this documentation shows that the two mains theoretical objectives of the "handbook" were met concretely: to train an *āšipu* and to produce a scholar, an *ummânu*. This curriculum is particularly well documented through the example of Anu-ikṣur.

However training of scholars is actually what the libraries as well as the *Āšipu*'s handbook attest. Indeed, the methods for the acquisition of the less formal aspects of the exorcist-doctor profession, such as the actual care, are far more difficult to identify and, for Uruk, this study remains to be done.

Appendices

Appendix 1: The Āšipu's Handbook (SpTU 5, 231)[78]

Obverse

1. Incipit of the serie(s) of exorcism which have been recorded for learning and consulting. The whole name(s).
2. 'Kulla' gloss: to lay [...] the foundations of a house; [...] and 'Installation of a priest'.
3. 'Incantation of Apsu'; '*Ginutaqqû*' and 'Purification rituals of the gods'.
4. '*Kiʾutukku* (Incantation by Šamaš)'; '*Šuʾillakku* (Prayer of raised hands)' and '*Kimiltu* (The anger god(s))'.
5. 'Ritual of the months *Duʾuzu*; [...]; *Ululu*; *Tašrîtu*' and 'Royal rituals'.
6. [...] and '*Kataduqqû* (Physiognomic omens)'.
7. [...] and 'To set out the evil'.
8. [...] and 'Magic circles'.
9. [...] and 'All illnesses'.
10. [...] and 'Substitute statues'.
11. [...] enclosures' and 'Mouth washing [...] of the gods(?)'.
12. '*Ušhulgallû* (Evil sorcery)'; [...]; '*Ušburrudû* (To dispel sorcery)' and '*Namerimburrudû* (To remove a curse)'.
13. '*Kiʾutukku* (Incantation by Šamaš) [...]'; '*Ušburrudû* (To dispel sorcery)'; '*Namerimburrudû* (To remove a curse)'; 'Blast of wind' and '*Lamaštu*'.
14. 'All evil'; [...]; '*Šurpu* (Burning)'; 'To transform a bad dream to a good one' and 'Potency incantation'.
15. Woman not able to give birth, ..., "Baby" incantation.
16. Sick eye, ..., and *Būšānu*-disease.
17. *Sakikkû*, ..., and Incantations against all illnesses.
18. To halt a nosebleed, ..., and To stop diarrhea.
19. To cure a snake bite, ..., and *Samana*-disease.
20. "Foot of evil" from a house, to avert ... epidemic and To cause offerings to be received.
21. Ritual of the city, house, field, orchard (and) canal and Contribution to Nidaba.
22. To avert the rainstorm ... and Exorcism(?) of the steppe(?).

Reverse

1. To pass through the steppe, That enemy arrow not approach and Imprisonment.
2. To purify bulls, sheep and horses.
3. (Omen) decision by stars, birds, ... and goats, Oracular utterances of ...

[78]Translation by the author (Clancier 2009b for the complete edition). The most complete manuscript comes from Assyria (*KAR* 44), cf. (Geller 2000; Jean 2006, 63 *sq*).

4. '*Abnu šikinšu* (The stone whose appearance is)', '*Šammu šikinšu* (The plant whose appearance is)', 'Tablets of stones', 'Tablet of plants', 'Necklace$^{(?)}$' and 'Amulet$^{(?)}$'.
5. Titles of all$^{(?)}$ the series of exorcism of Esagil-kin-apli, son of Asalluhi-mansum,
6. *apkallu* of Hammu-rabi, King of Babylon, descendant of Lisia, *pašīšu*-priest of Ezida.
7. For learning and consulting of actions of touching the reed established by Ea, *Kikiṭṭu*-ritual, *Šerkugû*-ritual.
8. all existing against evil spells, *namburbi*, signs from heaven and earth.
9. The totality of the wisdom and secrecy of the *kakugallūtu*.
10. ... projects of heaven and earth, the secret of *Apsû*, and the non-canonical incantations.
11. ... paralysis, The *rimûtu*-paralysis, The *sagullû*-illness, The symptoms, The *kissatu*-illness and ....
12. Cures of epilepsy, (The demon) Master-of-the-roof, The hand of a god, The hand of Ištar and The hand of a spectre.
13. The evil *Alû*-demon, *Lilû*, *Arad-Lilî*, Evil spirit, The hand of the oath-breaking and The hand of a man.
14. And the cures of the totality of illness, a bout of fever and the cures for women,
15. until you obtain knowledge of all *išippūtu*, you come to know the secret,
16. then learn "The creatures of the steppe", the glosses and (the texts in) Emesal,
17. you will learn to do research in the rituals in Sumerian and Akkadian,
18. ..., *Azad*, *Ušuš*, *Enūma Anu Enlil* and *Šumma ālu ina mēlê šakin*,
19. to devise and exchange (for) an agreement.
20. ... valiant$^{(?)}$, ..., wise, erudite, the gods of *āšipūtu* will give extended understanding.
21. May the god, his protector, be good with him! His name will be pronounced until the distant day.

Colophon

22. In accordance with the original tablet, written and collated ... Rimut-Anu

Top Edge

1. ... Šamaš-iddina, descendant of Šangu-Ninurta: Uruk, month *Tašrîtu*, day ..., year ... Darius King.

Appendix 2: Genealogies of Šangu-Ninurta and Ekur-zakir

Šangu-Ninurta

Nadin

Šamaš-iddin
no date preserved

Anu-ikṣur	*Rimut-Anu*
no date preserved	Darius II (423-405)

Anu-ušallim
no date preserved

Ekur-zakir

Ištar-šum-ereš
no date preserved

Iqiša
from 322 to 316

Ištar-šum-ereš
from 318 to 316

Ša-Anu-iššu
229

References

Abrahami, Ph. 2003. A propos des fonctions de l'*asû* et de l'*âšipu*: la conception de l'auteur de l'hymne sumérien dédié à Ninisina. *Journal des Médecines Cunéiformes* 2: 19–20.
Clancier, Ph. 2005. Les scribes sur parchemin du temple d'Anu. *Revue d'Assyriologie* 99: 85–104.
Clancier, Ph. 2008–2009. Medical tablets of the *āšipu*'s libraries. Geography of knowledge in Assyria and Babylonia Project. http://oracc.museum.upenn.edu/cams/gkab/corpus/
Clancier, Ph. 2009a. *Les bibliothèques en Babylonie dans la deuxième moitié du Ier millénaire av. J.-C*, Alter Orient und Altes Testament 363. Münster: Ugarit Verlag.

Clancier, Ph. 2009b. *SpTU 5, 231 [Ašipu's Handbook]*. Critical edition and English translation by P. Clancier. CC BY-SA 3.0 license. CAMS. http://oracc.museum.upenn.edu/cams/gkab/corpus/. (*SpTU* 5, 231).

Finkel, I. 1988. Adad-apla-iddina, Esagil-kin-apli, and the series SA.GIG. In *A scientific humanist. Studies in memory of Abraham Sachs*, ed. E. Leichty et al., 143–159. Philadelphia: University Museum.

Finkel, I. 2000. On late Babylonian medical training. In *Wisdom, gods and literature*, ed. A. George and I. Finkel, 137–223. Winona Lake: Eisenbrauns.

Geller, M. 1997. The last wedge. *Zeitschrift für Assyriologie* 87: 43–95.

Geller, M. 2000. Incipits and rubrics. In *Wisdom, gods and litterature*, Studies in Assyriology in honour of W.G. Lambert, ed. A. George and I. Finkel, 242–258. Winona Lake: Eisenbrauns.

Geller, M. 2007. Médecine et magie: l'*asû*, l'*āšipu* et le *mašmâšu*. *Journal des Médecines Cunéiformes* 9: 1–8.

Gesche, P.D. 2001. *Schulunterricht in Babylonien im ersten Jahrtausend v. Chr*, Alter Orient und Altes Testament 275. Münster: Ugarit Verlag.

Heeßel, N. 2000. *Babylonisch-assyrische Diagnostik*, Alter Orient und Altes Testament 43. Münster: Ugarit Verlag.

Hoh, M. 1979. Die Grabung in Ue XVIII 1. In *UVB XXIX-XXX*, ed. J. Schmidt, 28–35. Berlin: Gebr. Mann Verlag.

Jean, C. 2006. *La magie néo-assyrienne en contexte. Recherches sur le métier d'exorciste et le concept d'āšipūtu*, State archives of Assyria studies 17. Helsinki: Helsinki University Press.

Jursa, M. 2005. *Neo-Babylonian legal and administrative documents. Typology, contents and archives*, Guides to the Mesopotamian textual record 1. Münster: Ugarit-Verlag.

Labat, R. 1951. *Traité akkadien et diagnostics et pronostics médicaux*. Paris: Brill.

Lambert, W.G. 1957. Ancestors, authors and canonicity. *Journal of Cuneiform Studies* 11: 1–14.

Leichty, E. 1970. The omen series Šumma Izbu, TCS 4, Locust Valley.

Oelsner, J. 1986. Materialen zur Babylonischen Gesellschaft und Kultur in Hellenistischer Zeit. *Assyriogia* 7. Budapest.

Pedersén, O. 1998. *Archives and libraries in the ancient near east, 1500–300 B.C.* Bethesda: CDL Press.

Robson, E. 2008. Mesopotamian medicine and religion: Current debates, new perspectives. *Religion Compass* 2(4): 455–483.

Schmidt, J. 1972. Grabung in Ue XVIII 1. In *U.V.B. XXVI-XXVII*, ed. J. Schmidt, 56. Berlin: Gebr. Mann Verlag.

Stol, M. 1993. *Epilepsy in Babylonia*, Cuneiform monograph 2. Groningen: Styx.

Van Buylaere, G. 2009. Sign-writing exercise. *Geography of knowledge in Assyria and Babylonia project.* CC BY-SA 3.0 license. CAMS. http://oracc.museum.upenn.edu/cams/gkab/corpus/. (*SpTU* 5, 277).

Waerzeggers, C. 2003–2004. The Babylonian revolts against Xerxes and the 'End of Archives'. *Archiv für Orientforschung* 50: 150–173.

Westenholz, A. 2007. The graeco-babyloniaca once again. *Zeitschrift für Assyriologie* 97: 262–313.

Weidner, E. 1966. Ein Kommentar zu den Schlangen-Omina. *AfO* 21: 46 et pl. IX–X.

Part II
Critical Approach

Chapter 4
Does a Master Always Write for His Students? Some Evidence from Old Babylonian Scribal Schools

Christine Proust

4.1 Introduction

Ancient Mesopotamia has given us more than 2,000 clay tablets containing mathematical texts. For the large majority of them (more than 80 %), these tablets were written in the Old Babylonian period, that is, between 2000 and 1700 B.C.[1] The cuneiform mathematical corpus has been classified by Neugebauer and Sachs as follows: "*the mathematical texts from ancient Babylonia fall into two main categories: 'table-texts' and 'problem-texts'*".[2] Later archaeological and historical researches has shown that the "table-text" category essentially covers school texts, that is, exercises written by young scribes during the elementary education. School tablets have often lain forgotten in museum storerooms and have only been

[1]**Note on the dates**: All the dates given in this chapter are Before the Common Era. The dating follows "middle chronology", according to which the dates of Hammurabi's reign are 1792–1750. This chronology, although contested (Gasche et al. 1998), is the most widely used by historians of the Ancient Near East. **Note on the sources**: the information on the tablets quoted in this chapter (physical description, references, photographs, copies, transliterations, translations) are available on the Cuneiform Digital Library Initiative website (CDLI http://cdli.ucla.edu). The statistics come from the CDLI website database, completed by data from my own database which includes unpublished tablets. To April 2013, the CDLI has identified about 1940 mathematical tablets, which represents a relatively complete inventory of mathematical sources known today. We must, however, add to this total further unpublished tablets kept at the Oriental Institute of the University of Chicago (approximately 200 school tablets from Nippur), and those unearthed at Mari (140 school tablets), so a total approaching 2,300 tablets. Of these, approximately 1900 date from the Old Babylonian period.

[2]Neugebauer and Sachs (1945, 1).

C. Proust (✉)
Laboratoire SPHERE UMR 7219, CNRS & Université Paris Diderot,
rue Thomas Mann 5 75205, Paris, France
e-mail: christine.proust@univ-paris-diderot.fr

A. Bernard and C. Proust (eds.), *Scientific Sources and Teaching Contexts Throughout History: Problems and Perspectives*, Boston Studies in the Philosophy and History of Science 301, DOI 10.1007/978-94-007-5122-4_4,

published systematically in recent years.[3] Today they represent a considerable part of the known mathematical documentation. The vast majority of the known elementary school tablets come from Nippur, a site that has been excavated under relatively controlled conditions by American expeditions from Universities of Pennsylvania and Chicago. Other mathematical texts, those Neugebauer called "problem texts", can be described in a very approximate way as "erudite texts", as opposed to the texts written by young pupils. The context in which the tablets containing the erudite texts were produced is often difficult to identify due to the fact that their provenance is generally unknown. Indeed, the majority of them were bought from antiquities traders by European and American museums and collectors in the 1920s and 1930s. However, some groups of tablets which come from legal excavations do benefit from a relatively clear archeological context. Examples of this are groups found at Tell Harmal and Tell Haddad in the kingdom of Eshnunna (Diyala Valley, in the north of Mesopotamia), and at Susa in Elam, in the west of modern Iran.[4] The fact that interests us here is that, when their provenance is known, the tablets containing erudite texts were found along with elementary school tablets. The same is true for literary tablets, which are generally found along with elementary lexical exercises.

As a consequence, cuneiform mathematical texts from the Old Babylonian period are clearly the result of the activities of scribal schools and, in this sense, emanate incontestably from a "teaching context".[5] The authors of mathematical texts in the Old Babylonian period had probably functions of masters in scribal schools. Jens Høyrup describes them as follows:

> The mathematical texts are school texts. [...] their authors [...] were teachers of computation, at times teachers of pure, unapplicable computation, and plausibly specialist of this branch of scribal education; but they remained teachers, teachers of scribal school students who were later to end up applying mathematics to engineering, managerial, accounting, or notarial tasks. (Høyrup 2002, 8)

As the mathematical texts were written by masters does it follow that they were always written for teaching purposes? A positive response is more or less explicitly assumed in most of the publications devoted to cuneiform mathematics. Mathematical texts are generally considered as evidence of pedagogical activity, meaning they are assumed to have been written either by students or for students. My goal is to show the simplistic nature of such a narrow alternative.

First of all, it should be noted that the context of scribal schools is specific to the Old Babylonian period, and that what is claimed here cannot be applied identically to older periods, notably the Sargonic period (2340–2200) and the Third Dynasty

---

[3]Veldhuis (1997), Robson (2001, 2002), Friberg (2000), and Proust (2007, 2008b).

[4]Neugebauer (1935–1937), Neugebauer and Sachs (1945), Thureau-Dangin (1936), Goetze (1951), Bruins and Rutten (1961), and Gonçalves (forthcoming) who provides a complete bibliography on Diyala's material.

[5]This general context of mathematics is recognized since long time: "[Problem texts] are school products intended to illustrate the rules for dealing with problems which are properly called 'algebraic'."(Neugebauer and Sachs 1945, 1).

of Ur (2100–2000), nor to more recent periods, notably the Hellenistic period (323–63). These archaic and later periods have given us a small mathematical corpus produced under quite different contexts, which we will not address in this chapter.[6]

Turning now to the Old Babylonian corpus, the problems posed to historians in the contextual analysis of elementary school texts are not the same as those posed by the erudite texts. The elementary school texts are by definition the product of teaching activities. They are generally perceived to be simple, easy and even relatively puerile texts, created for pedagogical purposes. It is the misleading synonymy of "elementary", "simple" and "puerile" that I take issue with. The erudite texts, for their part, are generally understood to be expressing a type of didactic discourse the masters addressed to their students. The assumption that a mathematical text written by a master is a teacher's guide, may lead to misunderstanding the meaning of certain texts.[7]

Moreover, what is exactly a "scribal school" (*Edubba* in Sumerian) in Old Babylonian period is the subject of debate among historians.[8] Without entering into this discussion, let's just underline the fact that the practices of teaching seem to have been very different from one city to another. For example, in some major centers such as Nippur, education has probably been developed in a relatively large scale, in formalized institutions. However, the education may have been limited to the family activities in others cities.[9] Schools were independent from the clergy at Nippur, but priests may have played an active role in others.[10] In other contexts, such as the mid Assyrian merchants, literacy and numeracy were acquired more by the practice of the profession than by teaching in a separate institution.[11]

In this chapter, I am going to focus on three groups of Old Babylonian period tablets, each group will be the subject of a section. The first is a group of elementary school tablets from Nippur (Sect. 4.1), numbering around a thousand tablets from the same provenience, and having the same dating. The teaching situation in which these writings were produced is relatively well known. It is documented in Sumerian literary texts, in some archeological data and in the school texts themselves. Many studies of this rich and varied documentation are available to historians today.[12] One of the most interesting aspects of the Nippur tablets is precisely that they form a remarkably homogenous, coherent and structured group. In Sect. 4.1, I endeavor

[6]In (Bernard and Proust 2008) there are comments on the various contexts through the evocation of the history of reciprocal tables-texts that have passed through millennia without great modification, but whose use has changed considerably over time.

[7]I think this is the case for the interpretations that were given for texts such as CBS 1215, which deals with the inversion method of regular numbers, Plimpton 322, which deals with Pythagorean triples, or series texts (see Britton et al. 2011; Proust 2012).

[8]For an up to date presentation of the debate and for related bibliography, see (Michalowski 2012). I rely on Michalowski's analysis, who stresses the diversity of teaching situations.

[9]Tanret (2002).

[10]Charpin (1986, 420–486).

[11]Michel (2008).

[12]See the bibliography given in footnote 3.

to show how each piece of text fits within the whole and how the meaning of a text is only understandable in relation to other texts. I show that, far from being puerile texts, these school writings reveal a highly-sophisticated system, probably developed originally for purposes other than education.

The second group studied in (Sect. 4.2) is composed of procedure texts (lists of problem statements followed by their resolution) and catalogues (lists of problem statements with no indication of their resolution). At first sight, the written style of these texts evokes that of a master addressing his students. To a certain degree they could have a similar function to that of a school textbook. However, I will show that teaching was probably not the only objective pursued by the authors of the catalogues.

The third group is the series texts (Sect. 4.3), which are lists of thousands of problem statements, written on a succession of tablets and numbered like pages in a book. I show that in this case, one can seriously doubt that such texts were written for teaching purposes. In writing these vast series texts, the scribal school masters were visibly not addressing their students, but their peers.

Each of these three groups shows a certain internal consistency, which allows the texts to be placed in a rich textual environment. The meaning of each text is therefore enhanced by other texts in the group it belongs to. Analysis of these groups allows us to show that an elementary school text was not always puerile; that a text written for teaching could at the same time serve other purposes and that a text written by a master did not always have an educative objective. The contrasting analysis of these three groups thus allows me to highlight the contextual diversity of both the production and the usage of the texts, all of which nevertheless deriving from Old Babylonian-period scribal schools in southern Mesopotamia.

4.2 School Tablets and School Texts

In this section I would like to draw attention to the fact that the school *tablets*, that is, the documents produced by student scribes, do not always contain *texts* created for learning. The following examples show that, in certain cases, the texts clearly reflect pedagogical intent, but in other cases, the texts were originally developed for other purposes.

Before addressing the writings of young scribal school students, I should briefly present these schools.

4.2.1 Scribal Schools

Although the existence of scribal schools was identified by Hermann Hilprecht[13] at the beginning of the twentieth century, studies have only recently focused

[13] Hilprecht (1906).

specifically on the description of the teaching itself and the consequences that precise knowledge by historians of the organization and ideology of scribal schools could have on the interpretation of texts. Work carried out by Niek Veldhuis is probably the most innovative on this point. Veldhuis studied Nippur school tablets and has been able to recreate the elementary school curriculum followed by young scribes in this city's schools.[14] Moreover, Sumerian literature, evidenced mainly at Nippur, brings lively and detailed accounts of life in the schools.[15] Symmetrically, some literary sources are themselves seen in a new light through the relatively accurate knowledge of the teaching context from which they come. So Veldhuis proposes a new reading of the collections of proverbs which, according to him, are not only "wisdom texts" as generally stated, but are part of the training curriculum for scribes, being used above all in learning the Sumerian language. It should be remembered that in the Old Babylonian period, Sumerian was the scholastic language, but it was no longer spoken by the population at large. According to Veldhuis, Sumerian lexical lists, proverbs and literary texts form a consistent whole, a kind of network through which the proverbs take all their meaning.

> In practice this means that we may try to look at lexical and literary texts as a synchronic corpus. The educational texts from Nippur -be they literary, lexical, or proverbial- belong together in a single educational system. They share a functional, physical, historical, and presumably cognitive background. Ideally, understanding of a single composition involves understanding of the whole corpus and the structure of that corpus. [...] This way of looking at texts is hardly new in Assyriology. It finds a parallel in the so-called archival approach to administrative texts. (Veldhuis 2000, 389).

A similar analysis was developed by Herman L. J. Vanstiphout[16] on a literary composition presenting itself as a hymn to Isin king Lipit-Eshtar (reign 1934–1924). The meaning of this hymn becomes clear when one considers it was the first literary text addressed by young scribes at Nippur during their training. In fact the text presents a succession of grammatical structures for learning Sumerian verb forms. It is noteworthy that the tablets on which the extracts of this text are found often have the elongated form characteristic of supports used at the beginning of the advanced training level, which followed the elementary level.

An "archival approach", taking Veldhuis's expression,[17] can be applied to the mathematical texts and produce interesting results as it captures the originality of certain teaching concepts used in the scribal schools. For example, the surface

[14]Veldhuis (1997).

[15]Several authors have drawn attention to the fact that the literary texts used for training did not objectively reflect the reality of schools but rather reflected the ideology particular to the scribal milieu of Nippur (Michalowski 1987, 63; George 2005). Some literary texts, however, contain many details on the school curriculum that are confirmed by independently obtained data, and are therefore valuable sources of information.

[16]Vanstiphout (1979).

[17]This method consists in analyzing the text, not in isolation, but as an element in a collection of documents found at a precise archaeological *locus*. See on this subject (Veenhof 1986).

calculation exercises found at Nippur can only be understood if one takes into account their relationship to other mathematical school texts from the same site (see below).

The Nippur site was excavated in a systematic way by teams of American archeologists from the University of Pennsylvania for the first campaigns (1888–1900), then by teams from the Universities of Pennsylvania and Chicago for the latest (1948–1990). These expeditions led to the discovery of thousands of school text tablets, found mostly in the relatively small zone known as "Tablet Hill" where the scribal schools seem to have been located.[18] These "archives" date for the most part from the reign of Hammurabi's son, Samsu-Iluna (1749–1712). Where the archeological context of the school tablets is known, it shows that they were thrown away after use and recycled as construction material. These tablets are divided today between several collections kept in Philadelphia (the collection includes approximately 500 mathematical tablets), in Istanbul (312 mathematical tablets), in Jena (79 mathematical tablets), in Chicago (approximately 200 mathematical tablets), and in Baghdad.[19] These collections total over a 1,000 pieces, of which around 900 have been published. They are therefore quite representative of Nippur scribal school output in the field of elementary mathematics during the first half of the eighteenth century BCE.

Elementary school tablets from Nippur are easy to recognize as they have specific physical characteristics which differentiate them from other mathematical tablets. Four types of elementary school tablet have been identified at Nippur: those of type I are large, multi-column tablets; those of type II contain a master's example and the pupil's copy on the obverse and the transcription of a memorized text on the reverse; type III are small elongated tablets containing a short extract and type IV are small lenticular or square tablets.[20]

Another characteristic feature of elementary school tablets is that they contain stereotypical texts similar to those found in all Mesopotamian scribal schools. From studying these we can reconstruct the scribes' mathematics training curriculum, which, broadly, was made up of an elementary, an intermediate and an advanced level. The first two levels are relatively well known thanks to information provided by physical evidence, the most important being in relation to the typology of the tablets. The advanced level is much less documented (see Sect. 4.2). The elementary and intermediate stages of the mathematics curriculum are represented as follows (Table 4.1):

The metrological lists are the enumeration of measurements of capacity, weight, surface and length, in that order. The metrological tables enumerate the same items as the lists but give for each measurement a correspondence to an abstract number

---

[18]The best documented of the scribal schools at Nippur is "House F", whose contents were analyzed by (Robson 2001).

[19]These tablets are for the large part published (Proust 2007, 2008b; Robson 2001) or are on the way to being published.

[20]For more details, see (Veldhuis 1997; Robson 2001; Proust 2007, 2008b).

Table 4.1 Mathematics curriculum at Nippur (Proust 2007, 268–9)

Texts	Number of tablets were these texts are identified	Level	Tablet types
Metrological lists	187		
Metrological tables	161	Elementary	I, II, III
Numerical tables	417		
Exercises	35	Intermediate	IV

written in sexagesimal place value notation, characteristic of cuneiform mathematics.[21] The numerical tables are made up of reciprocal tables, multiplications, squares, square roots and cube roots; they are written entirely in a floating point notation. The exercises focus mainly on calculating reciprocals, multiplication and surfaces.[22]

Were all these texts designed only for teaching? The answer seems clear (and positive) to me for the calculation exercises, but less clear for the metrological lists and numerical tables.

4.2.2 *School Tablets Containing Didactic Texts*

Among the documents that correspond to the intermediate level, are found ten or so tablets containing a calculation for the surface of a square, carried out using the same model: a short text is written in Sumerian in the lower right-hand corner, and several numbers are placed one under the other in the upper left-hand corner. These type IV tablets have the characteristic cushion shape of intermediate-level school text tablets at Nippur. Tablet UM 29-15-192, a copy and transcription of which are given below is an example (Table 4.2).[23]

This tablet is striking in that the layout of the text is in two zones. On closer inspection we notice that the numeric data in the two zones are not of the same nature: the data in the lower right-hand corner are metrological (side and surface), written in accordance with the rules of the metrological lists; the data in the upper left-hand corner are made up of sexagesimal place value notation of the same type as those found in the numerical tables. More interesting still is that there is a correspondence between the numeric data in the two areas, and that this correspondence is exactly that which the metrological tables give. For

[21]A detailed description of the metrological lists and tables, as well as place value notation can be found in (Proust 2009b). An important feature of cuneiform sexagesimal place value notation is the fact that no mark indicates the order of magnitude of numbers, or, in other words, that the notation is "floating".

[22]See (Proust 2007, ch. 6 and 7).

[23]Tablet UM 29-15-192 comes from Nippur and is kept in Philadelphia; it is published in (Neugebauer and Sachs 1984, 251).

Table 4.2 Tablet UM 29-15-192 (Neugebauer and Sachs 1984, 251)

Transcription	Translation	Copy
20	20	
20	20	
6.40	6.40	
2 šu-si ib$_2$-sig$_8$	2 šu-si the side (of the square)	
a-ša$_3$-bi en-nam	What is the surface?	
a-ša$_3$-bi igi-3-gal$_2$ še-kam	Its surface is 1/3 še	

example, the metrological table of lengths establishes that the measurement 2 šu-si (approximately 3 cm), which we read at the bottom on the left-hand side, corresponds to the number 20 in sexagesimal place value notation. And this is exactly the same number as is written in the top left-hand corner. The product of the calculation 20 times 20, necessary for the evaluation of the surface, is provided in the 'by 20' multiplication table. The result is the number 6.40 which is placed under the number 20. The metrological tables establish that the number 6.40 corresponds to a surface of 1/3 še, which is that given in the text at the bottom. So the two number systems used in the two zones on the tablet refer to the two number systems in the metrological tables. The calculation of surfaces is based on the transformation of metrological data into "floating" sexagesimal place value notation to allow the multiplication to be carried out easily, then by reading the metrological table in reverse to obtain the result as a measurement of surface. Note that mental evaluations of orders of magnitude are necessary in the last stage of this process. Calculation using "floating" sexagesimal place value notation greatly facilitates the passage of linear dimensions into bi-dimensional elements; the calculation of the product would be extremely problematic if units of measurement were involved because of the irregularities of metrology. There is, therefore, in the calculation practiced by the scribes, no conversion from one unit to another, contrary to what the reading of the tablet alone, isolated from its contextual environment, may suggest.[24]

But the reach of these results goes much further: to understand the methods taught to young scribes is also to understand the fundamentals on which the scholars

[24]This last interpretation, based on the idea that surface and volume calculations use unit conversions, is largely referred to in the publications on this subject. I have explained in detail the reasons for my disagreement with this interpretation in (Proust 2008a). A recent example where the surface exercises are explained, in an uncontextualized way, by the conversion of units is found in (Robson 2008, 8–12).

relied. As a result, the elementary school texts provide historians today with the opportunity to interpret advanced mathematical works not using their own algebraic tools, but rather with methods forged by the scribes themselves.

Tablet UM 29-15-192 is a school text in the sense that it was written by an apprentice scribe as he learned mathematics. In addition, its content has a clearly didactic function: the text is designed to teach precise mathematical knowledge. As the layout of the text on the tablet shows, the learner is guided by a master whose intentions are perceptible: to explain the relationship between sexagesimal place value notation and metrological data in surface calculations, and using this relationship in a real calculation.

4.2.3 *School Texts are not Puerile Texts*

If we now consider metrological tables (or multiplication tables), the analysis is a little different. As indicated above, the versions of the texts found at Nippur come from elementary school tablets. But were these tablets created especially for teaching? Several clues seem to indicate that this is not the case. On one hand, the oldest reciprocal tables, dating from the end of the third millennium were written on tablets different in appearance to the Old Babylonian tablets. Indeed, they are of much higher quality: refined clay covered of fine writing. They do not appear to have been produced by inexperienced young apprentices, but rather by expert hands. The oldest versions of the tables were not, perhaps, written for elementary teaching. On the other hand, metrological and numerical tables are also documented as being written on valuable supports, very different from those used in teaching, as are type I, II and III tablets at Nippur (see Table 4.1 above). An example is the beautiful AO 8865 Prism kept at the Louvre and containing metrological tables, tables of squares, square roots and cube roots.[25] The prism is transpierced by an axial hole, around which the prism can be rotated to change the view; the tablet seems therefore to have the function of a reference document. It is clear, moreover, that the tables, which were assimilated by the young scribes at Nippur at the beginning of their mathematical training, also served as data sets that would be brought into use throughout their career. They are, therefore tools which were not necessarily initially made for teaching but that were used in teaching, a little like the trigonometry tables which modern schoolchildren carried in their schoolbag before the era of the first electronic calculators.

In fact, the lists and metrological tables probably originally had, at the time of their design, an essentially normative function. They were developed as part of policies of large scale standardization and unification of systems for weights and measures undertaken by the first centralized states in Mesopotamia in the

[25] Proust (2005).

second half of the third millennium.[26] These normative texts were produced by a bureaucracy with the objective of economic and social regulation. The normative texts seem to have been re-used in education, and set out by the masters in a particular way on the school text tablets (of types I, II and III at Nippur), to allow their assimilation and probably memorization by the future scribes.

These remarks show the ambiguity of the term "school text": a *tablet* could be written by a student scribe, without the *text* it contains being didactic, at least in its original intent. These distinctions between the text and its support on one hand, and between the creational context and the context of its use on the other, here echo those made by Anne-Marie Chartier (Chap. 2 in this volume) regarding the use of Psalters in learning to read in the seventeenth century. Like metrological lists, Psalters were not created with the intention of teaching, but were used for this purpose (*ibid.*).

Another ambiguity comes from the fact that the words "school texts" are generally applied by Assyriologists to all texts used for the learning of cuneiform writing and calculation. However this term carries the implication of the initial training of young people, sometimes children, who did not already know how to read or write. This was the case in Old Babylonian scribal schools. But the texts used in the Hellenistic period for learning cuneiform at temples in Uruk and Babylon were certainly not addressed to children taking their first steps in writing and arithmetic, but to young Aramaic speaking students, already fluent in written and spoken Greek, learning the difficult cuneiform script and the dead languages (Sumerian and Akkadian) it conveyed, with the intention of acceding to scholarly professions such as medicine and divination (see Chap. 3 by Philippe Clancier). Although the texts used in these two cases were often the same and certain elements of the Old Babylonian tradition were passed down from one generation to the next for more than the 1,500 years that separate the two periods, it is clear that in the Old Babylonian and the Hellenistic periods, the words "school text" referred to completely different actors and teaching practices.

This section has shown, through several examples, the importance of differentiating between the *tablets* and the *texts* that the tablets contain. It has also shown that "school" texts are neither necessarily "didactic" nor "puerile". The school text tablets from Nippur sometimes contain didactic texts (see the example of the surface calculation); sometimes contain normative texts of non-scholastic origin (see the example of the metrological texts). It is equally possible that some school texts, written originally for teaching purposes, could later become scholarly texts belonging to a cultural heritage rather than an educational curriculum (we can think of certain Sumerian literary texts). Finally, the same texts could have been used, at certain times, in the initial training of scribes when they were still children, and, at other times in the specialized teaching of young scholarly professionals (see the example of the savants of Uruk and Babylon in the Hellenistic period).

[26]For a detailed analysis of the normative character of the metrological texts, see (Proust 2009b).

4.3 Texts Written by Masters for Advanced Teaching

In the elementary school texts, the historian has a wealth of structured documentation having relatively clear archaeological context. For the "erudite" texts,[27] the situation is far from being as favorable, because, as is indicated in the introduction, most of the known erudite mathematical texts come from illegal excavations. Their dating and geographical provenance can only be assessed from indirect evidence. Lexical and epigraphical criteria allow Old Babylonian tablets to be clearly distinguished from these from Late Babylonian and Hellenistic periods. Within the large corpus of Old Babylonian texts, one can differentiate groups having different origins and dates. The first philologist to discern regional differences was Albrecht Goetze.[28] A linguistic study of Akkadian terminology and graphics allowed this scholar to make a distinction between northern and southern texts, early and later texts; and in some cases, he was able to define homogeneous groups probably originating from the same site. The "erudite" texts bear witness to several different mathematical traditions, the three most significant, but not the only ones, being those of southern Mesopotamia, the kingdom of Eshnunna and Susiana. According to Høyrup, who extended Goetze's classification (Høyrup 2002, ch. 9), the different communities of scribes cultivated specific mathematical interests and developed different methodological approaches. The diversity of practices is also evident in the typology of the tablets, the language (Sumerian and Akkadian) and the writing techniques used. The examples studied in this chapter probably all come from the southern Mesopotamia (from Ur to Sippar) and as a result I will focus principally on information on this area.

A first series of physical clues comes from the study of literary tablets. A typological analysis of the tablets from scribal schools at Nippur was carried out by S. Tinney.[29] This author differentiates tablets written in a single column (type S), corresponding to the beginning of the advanced curriculum, from tablets written in multiple columns (type M). Although based on literary documentation, this typology probably has meaning for mathematical texts because the artifacts produced by the schools show that the type of tablet depends on the level of teaching and the didactic situation, not the genre of text. The typology of tablets used at Nippur was not always adopted by scribes in schools in other cities. However, it does provide clues which, used with caution and cross-referenced with other information, can help clarify contextual elements of documents of unknown provenance. This section will be devoted to texts written on type S tablets, and Sect. 4.3 to texts written on type M.

[27]I use the term "erudite" here in contrast to "elementary school" (see the introduction of this chapter). This term refers to teaching texts destined for advanced student or pieces of pure erudition.

[28]Goetze (1945).

[29]Tinney (1998, 46, 1999).

Table 4.3 Concordance

	Museum number	Category of text	Type of tablet	Origin
A	YBC 4663	Procedure text	S	Antiquities market
B	YBC 4662	Procedure text	S	Antiquities market
C	YBC 4657	Catalogue	S	Antiquities market

The three texts examined in this section were selected partly because they are good examples of Old Babylonian period mathematics teaching texts and partly because they are closely interconnected. For ease of reference, the three tablets considered in this section are lettered (A, B and C). The concordance with the museum (here, the Yale Babylonian Collection) inventory numbers is as follows (Table 4.3):

4.3.1 Two Procedure Texts: The Masters Teach

Tablet A (YBC 4663) is very elongated and inscribed in a single column; the text ends on the reverse with a line, followed by a large blank space (see plate, Fig. 4.1). The origin of the tablet is unknown, but Goetze classified it in the southern groups.[30] It contains a suite of eight solved problems on the digging of a trench, each problem is written in a section delimited by horizontal lines. The parameters of problems 1–6 (data and unknowns) are the dimensions of the trench (length, width and depth), its base, the volume of earth extracted, the number of workers necessary to dig the trench,[31] each worker's daily task, that is, the volume of earth assigned to them to be dug each day, their daily wage, their total wages (wages are expressed in weights of silver). All these parameters are linked by the simple relationship:

$$\text{total wages} = \text{daily wage} \times (\text{length} \times \text{width} \times \text{depth})\,/\text{daily task}$$

The first problem gives values for the following parameters: the length, the width, the depth, the task (the volume assigned each day to each worker), the workers' daily wage. It requires the calculation of: the base, the volume, the number of workers necessary and the total wages. The resolution procedure is explained. The next problems (2–6) are variations that consist in carrying out circular permutations between given and unknown parameters. Problems 7 and 8 also offer variations of the same situation, but they are more complex as the data relate to linear

[30] Goetze (1945, 148).

[31] It is assumed here that, by default, the work lasted 1 day: other texts lead to the belief that the work lasted 9 days (see YBC 4657 and YBC 4662 quoted later and Neugebauer and Sachs 1945, 74), and therefore the "number of workers" is in fact the number of workers × days (30 workers for 9 days).

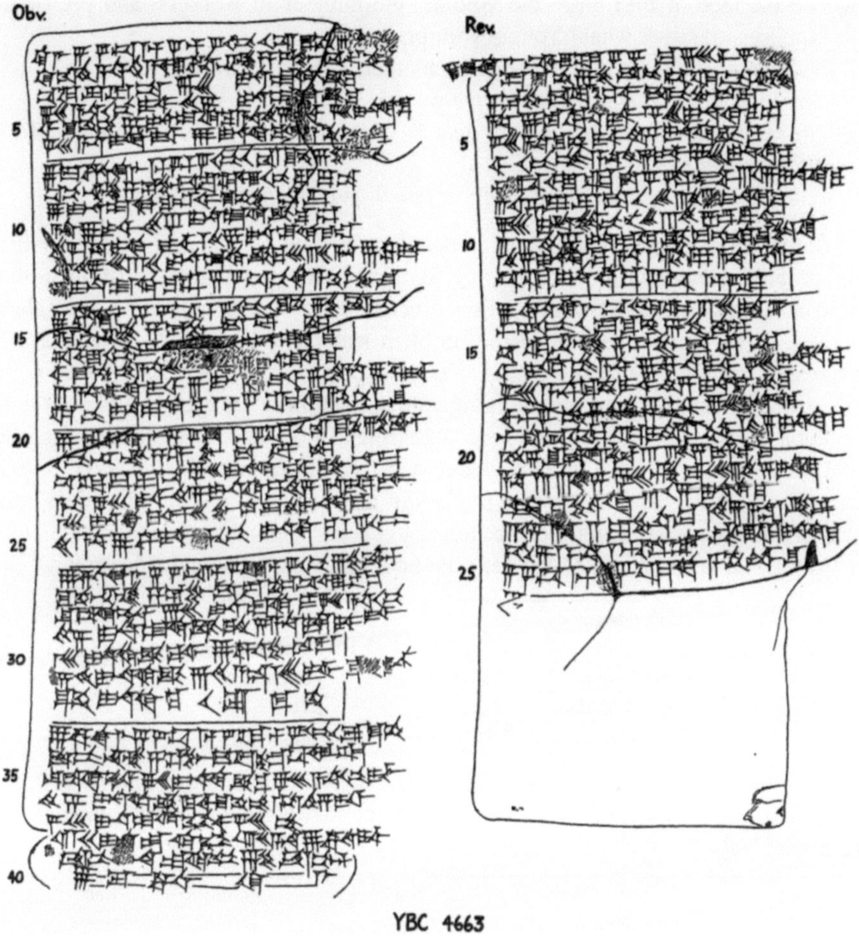

Fig. 4.1 Procedure text, type S tablet (YBC 4663, Yale University) Copy Neugabeur and Sachs 1945, plate 7

combinations of length and width, leading to what we call today a quadratic problem. The numeric values in the eight problems are always the same. The group forms a coherent structure.

Meticulous examination of the text shows how the resolution of the problem employs the arithmetical tools taught in the elementary levels. Consider for example the first problem statement, for which the translation is as follows[32]:

1. A trench. 5 ninda is the length, $1\frac{1}{2}$ ninda the width, $\frac{1}{2}$ ninda its depth, 10 gin_2 the volume of assignment (for each worker), 6 še (silver) [the wages of a hired man].

[32]I use standard font for the words translated from Sumerian ideograms, and italics for the words translated from Akkadian.

2. The (surface of the) base, the volume, the number of workers, and the (total expenses in) silver what? You, in your procedure,
3. The length and the width multiply each other. *This will give you* 7.30.
4. 7.30 *to its depth raise. This will give you* 45.
5. The reciprocal of the assignment detach. *This will give you* 6. *To* 45 *raise. This will give you* 4.30
6. 4.30 *to the wages raise. This will give you* 9. *Such is the procedure.*

Before examining this short text in detail, a few comments are in order. Firstly, the problem statement is written using Sumerian ideograms (standard font in the translation above), and the resolution procedure is written in Akkadian (italics above). Then, we notice that in the problem statement (lines 1–2), the data are expressed as concrete measurements, with units of measurement in accordance with the standards provided by the metrological lists, but only sexagesimal place value notation appears in the resolution procedure (lines 3–6). These observations suggest that, as in the case of the surface calculation referred to on tablet UM 29-15-192 mentioned above, there has been a process of conversion of metrological data into sexagesimal place value notation according to the correspondence established in the metrology tables. It is indeed the case, as shown by the correspondences below.[33]

Length:	5 ninda	→5	(table L)
Width:	$1\frac{1}{2}$ ninda	→1.30	(table L)
Depth:	$\frac{1}{2}$ ninda	→6	(table Lh)
Daily task:	10 gin_2	→10	(table S)
Wage:	6 še	→2	(table P)

The calculation, that is to say, the sequence of multiplications and inversions described in the procedure, is carried out on the floating sexagesimal place value notation.[34]

Base:	$5 \times 1.30 = 7.30$	→$7\frac{1}{2}$ sar	(table S)	line 3
Volume:	$7.30 \times 6 = 45$	→45 gin_2	(table S)	line 4
Workers:	$45/10 = 45 \times 6 = 4.30$	→4(geš$_2$) 3(u)	$4 \times 60 + 30$	line 5
Total wages:	$4.30 \times 2 = 9$	→9 gin_2	(table P)	line 6

The final step, omitted in this section but present in other sections of the tablet, is to convert the sexagesimal place value notations into concrete measurements by using the metrological tables in reverse (see Sect. 4.1).

[33]One can refer to the metrological tables in their entirety on the CDLI website (Proust 2009b). The arrows represent the "reading" of the tables, that is to say the correspondence established by the metrological tables. These are referred to, in brief, as follows: "table L" means the length measurement table, "table Lh " means the height measurement table; "table S" means the surface measurement table; "table P" refers to the weight measurement table. For more details on the calculation of volumes (units, using the height table and surface tables for volumes), see (Proust 2008a).

[34]The following equalities must be understood modulo a factor 60^n, n any positive or negative integer.

The interpretation above focuses on the arithmetical calculations. But another, more geometric aspect is reflected in the Akkadian vocabulary of the procedure, notably through the use of several different terms for multiplication (that I've translated by the verbs "to multiply" and "to raise"). These allusions to geometric figures show what the meaning is of each of the operations. This demonstrative aspect of Babylonian mathematical procedures, perceptible in the vocabulary chosen by the scribes, was discovered by Høyrup and developed in particular in (Høyrup 2002).

Analysis of the above calculations shows that the text's authors were steeped in the elementary knowledge taught in scribal schools. It is clear that metrological tables were used to carry out the calculations. This usage is confirmed by the fact that the correspondences between measurements and sexagesimal place value notations recorded on tablet A coincide perfectly with the metrological tables. One can add that, as noted above, metrological and numerical tables of the type found in abundance at Nippur have also been found in most sites in the Near East that housed ancient scribal schools. Tables identical to those found at Nippur have, for example, been recorded at Ur, Uruk, Larsa, Sippar, Kish, Ishchali, Mari, and even Ugarit. This suggests therefore that the computational tools that can be reconstructed from the Nippur documents were widely used elsewhere.

It is noteworthy that these short, simple problems provide an opportunity to use, one after the other, direct and reverse reading of all the metrological tables, carry out surface and volume calculations, multiplications and inversions. The entire repertoire of elementary calculation is presented in a systematic way. This didactic progression presents a certain analogy to the repertoire of verb forms presented in the hymn to Lipit-Eshtar evoked above. This suggests that Tablet A was probably used in the very beginning of an advanced level of learning. This stage corresponds, in the field of literature, to learning the basic structures of Sumerian. In the field of mathematics, this stage seems to correspond to learning techniques used in solving problems on surfaces and volumes.

Several tablets closely related to the one we have just examined provide a rich information on textual environment. Indeed, the Yale collection contains another procedure tablet (designated B in the following), very close to Tablet A in its inventory number, its appearance, its content and its structure. Like tablet A, tablet B is a type S, whose problems relate to a trench of the same dimensions, and with the problem statements written in Sumerian and their resolutions in Akkadian.

4.3.2 *A Catalogue: The Masters Classify, Order and Arrange*

Another tablet, C (YBC 4657 – see Fig. 4.2), contains the problem statements found in procedure texts A and B and in the same order. Tablet C, also a type S, is kept at Yale and was probably bought at the same time from the same dealer. The tablet contains a list of 31 problem statements written in Sumerian, all of them dealing with the excavation of a trench. Contrary to A and B, it does not show the resolution procedures, but only the answer to the question.

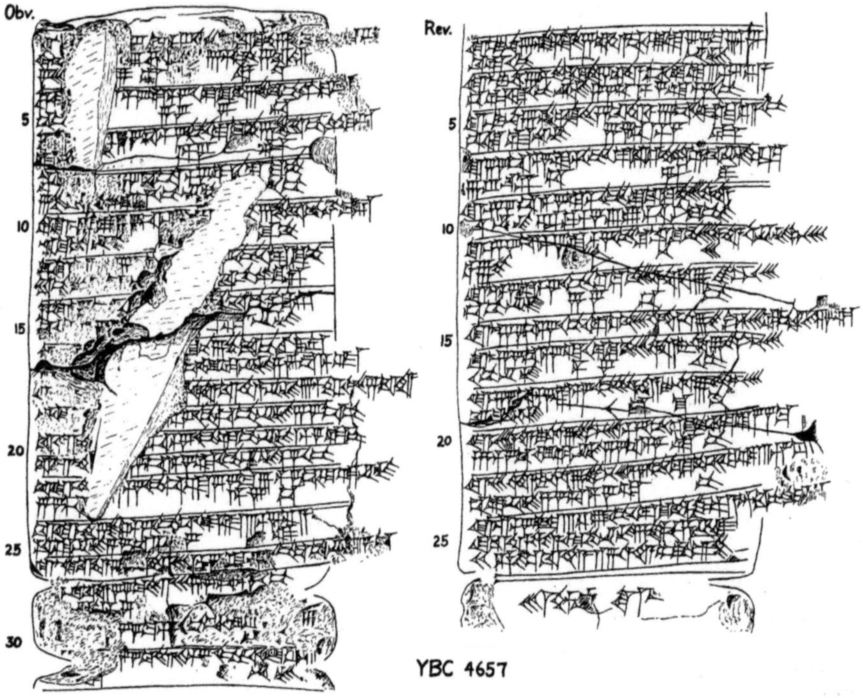

Fig. 4.2 Catalogue, type S tablet (YBC 4657, Yale University)

As noted in the introduction, the lists of problem statements are called "catalogues" by specialists.[35] Currently ten catalogues of mathematical problems are known to us: the eight kept at Yale are of unknown provenance; the two at the Louvre come from Susa. The catalogues have the following peculiarities: they are lists of problem statements with no indication of how to resolve them; they are nearly entirely written in Sumerian ideograms, they are thematically homogenous and they finish with a colophon giving the number, and sometimes, the theme of the problem statements. The thematic unity of the catalogues is underlined by the fact that the theme is indicated at the beginning of each section by the use of a kind of key word in the role of a tag: trenches (ki-la$_2$), fields (a-ša$_3$), canals (pa$_5$-sig), weights of stone (na$_4$), and by the fact that this tag is often repeated in the colophon.

Tablets A, B and C are therefore very closely related and probably come from the same site.[36] Their characteristic typology leads them to be linked to the same level of mathematics training. Why then do we find, in the same context, two different

[35]The name comes from Jöran Friberg and from Jens Høyrup (2002, 8 including note 13).

[36]The three texts are published together in the same section by Neugebauer and Sachs (1945, 66 ss.), who underline: "These three texts form a closely knit group." (ibid, 73). However, the question of whether procedure texts derive from the catalogues (or vice versa) is not addressed in

categories of texts for the same problems, one with resolution procedures and one without? What is the relationship between the procedure texts and the catalogues? Were the procedure texts derived from the catalogues or the catalogues derived from the procedure texts? And what does this relationship tell us about the respective functions of both categories of texts?

A first hypothesis is that the catalogue was a type of guide used by a master and the procedure texts derived from it: the problem statements were taken from the catalogue, then solved, for example, by a student. It would be a relationship between a "master text" and extracts as often found in the school archives for Sumerian literary texts (see, for example the case of the hymn to Lipit Eshtar quoted above and the corresponding bibliography).

Another hypothesis emerges when examining the overall structure of the texts. Catalogue C is in fact composed of several juxtaposed groups, as highlighted by Neugebauer and Sachs (1945, 74).

The first eight problem statements in catalogue C form a coherent group, which is exactly that found on tablet A.

The next problem statements in catalogue C (9–12) form a second group made up of variants that involve only the dimensions of the trench (length, width, depth, base and volume). After producing four problem statements with circular permutations of these parameters (9–12), the scribe introduces, into the data, linear combinations of length and width (13–14), then the sum of the base and the volume (15–16), and at the same time a linear length and width combination on one hand, and the sum of the base and the volume (17–18) on the other. This second group forms, like the first, a systematic, coherent whole that increases progressively in difficulty and finishes with quadratic problems.

The next three problem statements (19–20) are variants using the same parameters as the second group, but they lead to the use of different procedures (they come down to linear systems).

The fourth group (22–28) amounts to problem statements similar to those of the first group, but involve a large number of parameters: the dimensions of the trench, the daily wage, the daily task, the number of workers and, what is more, the number of working days. These two groups form a body of problems of increasing complexity, describing a wide range of linear problems. This is the coherent whole found on tablet B.

The two problems after that lead to quadratic systems similar to those at the end of the first group. The last problem statement is identical to the first (1), but the numerical data differ from the rest of the tablet as the trench has a square section. These three sections (29–31) are somewhat disparate, and do not appear to form a coherent group.

The tablet ends with a colophon which indicates: "31 trench problems".

Neugebauer and Sachs 1945 nor indeed is the issue of the possible difference in function between procedure texts and catalogues.

Table 4.4 Problem groups in catalogue C

Groups	Section of C	Data	Mathematical content	Corresponding procedure text
1	1–6	Trench dimensions, wages, number of workers	Linear problem	A
	7–8	Idem including: linear combinations of length and width	Linear- quadratic system	
2	9–12	Trench dimensions	Linear problem	Unknown
	13–14	Idem including: linear combinations of length and width	Linear- quadratic system	
	15–18	Idem including: sum of the base and the volume	Linear- quadratic system	
3	19–21	Trench dimensions including: linear combinations of length and width	Linear system	B
4	22–28	Trench dimensions, wages, number of workers, length of time worked	Linear problem	
5	29–30	Trench dimensions, wages, number of workers, length of time worked including: linear combinations of length and width	Linear- quadratic system	Unknown
6	31	Trench dimensions, wages, number of workers	Linear problem	Unknown
	Colophon	"31 trench problems "		

4.3.3 Procedure and Catalogue Texts

In Table 4.4 below the different groups of problems in catalogue C and corresponding A and B texts are summarized[37]:

All catalogue C problem statements are homogeneous from a thematic point of view: all focus on the problems of a trench, all begin with the same formula "a trench" (ki – la$_2$), the dimensions of the trench are the same in all the texts (except the last problem statement). But over the 31 problems there is no regular progression in the methods brought into play, nor systematic classification. However, there are clearly differentiable subgroups that are highly coherent and show an internal progression. Problems 1–8 in group 1 (see Table 4.4), found in the tablet containing A procedures, show a clear didactic progression. It is the same for Group 2, for which associated procedures have not been found. Problems 19–28 of groups 3 and 4, found in the tablet containing the B procedures, form a mathematically homogenous whole and focus on linear situations.

[37]For more details, see (Neugebauer and Sachs 1945, 74; Proust 2012).

The relationship between the catalogue and the procedure texts is therefore probably the opposite of what was assumed above: the catalogue is a kind of compilation made up of several pieces of pre-existing procedure texts. The function of such catalogues in this case remains unclear. It could, for example, be related to the establishment and institutionalization of an advanced mathematical training curriculum. It is possible that setting down this curriculum in writing went hand in hand with the development of the first libraries.[38] This would be a phenomenon similar to that described by Philippe Clancier for the "Manual of the Exorcist", in, it is true, a quite different context (see Chap. 3).

In this chapter, we have seen a sample of texts clearly linked to education, but in different ways for the procedure texts and catalogues. These are mainly texts written on type S tablets, i.e. tablets, which at Nippur and probably throughout southern Mesopotamia in the Old Babylonian period, were used in a stage between basic education and professional scholarliness. Some of these texts appear to result from the training of students in problem-solving (procedure texts A and B). Others, the catalogues, could be compilations bringing together a vast repertoire of problems drawn from several different procedure texts. These compilations could have been developed by masters to streamline the organization of the mathematical training curriculum, and maybe to classify and archive educational texts, thus constituting the first libraries.

Teaching, classifying and archiving figure among the possible objectives pursued by the authors of the procedure texts and catalogues analyzed here. The example of the "mathematical series texts" presented in the third part testifies that the scribal school masters could also develop projects not directly linked to their teaching activities.

4.4 Texts Written by Masters for Their Peers

Mathematical series texts (i.e. written on several numbered tablets – see introduction) form a small corpus comprising around 20 tablets. The majority of them are at Yale University, but examples are also kept in Berlin, Paris and Chicago.[39] These tablets contain long lists of problem statements (more than 1,500 in total). Their geographical provenance is unknown, and there is no archeological contextual data relating to them. The corpus of mathematical texts arranged in series is remarkably homogenous; it presents the following characteristics: (1) they are multi-column tablets; (2) the tablet numbers in the series are recorded in the colophon; (3) the colophon also contains the number of problem statements written on the tablet;

[38] Glassner (2009).

[39] For the tablets at Yale and in Berlin see Neugebauer (1935–1937), I ch. 7 and III pp. 27–45; for those in Chicago, see Neugebauer and Sachs (1945), texts T and U; for those at the Louvre, see Proust (2009a).

(4) the writing consists only of Sumerian ideograms; (5) no indication for the resolution of the problems is given; (6) each list is on a single theme (e.g. fields, or canals, or bricks); (7) the style is extremely concise and the lists highly complex. Based on this brief description, one might think that the series texts are quite similar to the catalogues: features 3–6 listed above are indeed shared by both categories of texts. In fact, Neugebauer and Sachs considered that there was nothing to differentiate the series texts from other mathematical texts, and there was no justification in giving the series texts a specific name:

> Several of the texts which present large numbers of problems without giving answers bear colophons giving the tablet a serial number. This gave rise to the name "Series Texts" used in MKT for this whole group of tablets. We think it wise, however, to abandon this name because the new material makes it difficult to define the borders of this group. (Neugebauer and Sachs 1945, 37).

The function of the series texts, therefore, should be no different from that of the catalogues. For Neugebauer and Sachs, the series texts were repertoires of teaching exercises, an opinion that was regularly taken up later in the historiography. Here is what Neugebauer and Sachs say about the two series texts kept in Chicago:

> This text, like the following one, can best be compared to an extensive collection of problems from a chapter of textbook. It is obvious that a collection of this sort was used in teaching mathematical methods. They constitute a large reservoir of problems from which individual problems of any required type (say, speaking from modern point of view, of a certain category of quadratic equation) could be selected. (Neugebauer and Sachs 1945, 116).

I would like to emphasize here the differences between the series texts and the catalogues and demonstrate that these differences reflect that their authors were not pursuing the same objectives.[40]

The typology of the series tablets is a first clue to take into consideration. Unlike the catalogues, series texts are written on type M tablets (see Fig. 4.3). Another peculiarity of the series texts lies in the writing techniques used to describe the operations and the parameters; these techniques are evidence of virtuosity not seen in the catalogues. The structure of the problem statements is also quite unique. We have seen that in the case of the catalogues, the lists are essentially built on a set of circular permutations of the parameters. In addition, each problem statement is autonomous in the sense that it contains all the information necessary for the understanding and resolution of the problem. In the case of the series texts, the construction of the lists is only marginally based on circular permutations. The structure is based on a system of linked variations producing an arborescent structure.

To illustrate this process, let us take as example an extract of tablet AO 9071 (see Fig. 4.3). According to its colophon, this tablet is the seventh in a series and

[40] I shall confine myself here to very general remarks about the series texts. Entering into the detail of the texts largely goes beyond the scope of this contribution. For more information, see (Proust 2009a, 2012).

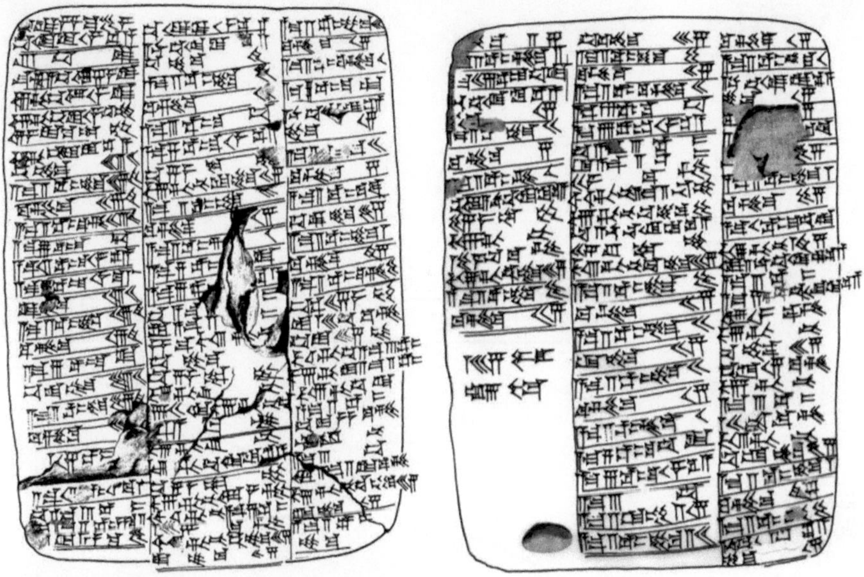

Fig. 4.3 Series text, type M tablet (AO 9071, Le Louvre Museum)

contains 95 problems on the dimensions of a rectangle; the unknowns are the length and width. Problem statement 59 reads:

I added: 45 (ba-zi-ma 45)

It is clear that such a problem statement is incomplete and does not contain all the information required to solve the problem. The necessary information is in fact dispersed throughout the previous sections. Once all the information has been recovered, the problem can be represented in modern language by the following linear system, where x is the length of the rectangle and y is its width:

$$\begin{cases} (3\times x+2\times y)\,\frac{1}{13}+x=40 \\ \{\langle\{[(x+25)+(y+1.30-x)+(x+y+35)]\,\frac{1}{11}+4x\}\,\frac{1}{7}+y\rangle \\ \quad\times 2\times\frac{1}{16}-(x+2y)+x+y+(3x-2y)\}\,\frac{1}{7}+(x+y)=45 \end{cases}$$

How can a problem statement so concise represent a problem so complex? The process is as follows: the problem statements are made up of four blocks, and each block undergoes changes. Block 1 (which here corresponds to the first equation in a linear system) undergoes several variations; for each variant in block 1, block 2 undergoes several variations; for each variant in block 2, block 3 undergoes several variations; for each variant in block 3, block 4 undergoes several variations. Thus we obtain a four-level arborescent structure. Information is grafted onto this tree-like

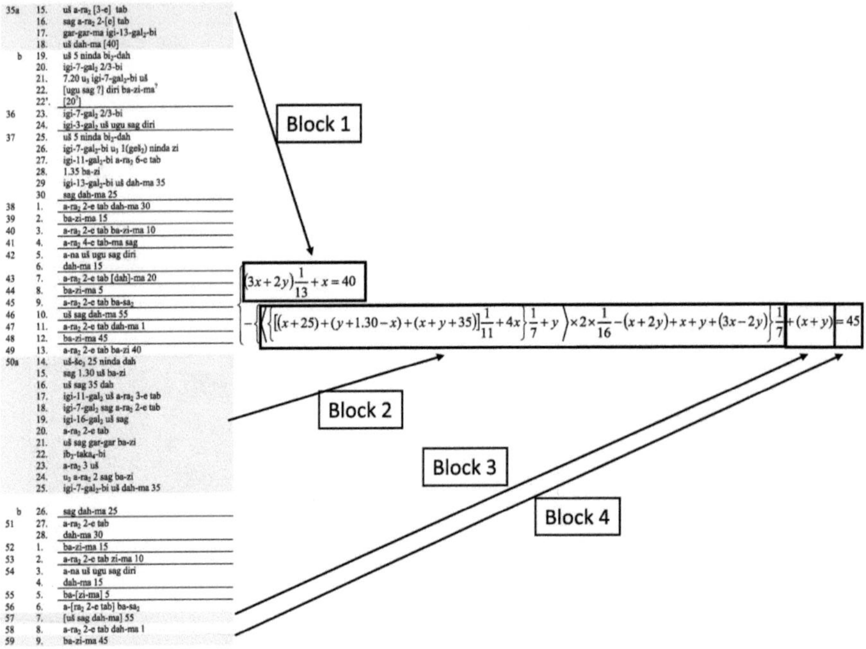

Fig. 4.4 Distribution of the information on series tablet AO 9071, #59

structure; its distribution accords the following principle: any information given in one level of the arborescence is omitted from statements in the levels which derive from it. As a result, when reading one section, the user has to go back in the text to look for information given in the previous sections. In the case of this problem statement, the information is found at four different points: in sections 35, 50, 57 and 59.

The following diagram shows how the information referred to in problem statement 59 is distributed in the text (Fig. 4.4).

Moreover, the text is compressed with various techniques, such as the omission of grammatical particles, or the representation of complex expressions by an ideogram. Finally, we obtain a highly concise and technical text requiring a lot of expertise from the reader.

Another major difference between the catalogues and the series texts concerns the mathematical content of the problem statements. The problem statements in catalogues are widely represented in cuneiform mathematics. The parallels between the two procedure texts A and B and the catalogue C are a striking illustration of this. Conversely, the problem statements set out in the series texts are, for the most part, unique in the known mathematical corpus. One can even wonder if some of them were ever intended to be solved. The example given above (#59 from AO 9071) is an example of this. No known Old Babylonian source bears witness to the existence of methods to reduce such a problem. Some simple (and very laborious) reductions are documented in some texts from Susa.[41] But, assuming that these reduction techniques were also carried out in Mesopotamia, they remain insufficient to overcome many of the problems found in the series texts. One can argue this is an *ex silentio* argument, and that in the future procedure texts relating to the series texts may be discovered. However, it is clear that some of the fourth or fifth degree problems, irreducible to lower degrees, found in series texts cannot have parallel in procedure texts as they cannot be solved.[42] It is as if lists were developed in the series texts according to formal rules to produce increasingly sophisticated problem statements, regardless of any educational application. These long lists were presumably an investigation into all the possible ways of creating problem statements. Varying the parameters seems to have a heuristic value, not a didactic function as in the procedure texts and the catalogues described in the previous sections of this chapter.[43]

Some considerations on dating complete this contrasting picture. Arrangement in series with colophons is relatively rare in the Old Babylonian period. This textual form appears quite late with omen texts. We can presume that the mathematical series texts date from the very end of the first Babylonian dynasty. This hypothesis was suggested by Høyrup, who comments that this genre of text could only have been developed after the maturation phase of the discipline.[44] The hypothesis of an even later dating, possibly the Kassite period, was put forward by Neugebauer

Together, these elements seem to indicate that the context of the series texts is not that of classic Old Babylonian scribal schools of Southern Mesopotamia. They can hardly be seen as repertoires of problems destined to be solved by students. They do show extensive reflection on mathematical writing techniques. Their purpose seems to be primarily to improve mathematical language in order to produce the broadest possible range of problems. Such thinking is most probably more connected to the activities of a community of scholars than to needs directly linked to education.

4.5 Conclusion

The different examples analyzed in this contribution show the diversity of situations in which mathematical texts were produced in scribal schools: some texts were written by beginner-level or more advanced apprentice scribes, others by masters;

[41] TMS 7 and 16 (Høyrup 2002, 85, 181).

[42] See for example sections #35 to 47 of the YBC 4668 series tablet (Neugebauer 1935–1937, I:455–6), which contain fourth degree problems, or sections #28 to 32 of YBC 4710 (Neugebauer 1935–1937, I:410–1), which contain fifth degree problems.

[43] It is interesting to note that the function of the sets of parameters in the collections of problems is not always the same. See for example cases described by Alexei Volkov (ch. 9).

[44] "the sophistication of many of the series texts – regarding mathematical substance as well as the pluridimensional variation of statements – shows them to belong to a more mature phase of scholastization than the Eshnunna corpus" (Høyrup 2002, 351).

some texts used in teaching were specifically created for this objective, others not; some tablets appear to be textbooks, but in fact meet objectives other than those of education. Thus "educational context" is a vague term, covering a wide spectrum of different situations, uses and objectives. The activities of masters included teaching and other objectives, such as communication between peers. These components are strongly interconnected, and yet they do not completely overlap. Developments in mathematics are the result of both the activity of teaching and interaction within a community of scholars.

In the Old Babylonian period, education went hand in hand with creative activity, supported by a very active milieu. Maybe this environment was more developed in some major centers (Nippur, Ur, Eshnunna, Kish, Susa for example) than in others. The most active centers were able to influence other, less creative ones, where the educational framework was less institutional. The existence of an erudite scribal milieu at Nippur is without doubt, and is likely elsewhere. A network of long-distance links between the scribes seems to have existed, as shown by the similarities of the content of school tablets found through the Ancient Near East. These links between scribes, even far apart from each other, appear to have been sometimes stronger than the short links between scribal schools and local administration.

Through the examples presented in this chapter, I wanted to draw attention to the trap presented by words like "scribal school" and "school text", whose implicit meanings heavily influence interpretations. Old Babylonian scribal schools were the places where the learning of cuneiform writing and arithmetic took place, but some of them were also intellectual centers. Some texts, written by the students themselves clearly reflect elementary teaching activities, others, written by the masters, bear witness to the activity of teaching, while others still show communication between scholars. These are the three examples which have been illustrated in this chapter. But not all currently-known mathematical texts fit easily with these examples. Many of them, either because their archaeological context has been lost or is poorly known, or because the archaeological data do not say enough, are more difficult to interpret. Were they written by advanced students or by masters? Who did they address? Do they contain innovations or are they just transcriptions of the master's "words"? Are they repertoires of exercises, or elaborations designed to generalize or justify mathematical results? The available data rarely allow answers to all these questions at the same time. The texts should be examined on a case by case basis. If we examine all the different types of evidence provided by the mathematical documents, for example the typology of the tablets, the layout of the text, the choice of notation for the metrological and numerical graphemes, the language and the terminology, and the writing techniques, we can find answers or at least partial answers to some of these questions. What I would like to emphasize is that these answers are extremely diverse. In only seeing repertoires of teaching exercises in erudite cuneiform mathematical texts, we lose sight of the originality and remarkable inventiveness reflected by these mathematical texts produced more than 4,000 years ago.

References

Bernard, Alain, and Christine Proust. 2008. La question des rapports entre savoir et enseignement dans l'antiquité. In *Didactique, épistémologie et histoire des sciences. Penser l'enseignement*, Science, histoire et société, ed. Laurence Viennot, 281–302. Paris PUF, collection (dir. D. Lecourt).

Britton, John P., Christine Proust, and Steve Shnider. 2011. Plimpton 322: A review and a different perspective. *Archive for History of Exact Sciences* 65(5): 519–566.

Bruins, Evert, and Marguerite Rutten. 1961. *Textes mathématiques de Suse*, Mémoires de la mission archéologique en Iran. Paris: Geuthner.

Charpin, Dominique. 1986. *Le clergé d'Ur au siècle d'Hammurabi*. Genève: Droz.

Friberg, Jöran. 2000. Mathematics at Ur in the old Babylonian period. *Revue d'Assyriologie* 94: 98–188.

Gasche, Hermann, Jack A. Armstrong, Steven William Cole, and Vahe G. Gurzadyan. 1998. *Dating the fall of Babylon. A reappraisal of second-millennium chronology (A Joint Ghent-Chicago-Harvard Project)*. Chicago: Oriental Institute of the University of Chicago.

George, Andrew R. 2005. In search of the é.dub.ba.a: The ancient Mesopotamian school in literature and reality. In *An experienced scribe who neglects nothing: Ancient near eastern studies in honor of Jacob Klein*, ed. Yitschak Sefati, Pinhas Artzi, Chaim Cohen, Barry L. Eichler, and Victor A. Hurowitz, 127–137. Bethesda: CDL Press.

Glassner, Jean-Jacques. 2009. Ecrire des livres à l'époque paléo-babylonienne: le traité d'extispicine. *Zeitschrift für Assyriologie und Vorderasiatische Archäologie* 99(1): 1–81.

Goetze, Albrecht. 1945. The Akkadian dialects of the old-Babylonian mathematical texts (ch. IV). In *Mathematical cuneiform texts*, ed. Otto Neugebauer and Abraham J. Sachs, 146–151. New Haven: American Oriental Series/American Schools of Oriental Research.

Goetze, Albrecht. 1951. A mathematical compendium from Tell Harmal. *Sumer* 7: 126–155.

Gonçalves, Carlos. Forthcoming. *A dozen mathematical tablets from Tell Harmal*. Springer.

Hilprecht, Herman Vollrat. 1906. *Mathematical, metrological and chronological tablets from the temple library of Nippur*, Babylonian expedition. Philadelphia: University of Pennsylvania.

Høyrup, Jens. 2002. *Lengths, widths, surfaces. A portrait of old Babylonian algebra and its kin*, Studies and sources in the history of mathematics and physical sciences. Berlin/Londres: Springer.

Michalowski, Piotr. 1987. Charisma and control: On continuity and change in early Mesopotamian bureaucratic systems. In *The organization of power, aspects of bureaucracy in the near east*, ed. Robert D. Biggs and D. Robert, 55–68. Chicago: The Oriental Institute of the University of Chicago.

Michalowski, Piotr. 2012. Literacy, schooling and the transmission of knowledge in early Mesopotamian culture. In *Theory and practice of knowledge transfer. Studies in school education in the ancient near east and beyond*, ed. Wolfert S. van Egmond and Wilfred H. van Soldt, 39–57. Leiden: PIHANS.

Michel, Cécile. 2008. Ecrire et compter chez les marchands assyriens du début du IIe millénaire av. J.-C. In *Muhibbe Darga Armağanı (Mélanges en l'honneur du professeur Muhibbe Darga)*, ed. Taner Tarhan, Aksel Tibet, and Erkan Konyar, 345–64. Istanbul: Sadberk Hanım Müzesi (Sadberk Hanim Museum Publications).

Neugebauer, Otto. 1935–1937. *Mathematische Keilschrifttexte I-III*. 2 vols. Quellen und Studien zur Geschichte der Mathematik, Astronomie und Physik. Berlin: Springer.

Neugebauer, Otto, and Abraham J. Sachs. 1945. *Mathematical cuneiform texts*, American oriental studies. New Haven: American Oriental Series/American Schools of Oriental Research.

Neugebauer, Otto, and Abraham J. Sachs. 1984. Mathematical and metrological texts. *Journal of Cuneiform Studies* 36: 243–251.

Proust, Christine. 2005. A propos d'un prisme du Louvre: aspects de l'enseignement des mathématiques en Mésopotamie. *SCIAMVS* 6: 3–32.

Proust, Christine. 2007. *Tablettes mathématiques de Nippur*, Varia Anatolica. Istanbul: Institut Français d'Etudes Anatoliennes, De Boccard.
Proust, Christine. 2008a. Quantifier et calculer: usages des nombres à Nippur. *Revue d'Histoire des Mathématiques* 14(2): 143–209.
Proust, Christine. 2008b. *Tablettes mathématiques de la collection Hilprecht*, Texte und Materialen der Frau Professor Hilprecht Collection. Leipzig: Harrassowitz.
Proust, Christine. 2009a. Deux nouvelles tablettes mathématiques du Louvre : AO 9071 et AO 9072. *Zeitschrift für Assyriologie und Vorderasiatische Archäologie* 99(2): 167–232.
Proust, Christine. 2009b. Numerical and metrological graphemes: From cuneiform to transliteration. *Cuneiform Digital Library Journal* 1. http://www.cdli.ucla.edu/pubs/cdlj/2009/cdlj09_001.html
Proust, Christine. 2012. Reading colophons from Mesopotamian clay-tablets dealing with mathematics. *NTM Zeitschrift für Geschichte der Wissenschaften, Technik und Medizin* 20(3): 123–156.
Robson, Eleanor. 2001. The tablet house: A scribal school in old Babylonian Nippur. *Revue d'Assyriologie* 95: 39–66.
Robson, Eleanor. 2002. More than metrology: Mathematics education in an old Babylonian scribal school. In *Under one sky. Astronomy and mathematics in the ancient near*, ed. John M. Steele and Annette Imhausen, 325–365. Münster: Ugarit-Verlag.
Robson, Eleanor. 2008. *Mathematics in ancient Iraq: A social history*. Princeton: Princeton University Press.
Tanret, Michel. 2002. *Per aspera ad astra. L'apprentissage du cunéiforme à Sippar-Amnanum pendant la période paléo-babylonienne tardive*, Mesopotamian history and environment. Gand: Université de Gand.
Thureau-Dangin, François. 1936. Textes mathématiques babyloniens. *Revue d'Assyriologie* 33–2: 65–84.
Tinney, Steve. 1998. Texts, tablets, and teaching: Scribal education in Nippur and Ur. *Expedition* 40: 40–50.
Tinney, Steve. 1999. On the curricular setting of Sumerian literature. *Iraq* 61: 159–172.
Vanstiphout, Herman L.J. 1979. How did they learn Sumerian? *Journal of Cuneiform Studies* 31: 118–126.
Veenhof, Klaas R. (ed.). 1986. *Cuneiform archives and libraries. Papers read at the 30ème Rencontre Assyriologique Internationale*, Publications de l'institut néerlandais d'Istamboul. Leiden: Nederlands Historisch-Archeologisch Instituut te Istanbul.
Veldhuis, Niek. 1997. Elementary education at Nippur, the lists of trees and wooden objects. Ph. D. dissertation, University of Groningen.
Veldhuis, N. 2000. Sumerian proverbs in their curricular context. *Journal of the American Oriental Society* 120: 383–399.

Chapter 5
In What Sense Did Theon's Commentary on the *Almagest* Have a Didactic Purpose?

Alain Bernard

5.1 Introduction

Among mathematical texts from Antiquity, Theon's commentary on Ptolemy's *Almagest*,[1] together with his two commentaries on the *Handy Tables*,[2] are rare examples of works that can be credited as having an *explicitly* 'pedagogical

I thank Bernard Vitrac, Matthieu Husson and Ioanna Skoura for their helpful suggestions and corrections on the last-but-one version of this text.

[1]The *Almagest* is the modern name, transliterated from Arabic, for one of the works of the second century AD Alexandrian astronomer Claudius Ptolemy (Feke and Jones 2010). This treatise is one of the major works in ancient Greek theoretical astronomy. His historical importance is partly due to the fact that it has served as the basis for astronomical commentaries, criticism and corrections from Antiquity to the early modern period at least. The first known complete commentaries on the text are due to Pappus of Alexandria (beginning of the fourth century AD) and, later on, to Theon of Alexandria (end of fourth century AD). The latter commentary originally had thirteen books, book XI is lacking and book V was retrieved in the scholia to the *Almagest*, (Tihon 1987, 1992, 120–1). Books I-IV have been edited by Adolphe Rome (1936, 1943). For the other books the only edition is Camerarius's (Cam. 1538). The reference to the Greek text (*in Alm*) is to Rome's edition. In general, these editions rely on manuscripts that are not autographs, which are probably lost forever: this means that the copies, on which the editions in question are based, are much later than the original writings. See, on this point, the general remarks of Ritter and Vitrac (1998, 1233–4).

[2]These are the so-called 'little' and 'great' commentaries on the *Handy Tables*. Both have been edited and translated in French by Anne Tihon (1978, 1991, 1999) and the reference here is to these editions (LC and GC). For a convenient summary of the content of these two commentaries see (Jones 1999b, 162–167).

A. Bernard (✉)
UPEC (IUFM) and Centre A. Koyré (EHESS-MNHN-CNRS UMR 8560),
Paris, France
e-mail: alainguy.bernard@gmail.com

A. Bernard and C. Proust (eds.), *Scientific Sources and Teaching Contexts Throughout History: Problems and Perspectives*, Boston Studies in the Philosophy and History of Science 301, DOI 10.1007/978-94-007-5122-4_5,

purpose', for the simple reason that the prefaces to these three commentaries explicitly say so.[3] This simple fact raises some simple questions, such as: is it really a good teaching program on the *Almagest*? Does it make it clearer? Is it an efficient pedagogical artifact? Correlatively, what does it reveal about the way in which the *Almagest* was explained and taught in Theon's period? The very first historian to pose such questions was Adolphe Rome, as he was also the first to use, for this purpose, anthropological studies available in his own time.[4] Rome's interest and purpose were not, by any means, the study of the didactic features of Theon's (or indeed Pappus's) commentaries, which were only of marginal interest for him. This is thus the first explicit attempt to study the question.

If taken in isolation, however, the problem of Rome's approach is that it tends to be all too straightforward: Rome obviously took it for granted, not only that the commentary essentially had a didactic purpose, but also that it was the quasi-direct reflection of Theon's teaching program or even of his students' attitudes. Hence, some deficiencies he finds in Theon are interpreted as the reflection of the passivity of his students, as if we, modern readers, were present among Theon's audience and would be incensed at seeing that no one reacted to such or such a mistake made by the teacher.[5] This implicit idea is understandable, insofar as any modern historian commenting on a commentator is in an ambivalent position: he has to characterize the operation of the commentary as such, and to compare the commentary with his own understanding of the 'primary', commented text. If the latter has difficult content, as is the case here,[6] this necessary operation brings us very close to the 'primitive scene' of our first learning, when we were confronted with the discrepancy between our efforts to understand difficult subjects and the efforts of our teachers to facilitate this. It is then highly tempting, when the commentator we are studying seems to present himself as an able teacher, to react just as we would have done against our own former teachers: who has never been angry with a teacher, for good or bad reasons?[7] This problem is in itself very general: when it is question of learning and teaching, we unconsciously tend to

[3]Part II contains a translation of the preface to Theon's commentary on the *Almagest*. The prefaces to the two other commentaries are translated in an appendix (a French translation is also available in Tihon's editions). The fact that prefaces would explicitly address learners is not uniform among the prefaces to mathematical treatises written in Greek. On this matter, see the general overview in Vitrac (2008).

[4]Rome has general remarks on the pedagogical characteristics and shortcomings of both Pappus's and Theon's commentaries in his preface to the first volume of his edition (Rome 1936, LXXXV note 2). In addition, many of his footnotes contain remarks on the allegedly didactic dimension of the two commentaries.

[5]See, for example, Rome's notes on the end of Theon's commentary on chapter I.4 of the *Almagest*: *in Alm*. 398.6–400.15, especially (Rome 1936, n. 2 p. 398; n. 1 p. 400).

[6]The perception of 'difficulty' is of course highly dependent on the time and social context; but in the present case, there is little doubt that Theon's contemporaries considered the *Almagest* as a difficult text; see (Jones 1999b, 147).

[7]Anne Siety, in her suggestive essay (Siety 2001), develops the argument that teachers are – nowadays– very much the subject of fantasies that she describes as being specific to mathematics

identify ancient situations with the modern one which laid the ground for our own, elementary education: since the latter has become our second nature, we tend to take it as a universal experience and draw straightforward conclusions from this.[8]

Without denying the interest of Rome's approach, that presents a text such as Theon's as the immediate reflection of his teaching, I would like here to question the legitimacy of such a 'straightforward reading'. My purpose, on the other hand, is really not to engage in any empty (and posthumous) quarrelling with Rome's judgments on Theon as a teacher or on what Rome presented as the absence of a sense of criticism in Theon's students. As I already mentioned, Rome had the immense merit of investigating crucial questions for the very first time,[9] including this one. Moreover, his indications and intuitions, whatever their shortcomings, are certainly still valuable and worthy of being taken into consideration. The positive purpose is rather to show that his approach, to be really fruitful, would have benefitted from being more prudent. Later studies by historians of astronomy in the Hellenistic period and late Antiquity, that I will mention and summarize in the first part, have convincingly shown that further elements need to be taken into consideration if one wants to reach a balanced and convincing judgment on what Theon's text potentially shows us on his teaching. I shall try myself to put even more circumstantial elements in the balance that I think are relevant for the discussion and might guide us into further research on Theon's text.

As for the plan of the present study, I shall begin with a brief recapitulation of the reasons for which historians have been led to a more prudent approach to the questions initially raised by Rome, mainly on the basis of a larger set of documentary evidence than Theon's texts alone (part I). Then I will offer some remarks on the preface of Theon's commentary to the *Almagest* (part II), on the basis of which I will try to show Theon's dependency on the broader culture of commentary (part III) and then analyze, as a consequence, his position vis-à-vis Ptolemy's treatise as a model for astronomical studies (part IV). I will conclude, as mentioned above, with the tentative definition of further directions for research on these issues and on the importance of such an approach for the very definition of 'mathematical studies' in late Antiquity.

---

teaching. Interestingly, she describes these fantasies as essential 'mythical' in nature, therefore calling to a long history. See part I.3 'le prof de maths, cet inconnu'.

[8]This possible confusion lies behind the historiographical issues raised in the first pages of Anne-Marie Chartier's contribution, concerning the interpretation of former aids to reading such as 'psalters'.

[9]Other questions in which he was a recognized pioneer are the careful study and edition of geometrical figures and the special interest he paid to the design and building of astronomical instruments. Moreover, Rome never saw his notes as anything other than preliminary remarks ('un premier débroussaillage', Rome 1936, LXXXV).

5.2 Some Reasons to Be Prudent in Interpreting Theon's Claims About His Didactic Purpose

Our first reason for being prudent is the fact that teaching was not only a fact, but also (if not primarily) a key *value* in Antiquity: the very notion that an educated person should be the 'teacher of others', not in a narrow, technical sense of course, but in the general meaning that he should inspire the conduct of others through his actions, his behavior, discourses or writings, is a widespread one. It is even constitutive of ancient *paideia* and of the very notion of language.[10]

In Theon's particular case, we have two reasons to believe that this aspect has to be taken into account. The first is that we know that his commentary is partly dependent on another commentary on the *Almagest* written by Pappus of Alexandria less than half a century before him: unfortunately, only Pappus's comments on the fifth and sixth books of the *Almagest* are extant today,[11] so a direct comparison between Pappus's and Theon's respective introductions is impossible. If it had been available, though, we might plausibly have discovered many similarities between Pappus's and Theon's presentations of their own work, especially since Pappus elsewhere develops considerations on the teaching of mathematics that are similar, to some extent, to what is found in Theon.[12]

The fact that there were 'cultural' reasons for presenting oneself as a teacher, or literary *topoi* borrowed from others that nurtured, so to speak, such general claims, does not imply that the authors *pretended* to be teachers, but actually were not.[13] While it certainly opens the way towards the possibility that some authors were *interpreted* as being essentially teachers, Homer being the principal example,[14] it only means that, in the case when the commentators were actually teaching, they had to present the activity in a partly conventional and partly original way. This remark very much applies to Theon, as we shall see later.

The second reason for having doubts is that the *Almagest* itself is presented as a kind of pedagogical artifact – in the broad sense that it is deeply meant and constructed to enable one to proceed into the conversion of the soul that

[10]Marrou (1965), Kaster (1988), Hadot (2006), and Sluiter (1999, 175–7). I borrow this general remark from B. Vitrac's enlightening talk, presented during one of our workshops, on the reputation of Euclid's *Elements* as a 'pedagogical artefact'.

[11]The two books were the first edited by Rome (1931). The dependence of Theon's commentary on the sixth book to Pappus's one is without question, but it does not amount to mere plagiarism. On this question see (Rome 1936, LXXXIII–IV); I also refer the reader to Ioanna Skoura's forthcoming PhD thesis (Univ. of Athens) on the question.

[12]This point is made in (Jones 1999b, 167).

[13]Borrowing the terms proposed by Anne-Marie Chartier in her own contribution (cf. §3.1 'the different levels of contextualization'), the fact that we *only* have access to the cultural environment does not imply that the didactic context and learning situation are inexistent.

[14]Sluiter (1999, 176–9).

Ptolemy advocates in his preface.[15] Modern readers tend to presuppose, because it has become a kind of standard *topos* of historiography, that Ptolemy's treatise *is essentially* a theoretical work dealing with astronomical models for the movement of the sun, the moon and the five planets. This description of the work is of course an apt one as far as the basic content is concerned,[16] but ignores the fact that the *Almagest* is first and foremost a philosophical treatise meant to bring the reader into a philosophical choice of life, this choice implying that a specific kind of study should be associated with it, namely 'mathematical' studies in the particular and precise meaning given to this by Ptolemy.[17] Moreover, Ptolemy's treatise does not present itself as self-contained but as essentially dependent on previous works and research, so that it is itself a commentary on these previous works and especially Hipparchus's research.[18] To comment on a text such as the *Almagest*, therefore, cannot have meant simply converting a kind of ancient "research treatise" into a convenient teaching device: it rather required from the commentator that he would define the position of his work with respect to the source not just as the *subject-matter* for teaching, but as its *model*. We will later elaborate on this point of view in more detail (part IV); for the moment, we would like to emphasize that this aspect implies that Theon, as a commentator, might *present himself* as a teacher just because he is conforming to Ptolemy's model. This phenomenon is in no way isolated or rare in Antiquity and it necessarily makes any straightforward conclusions drawn from the explicit self-promotion of commentators on their possible teaching questionable.

A further reason why Rome's straightforward reasoning is problematic is that it really relies on the wrong material as far as the students are concerned. On the favorable (and not implausible) hypothesis that Theon's commentary *does* somehow reflect the content of his teaching, we cannot just judge from our own reactions as readers what may have been the reactions, expectations or exercises of students in Antiquity, for all these depend on particular educational settings and procedures that we need to know *first*, at least in theory, in order to judge whether or not the 'teaching material' was suited to them. Rome himself comes close to this idea when he proposes to compare the probable setting of education in late Antiquity with the way in which astronomical treatises were studied in pre-colonial Morocco.[19]

---

[15]See Taub (1993, 30–37), Feke and Jones (2010, 208–9), and Bernard (2010a, 502–512).

[16]Although I would question the use of the essentially modern notion of 'model' to describe Ptolemy's enterprise. This is probably a better approximation to what the ancients had in mind when dealing theoretically with the movements of the planets (see Rochberg 2004, 283–286), but it tends to apply an idea of complexity and fundamental discrepancy (between reality and its models) that seems to me to be alien to ancient thought. On this point see (Bernard 2010c).

[17]For the essentially practical side of ancient philosophy, see the classic work (Hadot 2006) and, as far as Ptolemy is concerned, see the references given in note 15 above.

[18]*Alm.* I 8.1–16.

[19]His reference (Rome 1936, LXXXV note 2) is to (Renaud 1931). Unfortunately, this suggestive comparison only serves to comfort Rome in his negative judgment on Theon and the kind of education it represents.

Moreover, the students' reactions, beyond the oral interactions that we cannot attain in the present case for lack of any narrative on how astronomy was concretely studied in that period, are usually made manifest by certain kinds of marginal notes either within the commentaries on the *Almagest*, or in the *Almagest* itself. If we were to judge, therefore, their attitude and point of view on the basis of the relevant material, this could not be on the basis of Theon's text itself, but on the basis of other kinds of documents actually produced by students. This fundamental remark is amply confirmed by other studies on the same volume, for which such documents and indices are indeed available.[20]

As far as ancient commentators on the *Almagest* are concerned, much work has already been done to explore such documents and to show that some of them indeed bear plausible indications that both the *Almagest* and, in some cases, the commentaries on the *Almagest*, were the subject of courses, for which we have kept notes in certain cases.[21] The documentary evidence that has been taken into account for such studies is, on the one hand, astronomical papyri written in Greek, the extensive study of which has been completed by A. Jones.[22] Their content shows that the methods of Babylonian astronomy are largely reproduced within these papyri, though with significant differences probably due to the fundamental shift in the milieu in which astronomical methods were cultivated and transmitted. Thus, A. Jones argues plausibly that, in the typical framework of the Babylonian communities working in temples,[23] the 'procedure texts' that come closest to what are later known as mathematical commentaries should not be understood primarily as a means of instruction but rather "aide-mémoires" for procedures that were essentially taught orally. In this respect, commentaries on the *Handy Tables*, like Theon's or others that were written before him, represent a new kind of aid essentially meant for astrologers wanting to appropriate themselves the means of prediction afforded by Ptolemy's kinematic tables. The tables themselves are then a received text (not one that one should learn to reconstruct in each case) and the basic problem, to which Theon's or other similar commentaries answer, is then to explain how to *use* such tables or, at best, correct them if they are false or badly transmitted. The profile of 'students' for such a course that underlies this shift is the one for which Greek papyri were probably meant already, that is an independent astrologer. What 'teaching' means, in this respect, basically includes some *reading* activity and such commentaries are then partly regarded as means for self-teaching. In any

[20]See Proust's and Clancier's contributions in particular. Much similar material can be found for the study of astronomy in Arabic Middle Ages, as F. Charette's contributions to our workshops have amply shown.

[21]Various synthesis articles can be mentioned in this respect (Tihon 1992; Jones 1994, 1999b; Pingree 1994).

[22]Jones (1999a). This work in fact expands on earlier works like those of O. Neugebauer (Jones 1994, 26–8).

[23]On this see (Jones 1994, 1999b, 155) and Rochberg's enlightening synthesis on the same question as far as Late Babylonian and Hellenistic astronomy is concerned (Rochberg 2004, 209–236). See also P. Clancier's contribution in the same volume.

case, these papyri indirectly provide much information on the general background, against which the study of the *Almagest* or its practical outcome, the *Handy Tables*, could be of general interest for the ancients.

The second set of available documents, on the other hand, are directly found within the manuscript tradition of the *Almagest* and of its commentaries, either as isolated pieces or as *scholia*, addenda or *prolegomena* to the aforesaid treatises.[24] Before we sum up what can be learned from this material, we should remember that none of these treatises, including the commentaries like that of Theon, is available in the form of autographs: the latter are unavailable and, most probably, definitely lost. What we have, by contrast, are copies made long after the period when the authors worked.[25] There is therefore a fundamental gap between the two kinds of documents we have at our disposal, which is due to the nature of the textual transmission.

Coming back to the studies on the manuscript tradition of these treatises, what emerges from them is a diverse picture, for these documents do not stem from the same periods or milieus, although the circle of late Neoplatonist commentators is obviously very well represented among the authors of scholia and courses on the *Almagest*.[26] Some of the treatises were subject to official edition by their author and might be interpreted as 'manuals', loosely understood as *aids* to the reading of the *Almagest* – and there are proofs that they were actually used as such in later courses.[27] Some *scholia*, by contrast, seem to be based on notes directly taken during a course on the *Almagest*, as is shown by the specific state of the text and the vocabulary used.[28]

Some *scholia* do reflect the fact that revisions were prepared for commentaries, like Theon's, either by Theon himself or some of his student(s), but were not

---

[24]A useful review of the available material and its significance is found in Tihon (1992), which deals with the specific problems raised by the edition and/or study of such documents. Some of these testimonies are still under scrutiny and in the course of being edited, like the so-called *Prolegomena* to the *Almagest* now being studied by Acerbi et al. (2010).

[25]Thus, the manuscripts Vat gr 1594 or Par. Gr. 2389, which are among the oldest used for Heiberg's edition of the *Almagest*, dates back to the ninth century and eleventh century AD respectively– at least seven centuries away from Ptolemy's own period! The Laur. Pl. 22.18n, the main basis for Rome's edition of Theon's commentary on the *Almagest*, is dated from the ninth century, some five centuries away from Theon himself.

[26]See the useful synthesis of (Pingree 1994) for an overview. See also the final remarks to A. Tihon's edition and translation of scholia to the *Handy Tables*, (Tihon 1973, 105–8) and the recent remarks of F. Acerbi et al. on the milieu in which the so-called '*Prolegomena* to the *Almagest*' were compiled (Acerbi et al. 2010, 64–66 and 68).

[27]Tihon (1976, 1992, 118–120). Note, however, that the notion of 'scholastic manual' should probably be used with some care, since it is essentially a modern notion and even for the contemporary period it appears to be problematic (Choppin 2008). That edited texts would be *essentially* meant for teaching is therefore a doubtful equivalence.

[28]Such is the case of the so-called *prolegomena*, cf. on this (Acerbi et al. 2010, 64), recapitulating and confirming on this point Mogenet's previous remarks about the nature of this work (Mogenet 1956). Tihon (1992, 123–4) remarks that the *Great Commentary* might be classified in the same category and reflects the incomplete state of the text; on this point see also (Jones 1999b, 169–170). Pingree (1994, 80f) proposes interpreting a set of scholia to the Vat. Gr. 1594 in the same sense.

systematically incorporated in all the versions that circulated afterwards. One interesting case of this phenomenon is the 'corrected' calculation for an eclipse for the year 364 CE, which has been found to correspond to an awkward and probably incomplete revision of a calculation provided by Theon in the sixth book of his commentary to the *Almagest*, on the basis of some data taken from the *Handy Tables* and Theon's commentary thereon.[29] This is probably what comes closest to the possible work of a student and/or companion of Theon, but this does not show any attempt to criticize him, rather it is an attempt to modify a difficult calculation.

Some remarks on these studies that pertain to our present subject are in order. First, most of the studies evoked above concern commentaries directly related to the *Handy Tables*, which are by nature directed toward practice and the natural needs of astrology.[30] This is partly a bias implied by the nature of the text edited so far[31] and by the difficulty of interpreting the rich material provided by the *scholia* to the *Almagest*,[32] although some of it has been explored already. With these caveats in mind, the richness of *scholia* relating to practical aspects confirms what the papyri edited by A. Jones had already shown.

Second, the studies in question, quite logically, do not bear on Theon's circle and teaching specifically, but stem from the study and edition of astronomical and astrological literature at large. As one might expect, the result is that we get many testimonies for anonymous commentators, who are sometimes difficult to identify.[33] What can be hoped for at best, as mentioned by A. Tihon, is to glean some clues on the rich activities of the schools of late Antiquity at large, to which we have no other access.[34] But on the precise way in which Theon and his circle of students and companions worked, we know very little and the little we do know from the above mentioned studies remains fragmentary and speculative, in absence of an explicit *testimonia* like those we have for Galen's methods, for example.

Finally, and interestingly, let us remark that the more a text is edited, the more difficult it is to be sure that it once related to any kind of 'in situ' teaching. A set of fragmentary *scholia*, sharing the unity of composition typical to a course and characterized by a specific vocabulary evoking oral teaching,[35] is a better testimony than a fully edited commentary, the purpose of which is, at best, self-teaching.

---

[29]Tihon (1976–1977, 51–2).

[30]On this see (Jones 1999a, 5, b, 156–168) and his general analysis summarized above.

[31]Theon's commentary on the *Almagest* is only partly edited, partly due to the catastrophic conditions in which Rome had to work during the Second World War and the priority given to the edition of the *Handy Tables*, cf. (Tihon 2011; Mercier 2011).

[32]Cf. the general remarks in (Tihon 1992, 120–1) and the example of Pingree's inquiry about one such subset of scholia (Pingree 1994, 80f).

[33]One famous case is the controversy about the attribution of the so-called *Prolegomena to the Almagest*, once attributed by Mogenet to Eutocius of Ascalon, then attributed by Knorr to a certain Arcadius. For an updated discussion, see (Acerbi et al. 2010, 65–6).

[34]Tihon (1992, 137).

[35]Tihon (1973, 103–8).

With these general precautions in mind, we can now turn to the last reason to be suspicious of any straightforward interpretation of Theon's activity, which is the precise way, by which he presents his commentary. While he indeed addresses students in it, it must be noticed that they are addressed in a very particular and sophisticated way. Even when taken at face value, these declarations portray a subtle picture of their various expectations, as we shall now see.

5.3 The Preface to Theon's Commentary: Translation and First Remarks

In order to make the various levels of interpretation we propose on Theon's enterprise clearer, we will comment on some aspects of the preface to his commentary to the *Almagest*.[36] The notes justify the translation adopted for several crucial expressions; in particular, they point to the many allusions that they contain to parallel passages in the *Almagest* itself. For more clarity, such parallel passages have been also underlined with dots.

[a] Of Theon of Alexandria, in the edition made by him, commentary on the first book of Ptolemy's *Composition*.

[b] Having been continuously solicited by my listeners,[37] Epiphanios my boy,[38] to suggest explanations on what seemed to each of them difficult points of the mathematical composition[39] of Ptolemy, I thought it would be good to produce its commentary[40] and that I had to apply myself to such an enterprise for the sake of two <main purposes>, namely for the training of astronomers[41] and for the incitement of those who are interested in elements.[42]

---

[36]For convenience I differentiate the various parts of the present introduction by letters [a], [b]. The translations of the prefaces to the *Little* and *Great Commentary* on the *Handy Tables* are proposed in an appendix: they will occasionally serve as points of comparison in the course of my commentary.

[37]*Akroatai*. The term, for which we here keep a literal translation, is all too easily translated as 'student'. This translation is not illegitimate in one sense, but it tends to impose a modern notion on a series of ancient terms, each of which carry specific nuances, as we shall see.

[38]*Teknon*, which literally means 'child' or 'my child' but which is very often used in a metaphorical way to designate a distinguished student. See Pappus of Alexandria's similar address to Hermodoros at the beginning of book VII of his *Mathematical Collection* 'hermôdore teknon' (634.3).

[39]'Mathematical composition' is for *mathematikê syntaxis*, the traditional name for the *Almagest*.

[40]The expression used is *hypomnêmatismos*, which strongly recalls the verb by which Ptolemy himself characterizes his enterprise at *Alm.* I.8.6–16. The verb is used at other places in the *Almagest* (*Alm.* I.159.20, II.206.7, 338.22, 608.8).

[41]*Askêsis tôn astronomountôn*; the notion is recalled slightly later in Theon's commentary on the very first words of the *Almagest* by the mention of "ho ton astronomein epaggellomenon", namely "the one who pretends to practice astronomy" (*in Alm* 319.13).

[42]*Hê protropê tôn stoicheioumenôn.*

[c] Moreover, it is possible that those who happen to be lovers of the truth and inquisitive[43] might succeed in perceiving that we have included many demonstrations to some points about which previous commentators[44] had not thought of in any way, <at least> from the commentaries they have left. Indeed, by proposing themselves to put aside the allegedly clearest things, they in fact appear to have neglected the most difficult. And while Ptolemy himself explicitly says against them, right at the beginning of his treatise that "we are about to demonstrate everything geometrically",[45] they achieve most <of their comments> just as for the *Handy Tables*, that is by presenting only the bare procedures.[46]

[d] As for us, we took the greatest care, not only to examine everything <in the *Syntaxis*> through geometrical proofs, as far as possible, but we have also not neglected, in general, any of the points that are reputed to be difficult, even though we are not of the number who deal with these kinds of issues.

[e] And so as not to make the commentary too long,[47] we present everything word by word for the first book of the *composition*, by reason of its elementary character[48]; as for the following <books in the *syntaxis*>, we voluntarily left aside everything that appears to be clear even to those who can only grasp a few things at the same time, and as for the points which might be relatively difficult to understand by the same <persons>, we present their demonstrations again.

[f] Finally, we shall have the greatest obligation to those who will be able to correct the points that were investigated by us "not as usefully as it was possible <to do>",[49] as he <e.g., Ptolemy> says himself; and we do not think it shameful,

[43] *Philalêthôs kai zêtêtikôs*, these typical qualifiers appear not infrequently in Ptolemy and belong to the group of terms by which he designates the 'legitimate philosophers', that is, those who follow a legitimate method in their inquiry. Hipparchus and Hipparchus's methods, in particular, are often credited with similar epithets or adverbs: *zêtêtikos, philalêthos, gnêsios*. In the present case the expression seems to be directly borrowed from a remark by Ptolemy within this complex theory of the movement of the moon about the corrections needed on previous models (*Alm*. I.328.4: book 4, ch. 9).

[44] The term used is *hupomnêmatistoi*, in correspondence with the expression used above for Theon's own enterprise.

[45] *Mellontes hapanta grammikôs apodeiknuein*, which is an almost literal quotation of the concluding sentence, in book I, on the transition between the exposition of the 'general conceptions' and the 'particular demonstrations': Ptolemy says that before the first demonstration, he must make an abridged presentation of the theory of chords in a circle, and he adds "*hapax ge mellontes hekasta grammikôs apodeiknuein*", that is, "and for this we shall demonstrate once and for all each thing geometrically" (*Alm*. I.31.5 H). Thus, 'each thing' (*hekasta*) has become 'all things' (*hapanta*) and, of course, the sentence taken out of its context appears to be much more general than it is in reality.

[46] *psiloi ephodoi*, which is also the expression used in the introductory sentences of Theon's *Little commentary* on the *Handy Tables* (see the translation in appendix): there he says that he shall only present the mere procedures of calculation, without presenting any demonstration that might justify them.

[47] The term used is again *hupomnêmatismos*, and the whole formula strongly reminds one of the similar formulas used by Ptolemy and that are often associated with the idea of length (e.g., *Alm* I. 159.19–21).

[48] Lit. "*Dio to stoicheiôdê*": it is unclear, whether Theon refers to the 'elementary character' of the first book of the *syntaxis* itself, or to the style of his own explanation. The ambiguity might be voluntary.

[49] This is a literal excerpt from the final remarks of Ptolemy's explanations about the basic purpose and structure of the *Almagest* (*Alm*.I 7.25–8.16), in which he explains that for the sake of brevity, he

so great and divine is the promise,[50] if we appear to need the correction of others.[51] For we do not trust ourselves to be among those, that can examine every point in an indisputable way.

The sophistication of Theon's preface mainly stems from two characteristics, namely the allusion to the complex expectations of the audience, due to the composite character of the latter, on one hand; and the many, more or less explicit allusions to Ptolemy's words themselves, on the other.

Listeners (*akroatai*) are explicitly mentioned in part [b] and are designated as belonging to two possible categories, those who practice astronomy on one hand, and those who are "incited to study elements" –the precise meaning and extent of these two categories being unclear. A. Jones, as already mentioned, convincingly defends the idea, based on his examination of the astronomical papyri he edited, that Theon's listeners were the very same people, by or for whom these papyri were written. [52] Nevertheless, it seems clear that Theon uses the term to designate the people who would take the time to study Ptolemy's mathematical models for themselves. The two interpretations, however, are not incompatible, as I have argued elsewhere[53]: practicing ancient astrology did not only mean to earn one's living by casting horoscopes,[54] but also to submit oneself to a permanent and demanding training, which had an intrinsic value – Ptolemy's approach both to theoretical astronomy and astrology (*apotelesmatika*) is based on this common idea.

It is not clear either, whether these categories have to be taken as distinct or not. A. Jones clearly sees them as separate categories, the *astronoumenoi* being understood as 'mere practitioners' in astrology and the others as students in mathematics pursuing a curriculum in liberal arts.[55] But the formula designating the second category (*stoicheioumenoi*) is vague. Jones understands it specifically as "students who had worked through Euclid's *Elements* and were progressing to

will only present what the ancients had sufficiently investigated (*êkribômena*) and that he will only rework that that were "not generally determinated or not as usefully as it was possible <to do>".

[50]This is again an allusion to the passage in book IV, in which Ptolemy explains what the attitude should be of those who correct the ancients' determinations. *Alm.* I. 328.8–11.

[51]Again taken from the same passage in book 4, *Alm.* 328.11.

[52]See part I. Pingree 1994, 78 is of a different opinion: "Those students who studied the *Almagest* under [Pappus's and Theon's tutelage] were interested primarily in learning the details of Ptolemy's mathematical models, and no practical motive such as the practice of astrology." Pingree's justification for maintaining a strict division between the two putative kinds of publics is that "Astrologers operated with far less sophisticated material ; at best they learned how to manipulate the *Handy Tables*, a skill also taught by Theon" (*ibid.*) Personally I find Pingree's interpretation far less compelling than Jones's better documented one (cf. part I); moreover, the preface to Theon's *Little Commentary* leaves no ambiguity on the fact that he would expect one, in principle at least, to understand the geometrical proofs of the *Almagest*.

[53]Bernard (2010a).

[54]For a useful synthesis on the practice of ancient astrology in the Ancient Greco-Roman world, see (Barton 1994).

[55]Jones (1999a, b, 165).

the *Almagest* as a work of more advanced mathematical reasoning".[56] This is not impossible,[57] but still remains speculative: in its vaguest meaning, the term refers to those who are instructed in elements or basic principles (cf. LSJ), an idea that might refer to the progression of the *Almagest* as well.

Whatever the case, the notion of a composite audience with various purposes and abilities is reinforced by the following allusions in parts [c], [e] and [f]. In part [c], Theon alludes to the possibility, depending on the readers' knowledge and ability for critical comparison, that these 'more inquisitive' people might identify the difference between his own commentary and the others that were written before his own: while the earlier commentaries tend to bypass allegedly easy passages, he, on the contrary, will show the real difficulty of these 'easy' points. In part [e], he alludes to the progressiveness of his commentary, by taking as a criterion the limited ability to grasp certain issues, just as a teacher would do for the sake of graduating the difficulty of his explanations. In part [f], finally, he points out the fact that some of his listeners might be advanced enough to be considered as peers able to correct him.

These remarks, on one hand, explain why one should refrain from speaking uniformly about 'students' for the diverse audience and readership of Theon's preface. On the other, the straightforward idea that they *were indeed* students tends to be confirmed by the comparison with the commentary itself, as well as with Theon's other prefaces or other testimonies like those evoked above (part I). Even a superficial reading of the commentary shows that the commentary on the first book is indeed much longer than for the others, either because detailed explanations are given about the ways in which 'elementary' calculations should be conducted,[58] or because allusive points in Ptolemy's general explanations[59] give Theon the opportunity for geometrical digressions giving weight to some remarks made in passing by Ptolemy.[60] By extension, then, many of the commentary's other peculiarities might be interpreted as stemming from the didactic character of the commentary, as Rome suggests in several places.[61] In the preface to his *Little*

---

[56] *Ibid.*

[57] As pointed out by François Charette (personal communication), to check this would require, in principle at least, having philological or papyrological evidence for an organized curriculum leading from more elementary (?) treatises, like the *Elements* or the treatise included in the domain of 'little astronomy', to more advanced studies like that of the *Almagest.*

[58] See, for example, *in Alm.* 452.14–458.8 (on 'sexagesimal parts') or 458.9–462 or 579.11–580 (on multiplication and division). Cf. the summary of similar passages in the *Prolegomena*, (Acerbi et al. 2010, 57–59).

[59] Namely, the explanations given in chapters 2 to 5 of the *Almagest*, in which the main hypotheses of the treatise are presented. Theon's commentary on this relatively short part of the *Almagest* runs (in Rome's edition) to more than a hundred pages (*in Alm.* 334–447)!

[60] For a typical example of 'Theonine' digression, see his long commentary, drawn from Zenodorus, of *Alm.* 13.16–19: it runs to some 40 pages.

[61] A typical example is Theon's presentation of what we know as Menelaus's theorem, *in Alm.* 539.17–539.25 and (Rome 1936, 538 n. 2).

Commentary, Theon very clearly alludes to the inability of most of his students[62] to follow the operations on numbers as well as the corresponding geometrical demonstrations. For them, he says, he only provides the "mere procedures", that is, the way to calculate the positions of the sun and the moon without justifying them with geometrical demonstration, as it is done both in the *Almagest* and in the *Great Commentary* on the *Handy Tables*. The latter is also presented as a more advanced treatise that those of his companions "who have made some progress"[63] might be able to follow.

Hence, while the 'straightforward' interpretation mentioned above is not implausible to some extent, for there is nothing to prevent us from thinking that Theon was indeed teaching and that the commentary basically is "the redaction of his course" (Rome, *in Alm.* 317, note 2), it certainly might and should be nuanced. We have seen the first reason for such doubts: in the content of his preface(s) Theon alludes to a diverse audience and the terms used for its members do not bear a uniform meaning. Moreover, two supplementary reasons have already been mentioned and will now be discussed in more detail. The second is that presenting oneself as a teacher is a *standard* feature of many ancient commentaries (part III); the third, which depends partially on the second, as we shall see, is that the *Almagest* itself is presented and conceived as a support for learning and study (part IV).

5.4 Clues to the "Cultural Background": Collective Study, Commentary, Critical Comparisons and Corrections

Whenever one wants to interpret the didactic character of a text such as Theon's commentary on the *Almagest*, one straightforward possibility is to call on *any* element of historical context relating to ancient teaching procedures and check whether they could help us to better understand Theon's attitude and purpose. This is the way followed, within the present volume, by E. Caianiello on the interpretation of Fibonacci's *Liber Abaci* or, for mathematics in late Antiquity specifically, by S. Cuomo in her groundbreaking study on Pappus of Alexandria.[64] By contrast, my intention here is less holistic – not least because such an attempt would greatly extend the length of the present study. I would like to pinpoint only certain aspects of Theon's approach and suggest how they might be interpreted in a larger environment than the strictly textual one evoked in part II.

Thus, in his aforementioned article on astronomical commentaries, A. Jones proposed to nuance Rome's interpretation by remarking that Theon's introduction shows rather that "the situation seems to have been, not a formal class or lectures,

[62]Here the translation is more legitimate, since the expression used by Theon explicitly refers to "those who come to us so as to learn such a subject matter" (see Appendix 2 and the note).

[63]Again a transparent allusion to the *Almagest*'s wording.

[64]Cuomo (2000).

but the importuning of students reading the *Almagest* by their own lights."[65] We might in turn nuance this observation by remarking that 'importuning' carries a negative connotation that was probably alien to the ancient feeling. That listeners (*akroatai)* had to intervene at a very fundamental level in a master's classes, for the sake of *submitting* the very *questions* that might be discussed, can legitimately be regarded as the *standard* expectation, not only for 'students' in Antiquity, but for their 'teachers' as well. The very meaning of the notion of 'problem' (*problêma*), in the context of rhetorical performance, was precisely related to the procedure of 'challenging' (*proballesthai*) the teacher and orator with questions. In Philostratus of Lemnos, for example, we find that famous sophists of his time would blame students for *not* proposing such *problêmata* to them.[66] In the case of mathematics, as Jones himself suggests, Pappus's sophisticated argument directed against his female colleague Pandrosion might plausibly be interpreted as an encouragement to her students to persevere in their research.[67] Pappus's blame certainly does not lie with the *problem* itself, understood as the 'challenge' set by one of Pandrosion's students, but the relative ignorance of the students, who might have been able to identify the difficulty inherent to their construction by themselves, had they known more about ancient analysis and the use of 'given' in geometrical reasoning. The very fact that Pappus would narrate such an event shows that such a confrontation was a legitimate and rewarding way of presenting one's own achievements. This is confirmed by Plutarch's interesting text *On Listening*, in which he also criticizes the potentially disturbing interventions of students of philosophy within narrations by their teachers. [68] But Plutarch's argument is clearly not disapproving nor aimed at discouraging students from setting problems, but to discuss the right and most profitable way of doing so. His argument only confirms, therefore, that such a procedure was not only legitimate, but had itself to be done in a suitable way.

That Theon himself conforms to this "cultural model", which, at least, seems to correspond to the above mentioned procedures of sophistic and philosophical training, is further confirmed by the fact that he would immediately suggest that more capable students might be able to compare between his commentary and others. This of course presupposes that such students must know *in advance* both the *Almagest* and some of its commentaries.[69] This situation is analogous to that which seems to by underlying, one and a half century later, Proclus's commentary

---

[65]Jones (1999a, b, 165).

[66]Bernard (2003a, 132 and 133 n. 197).

[67]For a detailed discussion of this passage, see (Knorr 1989, 63–76) and my own discussion in (Bernard 2003a, b).

[68]*Peri tou akouein* = *De Audiendo*, 42d–43f.

[69]It is probably not by chance that part [c] of Theon's preface looks so much like a promotional self-advertisement, meant to attract students to him. In the 'liberalised' context evoked by Jones (*ibid.* 157: the Greco-Roman astrologer as an 'independent professional'), one had to compete for students. Organized fights to attract students to the 'courts' of renowned sophists, in Eunapus's time, were not uncommon. Cf. note 66 above.

on Euclid: he concludes his preliminary presentation by pointing out the difference between 'standard' commentaries on Euclid, and more philosophically inspired ones, like his own.[70] This presupposes certain knowledge of the corresponding literature from at least some of his listeners. In the case of Theon, this hypothesis is reinforced by the fact that his commentary sometimes anticipates the content of the *Almagest* as if it was already known in advance, at least by some of the listeners.[71] This preliminary knowledge plainly conforms to the basic presupposition of ancient rhetorical teaching, which is that the student of rhetoric should rely on his permanent dwelling into the bulk of ancient literature: many of the ancient preparatory exercises, like reading aloud (*anagnôsis*), listening and writing down (*akroasis*), paraphrasing (*paraphrasis*), would be conceived as means of becoming more and more familiar with the bulk of works and books 'canonized' as *classics*, mainly for their long-recognized value for study. This is also coherent to what might be expected from ancient teaching situations, that is, that it made use of books that might be consulted and compared.[72] Finally, the request to go beyond standard expectations conforms to the ancient "expectation about expectations", as mentioned above concerning Plutarch.

Another aspect of Theon's preface strongly reminds one of the general background of ancient culture, namely the notion that certain listeners would not only be expected to know both the *Almagest* and the commentary literature on it, but also to be able to *correct* Theon where he might have failed (cf. part [f] above). The introduction to the *Great Commentary* mentions by name two companions (*hetairoi*), Eulalius and Origenes, who seemed to have requested a critical comparison between the *Handy Tables* and the methods and tables of the *Almagest*.[73] We have also mentioned above the author of a defective adaptation of the calculation of an eclipse, who might have been an advanced student.[74] Thus, we might legitimately suspect that Theon's audience might really have included more advanced participants that he could have considered as his peers.

Similarly, the notion of correction (*diorthôsis*), by which Theon concludes his preliminaries as well as his introduction to the *Great Commentary* (see appendix), raises the possibility that the elaboration of the commentary was a collective work or related to a collective work with the most capable students. What we know is that the three commentaries probably underwent several stages of revision, as the system of

[70] *In Eucl.* 84.8–23.

[71] See, for example, *in Alm.* 329.19–330.2. This preliminary knowledge might also be interpreted in light of the ancient practice, which was formalized in late Neoplatonist circles, of introducing introductory ('isagogical') remarks (*prolegomena*) situating the contents of the treatise studied in a larger context. See (Mansfeld 1994, 1998).

[72] An interesting and classical example of the use of books within technical studies is Galen, who frequently makes explicit the way he would consult books, especially the various versions of Hippocrates's writings, and compose his own texts. On this see (Andorlini 2000).

[73] For a more precise description of the contents and purpose of Theon's *Great Commentary* on the *Handy Tables*, see (Jones 1999b, 166–7) and the edition of the text (Tihon 1991).

[74] Tihon (1976–1977, 50–52).

cross references reveals.[75] We really do not know, for lack of relevant testimony, the way in which Theon elaborated his commentary and how this work was eventually related to discussions with students. If we consider, as a plausible comparison, that he proceeded as Proclus is described to have done by his disciple Marinus in his famous hagiography of the master,[76] then we might speculate that some issues were discussed in common during meetings (*sunousiai*) and that Theon, like Proclus, would have *afterwards* written down a commentary which would essentially be the result of this discussion, including the remarks of the most "inquisitive" students.[77] Such "speculative scenery" would, in a sense, vindicate Rome's notion that Theon's commentary was "la rédaction d'un cours" – but in a sense probably different from what Rome understood by this: not as a manual, but as an *a posteriori* composition that did not necessarily have an immediate use or interest for teaching.

For the same reason, Rome's notion that Theon's students were passive might ultimately rely on a misinterpretation of the underlying teaching situation. If we again accept the possible (but unproved) analogy between Proclus's and Theon's procedures, we might well surmise that the final commentary was a kind of compromise between Theon's own interpretation of the text and various interventions of more or less advanced students that were included in the text, by way of a kind of implicit 'patronage'. Thus, what Pappus does in the third book of his *Collection*, namely to report explicitly the geometrical construction proposed by students, might have reflected a more general procedure that was not always explicit.[78] Similarly, what A. Tihon suggests about the author of the calculation of the eclipse of 364 CE goes very much in the same possible direction (cf. part I above and note 30). Within Theon's commentary on the *Almagest* (part of which probably relies on Pappus's own), this hypothesis would explain the variety of comments included in the text as well as the various mistakes that can be found here and there: just as Pappus, in the third book of the *Collection*, would conclude his non-demonstrative, but essentially suggestive criticism of the construction that was submitted to him by a remark calling for further research and criticism on the part of his readers,[79] Theon (or Pappus before him?) might have included several developments in the commentary, not for the sake of providing a 'definitive' commentary, but for the sake of raising

[75] Jones (1999a, b, 168–9), Tihon (1992, 126–7).

[76] *Vit. Procl.* 26.21-27.37 Saffrey.

[77] This should not, of course, exclude the possibility that preliminary notes were written down as a preparation for the course. Festugières (1950, 34–47) discusses the meaning of *hypomnêmatismos* as a term describing the unedited notes taken for an oral course (*dialexis*, *homilia*). His remarks confirm that the edited version would be an *a posteriori* production.

[78] We might compare with the way by which D. Barbaro, in the Renaissance period, included one letter from Antonio Maria Pazzi (Gessner 2010). However, in this case as in Pappus's, the reference to Pazzi is *explicit* in Barbaro's text, while there is no such reference to any student or collaborator in Theon.

[79] *Math Coll.*52.31-54.3 Hultsch, book III. For further analysis, see Knorr (1989, 63–76) and Bernard (2003a, b).

interesting and thought-provoking issues. The defective demonstration at the end of Theon's commentary on *Almagest* I.4,[80] for example, might be such an unachieved attempt to introduce a discussion that was not to be found in classical treatises. Rome's reaction in denouncing a "pseudo-proof" and the passivity of students toward this demonstration might therefore be essentially unfair: since *his* attention was drawn by this attempt and its eventual defects, Theon's comment might be said to have attained its main goal as well. Had he really been a member of Theon's audience and not a modern historian, Rome might have proposed his own correction instead of just criticizing the text and its author, following Plutarch's advice that one should listen to (or read) others with benevolence, whatever their success or failures, thereby alluding to the necessity of *diorthôsis*.[81]

Since we are not in a position to identify what might be the various drafts of Theon's work[82] and since we do not know the precise way in which Theon taught either, we are reduced to speculation. But the above parallels should encourage us to remain prudent with any rash judgment on Theon's or his students' attitudes. Thus, Pappus's quasi-contemporary procedure, which is really our closest clue on the relationship between master and student in this kind of undertaking, points towards an image in which correction, self-correction and benevolence remain the key issues, and conform to the values expressed at the end of Theon's preface (part [f]).

More generally, then, we are led to judge Theon's text and what it might reveal about the underlying teaching situation against the background of ancient 'mondes lettrés', to adopt C. Jacob's terminology, and more precisely the cultural sphere of intellectual practice and values that is characteristic of the small elite of educated people (*pepaideuomenoi*) who, in late Antiquity, participated in literary culture. But to study the commentaries or original productions of this period does not only presuppose, in principle at least, to investigate what these precise procedures and values were, as we have done above; moreover, the latter were the subject of representation and this 'self-representation' is in itself 'part of the game': what we know of these cultured procedures is both what they were and the way they were represented. As noted by C. Jacob,[83] this mixture of facts and norms of practice should be ideally attainted through the direct and reflexive observation of situations[84] that can be observed directly, just as modern anthropologists or sociologists do. By contrast, Antique or Medieval teaching situations are indirectly

[80] *in Alm.* 398.6–399-8. This faulty demonstration is discussed in Knorr (1996).

[81] See *Peri itou akouein* 39d–40f, 41–43 Belles Lettres, especially 40c: "For it is the easiest thing in the world to find fault with one's neighbor, and also a useless and inane proceeding unless it be applied in some way to correcting [*pros tina diorthôsin*] or avoiding similar faults." (transl. Lacus Curtius, Loeb Library 1927; parenthesis added).

[82] Even in the case of the *Great Commentary*, which looks more like an incomplete work (Tihon 1991) this remains a very difficult task.

[83] In his introduction to the collective volume 'les lieux de savoirs' (Jacob 2007, 23).

[84] See the references signaled by Jacob in this note 1, *ibid.* p. 23.

attained through written or iconographic sources, which constitute as many 'media' or 'prisms' through which the underlying situation is represented, and therefore elaborated and transformed, by a specific discourse. Thus, in the case of Pappus, the concrete situation to which his report and criticism refers is introduced by a sophisticated argument relying on an ancient philosophical debate on mathematical epistemology, which serves him to enhance both his and the students' position.[85] In the case of Theon, we shall see next (part IV) that Ptolemy's own words and values are heavily utilized to represent each member of the 'scenery' that is represented, or alluded to, within the preliminaries.

If we were, then, to define what Theon's *monde lettré* was, we would have to take into account the fact that Theon's work is indeed a *cultured commentary* on a 'classical' source and, as such, it carries what I. Sluiter rightly calls an idea of the 'didactic tradition'.[86] By this is meant a deeply ingrained set of cultural values that in practice led many ancient writers and especially commentators to emphasize by various means the didactic import and weight of their own contribution. In other words, they presented themselves as teachers because *this self-representation was in itself part of a consistent tradition*, that she calls the 'didactic tradition'. Ultimately, the commentators would project, so to speak, the value of this tradition onto the authors of the commented sources themselves, presenting them essentially as teachers. Contrarily to what she explicitly suggests, the explicit commitment in the 'didactic tradition' does not imply that all commentators were teachers. It rather implies that most commentators abide to this tradition and consciously transmitted its values and encouraged others to do so. In other words, just as we can (legitimately) take Theon's allusions to teaching at face value, as a kind of anthropological testimony on his own practice, we can also read it as a literary composition first and foremost, abiding to certain norms and transmitting a certain sense of tradition. In the case of Theon, this remark is particularly apt, as Ptolemy himself is presented as a capable teacher, who always observes the correct attitude expected from such a *persona,* as we shall now see.

5.5 Ptolemy's Model: In the Wake of the Master

As indicated by A. Jones in the case of Theon, the overwhelming impression given by his or Pappus's kind of commentary, is that "the *Almagest* was by the fourth century ossified as a 'Great Book', something that an intelligent youth could read for intellectual profit with the guidance of Pappus's and Theon's commentary".[87] Only the term 'ossified' should perhaps be nuanced here: for, on the contrary, one might argue that as it had become a 'classic work' and was studied as such, Ptolemy's

---

[85] Bernard (2003a, 132).

[86] Sluiter (1999, 203).

[87] Jones (1999b, 168).

treatise had *to be kept alive* and *meaningful* for an intelligent paraphrase exercise. For sure, something essential of Ptolemy's spirit died with late commentators, namely the notion that one should continue and complete astronomical observations and determination of parameters in the way Ptolemy and his predecessors had done. But this does not imply that, *as a classic*, the *Almagest* was not 'alive' in a sense for its readers. In fact, the important questions seem to be the following: how was it kept alive, and on what did the commentators focus? Answering this question first requires reverting to the interpretation of the many allusions to the *Almagest* contained in Theon's preface.[88]

First, we have seen that Theon designates his own enterprise with the very same word, 'memorandum' (*hypomnêmatismos*), by which Ptolemy himself characterizes the structure of his treatise. In *Alm.* 8.6–16 he says that "we have tried to note down (*hupomnêmasisasthai*) everything which we think we have discovered up to the present time".[89] The verb is again used in book II, in which Ptolemy announces, after a paradigmatic calculation, that he will not repeat it in order not to lengthen the 'explanatory part' of the treatise' (*ton hupomnêmatismon tês syntaxeôs*).[90] Again in the important transition beginning in book IX, Ptolemy recapitulates the previous chapter by saying that the main topics concerning the fixed stars have been presented (*hupomnêmatisaito*), adding "insofar as the phenomena [observed] up to now provide the means of progress in our understanding".[91] This role attributed to the 'presentation' or 'explanation', as an aid and suggestion toward a better understanding of the theory, is also the one suggested in the interesting concluding remarks of the treatise.[92] We cannot dwell here on the general meaning of both the verb and the noun in Ptolemy's treatise, the two probably having to be understood against the background of Ptolemy's general theory of imagination and memory.[93] But these quotes are enough to see that this notion is quite central to Ptolemy's project and has to do with the *concrete guidance* provided by him to enable one to make progress in the theory of the heavens. It is to be understood, in other words, as part of a "didactic" project – at least in the very special sense that it might *suggest* a special kind of thinking to the astronomer-to-be.

That Theon would take up the Ptolemaic term, which could also be understood in a very straightforward way as 'commentary', to designate his own enterprise, is probably no accident. It is interesting, in this respect, to make a comparison with Theon's commentary on the very passage, in which Ptolemy summarizes the basic

[88] These allusions are indicated, in our translation given above (part II) by underlined dots.

[89] Translation Toomer. The verb should be translated in a better manner, but it is not my purpose here to discuss this point. For the same reason, I will not propose here a uniform translation of the term, that I am incapable of giving at the present moment.

[90] An other occurrence with very close meaning and almost the same wording is found in book X, *Alm.* II 338.20.

[91] *Alm.* II 206.7–8.

[92] *Alm.* II. 608.2–10.

[93] On this question, I refer the reader to J. Feke's forthcoming paper (Feke 2012).

structure of the treatise[94]: Theon focuses mainly on the duality of Ptolemy's attitude, carefully following and presenting what the ancients had best worked out, "cleaving to the learner's attitude"[95] on one hand, and to the attitude of a careful commentator on the other:

> following the work of illustrious predecessors, thanks to their available treatises, [he wants] to pay attention to what was rightly found out and not ignore this, but also to supplement missing [aspects], as he indeed appears to do in his whole composition of the theory. On the one hand, he suppressed some parts as redundant, some others he corrected as not adequately formulated, some others he added as postulates, blaming no one in any vulgar way, but paying only attention to the subject matter itself, from which he only draws [his reasons for] confirmation or refutation.[96]

In a quite revealing manner, what Ptolemy importantly says about the lessons of time itself is mentioned only afterwards and in a mere paraphrase by Theon.[97] All in all, Theon typically seems to project onto Ptolemy his own values as a commentator, so that he can indeed identify *his* endeavor with Ptolemy's own.

Much in the same way, the perspicuous readers who might be able to compare Theon's work with previous commentaries and spot Theon's originality, are named using the same expression used in Ptolemy for those philosophers who follow, like Hipparchus, the 'right' theoretical method. To judge the validity of Theon's own method is thereby suggested to be equivalent to bring oneself into conformity with the ethical model repeatedly advocated by Ptolemy himself. Much the same can be deduced from the final words by which Theon invites his best readers, again quoting Ptolemy (*Alm.* 8.15), to correct him and continue his work.

Finally, the very content and originality of Theon's commentary are also characterized as basically following Ptolemy's indications. The first allusion to Theon's insistence on geometrical justifications is supposed to follow Ptolemy's passing remark preceding his treatment of chords in book I: by a slight distortion of this remark, Theon suggests that supplementing such demonstrations is, in fact, Ptolemy's global purpose in the whole treatise.[98] Similarly, his clever mention of the consecrated formula 'so as not to make the 'presentation' (*hypomnêmatismon*) too long' used by Ptolemy allows Theon to suggest that his mode of commentary, and the level of detail he does or does not provide, basically follows Ptolemy's own mode of presentation.

This basic orientation of Theon's commentary can be confirmed by many other examples. For example, his commentary on Ptolemy's long philosophical preface might be interpreted as basically following Ptolemy's very special (and

---

[94] *in Alm.* 325.1–17.

[95] *In Alm.* 325.4–5.

[96] *In Alm.* 325.7–13.

[97] *In Alm.* 325.14–17.

[98] See note 45.

philosophically justified!) rejection of philosophical dissensus.[99] Similarly and in a striking manner, Theon puts much emphasis on the ethical dimension of the study of the *Almagest* as being *the* basic idea developed in Ptolemy's preface, preferable to the modern vision conveyed by contemporary historians, who only present Ptolemy's work as a theoretical one in astronomy. The importance of the 'order of theorems' (*Alm.* 8–9) is again reflected in Theon's repeated emphasis on this issue.[100] I will come back in the conclusion to the question of how far we might push forward this kind of reading of Theon. For the time being, I conclude that Theon's position toward Ptolemy certainly cannot be merely understood as *only* characterized by a general will to *clarify* Ptolemy's arguments or make it easier for its reader and possible students. Such an orientation is obvious in Theon,[101] but such a characterization would be somewhat short-sighted. The basic outlook and declared purpose of Theon rather seems to *follow Ptolemy's model and method*, at least according to the idiosyncratic way, by which he pictures and understands this method. Even 'supplementing' his text with clarifications and proofs might be understood as being the continuation of Ptolemy's project.

5.6 Tentative Conclusion: Open Questions and Research Issues on Theon's Commentary

In conclusion, the two perspectives mentioned above at the end of part II do not contradict each other, although it is useful to keep them distinct to some extent. Rome's notion that Theon's text was 'la redaction d'un cours' might hit the right mark, provided we understand this as meaning that it might plausibly have been the *a posteriori* writing down of collective discussions involving students of various levels. The very act of turning these experiences into a coherent text, which implied presenting it as a legitimate product of ancient culture, might have given it the

[99] I have developed this argument in the chapter devoted to Pappus, Theon and Hypatia in Bernard (2010b, 426–9).

[100] For the very same reason, I would nuance Jones's elliptic judgment that "neither Pappus nor Theon shows much interest in or understanding of the large scale plan of the Almagest" (Jones 1999b, 165). One might of course always contest that a given commentator *really* understood the structure of such a complex treatise; but Theon, at least, certainly pays the utmost attention to the whole structure of the treatise and constantly, if not obsessively, reminds the reader of this. This might be regarded as an amplification of Ptolemy's own concern for the overall structure of his progression.

[101] Theon in fact uses formulae like 'in order to make this clear . . . ' and therefore calls to the value of *saphêneia*, clarity, which is typical of many treatises from late antiquity. Remarkably, Theon does not put forward this value in this introduction, as if it was not really at stake for him. What seems really important for him is *to supply proofs* where it is possible to do so – even in places in which this might appear useless (cf. part [c] above).

dignity and value of a 'memorandum' (*hypomnêmatismos*) on the *Almagest.* This, as we have seen, could basically have meant following Ptolemy's model, rather than just clarifying the content of his treatise. This project included Theon's self-promotion as being a capable teacher of astronomy.

All this does not mean that Theon did not actually "teach the *Almagest*", but more simply that to prepare and lead such a teaching on the one hand, and to write a commentary on the other, are two different activities, each of which having their own purpose and constraints. As for Theon's qualities as an 'actual teacher', so to speak, we would need more concrete proof and counterfactual arguments to judge them. Such proof exists but is still meager (cf. part I). Let us also remark that the analysis of the explicit self-representation of Theon as commentator (Jacob's "prisme") is not only important to take into account but is also the easiest to document, since it essentially requires a careful comparison with the source text. By contrast, to attain a faithful picture of his actual method of teaching and work is a much more difficult question and should not, in any case, be restricted to Theon (cf. the conclusions of part I).

One important conclusion of the present analysis is that the kind of imitation of Ptolemy practiced by Theon is both real and easily detected (cf. part IV) but also represents a clever distortion of the spirit of Ptolemy's project. For the latter, accumulated and instrumented observations supplemented those which already exist and that might lead to a change of parameters, is a key issue. Theon's imitation and distortion of Ptolemy go in another, characteristic direction: he does not go into adding new observations but new proofs and logical coherency or justification, on one hand, and the construction of observation instruments on the other. The "ethical spirit" and general purpose is therefore somewhat retained, but the basic intellectual endeavor is directed onto new subject matters. This move can hardly be underestimated for, in the long term, it probably played a role in the progressive constitution of 'mathematics' as a relatively well-defined field of interest.

Finally, this discussion points toward new directions of research. In conformity with the 'double perspective' underlined above, one might of course follow Rome's lead in looking into the core of Theon's commentary for hints about didactic strategies, the commentary being understood as the reflection, more or less, of an actual teaching program.[102] In spite of their disparaging character, Rome's remarks still contain many hints and useful suggestions. Moreover, the analysis might be further pursued over the four books he edited. But one might also understand the bulk of the commentary as following the spirit of Theon's preface, that is, as basically being an imitation of Ptolemy. The question is then to check, on the relevant subjects, the kind of amplification that Theon carried out on Ptolemy's work and what new coherency this brought to the *Almagest* itself.

[102] Such a reading was suggested, in one or our workshops, by Fabio Acerbi commenting on the interesting passage devoted, in Theon, to useful rules for the memorization of the so-called Menenalus theorem (cf. note 61 above).

Appendix: The Prefaces to Theon's Commentaries on the *Handy Tables*

1. Theon's preface to his 'Great Commentary' on the *Handy Tables*, on the basis of the Mogenet and Tihon edition and translation of it.[103]

[a] Of Theon of Alexandria, <memorandum on the *Handy Tables* of Ptolemy, first part>

[b] Since you have assigned us the task of going through in detail into the reasoned method[104] of the treatise of the *Handy Tables*, my companions Eulalius and Origenes, I have tried to write down a memorandum[105] by proceeding as far as possible in the most exacting way, for we also[106] thought it would add not a small contribution to the mathematical theory[107]; thus, I have undertaken the teaching of the proposed subject matter in five books, so that those who have already made some progress in mathematics[108] might be able to follow, and closely following the treatise of the *Syntaxis*.[109] In this way, indeed, what is received in each of the hypotheses[110] will be more easily remembered by them.

[c1] We will <thus> teach, through quasi-causal teaching,[111] the exposition[112] of the *Handy Tables* and the reason of the <underlying> calculations, so that, each

---

[103]Mogenet and Tihon (1985). This edition includes a French translation of the text.

[104]Lit. *logikê ephodos*. 'Ephodos' refers to the action of following something step by step, like a calculation in the present case.

[105]Here the verb employed is, as in the introduction of the *Almagest* and of Theon's commentary on the latter (cf. note 40), *hypomnêmatisasthai*.

[106]The privilege of having 'launched' the issue is awarded to Eulalius and Origenes, and the treatise is first presented as a response to it, then Theon adds his *own* opinion and approval of the project.

[107]In the context of the study of the *Almagest* and the *Handy Tables*, such an expression cannot but have a very heavy weight: it directly refers to the 'mathematical theôria' to which Ptolemy explains that he devotes himself in *Alm.* I 6.21–23.

[108]*En tois mathêmasi*, which might be understood in a looser way: in their studies. Tihon: 'dans les sciences'.

[109]*Pragmateia tês syntaxeôs*.

[110]*Hypothesis* has to be understood here in the 'Ptolemaic' sense, as the kinematic-geometrical basis from which the tabular computations are established and justified.

[111]Lit. *dia tês aitiôdous didaskalias,* that is a 'quasi-causal' teaching. This is probably a double allusion, first to the "knowledge of the causes" thanks to which errors might be corrected in a table (see below the end of the same sentence); and second, to the fact that understanding the computational method and its geometrical basis really amounts to understanding the true movements of the stars: Ptolemy insists heavily on this aspect in book III of the *Almagest* when he introduces the basic purpose of building a table of mean motion and anomaly for the irregular movement of the sun, see *Alm.* I 208.15–27, book 3, ch.2; at the same time, the *Almagest* is not a work on the physics of the Heavens, so that this might only be qualified of 'quasi-causal', as Theon has it.

[112]The Greek *ekthesis* is frequently used by Ptolemy in various and interconnected meanings, from the presentation of a complement argument to the bare outlook of tables. The meaning here can only be the way to present the tables, that is, how and why they should be constructed (cf. the title of II.6, II.8 or II.13 in the *Almagest*).

time we have any doubt about a writing mistake in the numbers that are disposed in the tables, we might easily correct them by the knowledge of the causes;

[c2] furthermore, we will also make comparisons, wherever the disposition of the tables presented <in the *Handy Tables*> or the explanations given in the same treatise differ from what is explained in the *Composition*, since he <i.e. Ptolemy> has arranged them according to their own respective treatment;

[c3] and in order not to make this memorandum too long, we have considered it was superfluous, on one hand, to complete it with geometrical procedures that we could already explain in a sufficiently clear manner by discourse alone, since we have already explained these procedures in a more geometrical manner once and for all in our remarks[113] on the *Composition*; but, on the other hand, wherever the subject matter was not suitable for such<verbal> explanations, we have presented them <i.e., the geometrical arguments> in such a way, that they might be more easily understood.

[d] As for you, my companions, if anything has escaped my attention on such important matters,[114] for I am only a man,[115] I beg your comprehension for no equivalent treatment, before us, has reached us.

2. Theon's preface of his 'Little Commentary' on the *Handy Tables*, on the basis of the text edited and translated (in French) by Anne Tihon.[116]

Of Theon of Alexandria, on the *Handy Tables*.

The more reasoned method <leading> the calculation of <the positions of> stars[117] through the *Handy Tables*, Epiphanios my boy, has been provided by us in the five books of another treatise.[118] But since, as far as the latter is concerned, most of those who come to us so as to learn such subject matter[119] do not only happen to be unable to follow competently the multiplications or divisions of numbers, but also to be completely uninitiated in geometrical demonstrations, we have tried for

[113]*Scholia.*

[114]Lit. 'of such a quality', *têlikouta.*

[115]This is probably a discrete allusion to Ptolemy's self promotion as striving for a quasi divine disposition of the soul, as a consequence of his choice of studying the movements of Heavenly bodies through mathematics. This opposition mortal / divine is well reflected in the famous epigram sometimes attributed to Ptolemy and that is indeed present in several copies of the *Almagest*: "Well do I know that I am mortal, a creature of one day. // But if my mind follows the winding paths of the stars// Then my feet no longer rest on earth, but standing by // Zeus himself I take my fill of ambrosia, the divine dish." *The Greek Anthology*, transl. Patton, Cambridge MA: Harvard Univ. Press, 1916, repr 1960, vol.3, book IX, epigram 577, p. 321.

[116]Tihon (1978).

[117]*Psêphoria tôn asterôn*, the notion of 'calculation'/*psêphophoria* is used frequently by Ptolemy and always only to position the sun, moon or planets through specific tables; the notion is therefore intrinsically related to the use of Tables. It might almost be translated as 'algorithmic procedure using tables or series of tables'.

[118]This alludes to the *Great Commentary.*

[119]Lit. *Mathêsis tês toioutês didaskalias*, <they come> for the learning what is thus taught.

them also to make a commentary as far as possible more methodical, presenting the bare procedures[120] so that the presentation of this subject matter might appear more clearly to them.

References

References to Ancient Texts

Alm. Claudii Ptolemaei Opera quae extant omnia. Syntaxis mathematica I. Libros I-VI (1898) II. Libros VII.-XIII. (1903), ed. Heiberg. Leipzig: Teubner.

Cam. 1538. Camerarius's edition of Theon of Alexandria's commentary on the *Almagest.* In *Claudii Ptolemæi Magnæ constructionis, id est perfectæ cœlestium motuum pertractionis lib. XIII [Ed. Simon Grynaeus]. Theonis Alexandrini in eosdem commentariorum lib. XI [cura Joachimi Camerarii].* Basel: Johan Walder.

GC Great Commentary of Theon of Alexandria on the Handy Tables. The reference is to Tihon's edition (Tihon 1993), pages and line numbers.

In Alm. Commentary of Theon of Alexandria on the Almagest. The reference is to Rome's edition (Rome 1936), pages and lines numbers.

In Eucl. Proclus diadochi in Primum Euclidis Elementorum Librum commentarii, ed. Godfried Friedlein. Leipzig: Teubner, 1873 (rééd. Georg Olms 1967).

LC Little Commentary of Theon of Alexandria on the Handy Tables. The reference is to Tihon's edition (Tihon 1978), pages and line numbers.

Vit. Procl. Marinus de Néapolis, *Proclus ou Sur le Bonheur = Vita Procli.* 2002. Text edited, translated and annotated by H.D. Saffrey and A.P. Segonds. Paris: Belles Lettres.

Modern Studies

Acerbi, F., N. Vinel, and B. Vitrac. 2010. Les 'Prolégomènes à l'Almageste'. Une édition à partir des manuscrits les plus anciens. Introduction générale-Parties I–III. *Sciamvs* 11: 53–210.

Andorlini, I. 2000. Codici papiracei di medicina con scoli e commento. In *Le commentaire, entre innovation et tradition*, dir. M.O. Goulet-Cazé, 37–52. Paris: Vrin.

Barton, T. 1994. *Ancient astrology*. Londres: Routledge.

Bernard, A. 2003a. Sophistic aspects of Pappus's collection. *Archive for the History of Exact Sciences* 57(2): 93–150.

Bernard, A. 2003b. Ancient rhetoric and Greek mathematics: A response to a modern historiographical dilemma. *Science in Context* 16(3): 391–412.

Bernard, A. 2010a. The historical significance of Ptolemy's Almagest for its early readers. *Revue de synthèse* 131, 6e série: n°4, 495–521.

Bernard, A. 2010b. The Alexandrian school. Theon of Alexandria and hypatia. In *The Cambridge history of philosophy in late antiquity*, vol. 1, ed. Lloyd P. Gerson, 417–436. Cambridge: Cambridge University Press.

Bernard, A. 2010c. Review of (Rochberg 2004) in *Revue de synthèse* 131, 6e série: n°4, 615–617.

[120]Lit. *psiloi ephodoi*. See note 46.

Choppin, A. 2008. Le manuel scolaire, une fausse évidence historique. *Histoire de l'éducation* 117: 7–56.
Cuomo, S. 2000. *Pappus of Alexandria and the mathematics of late antiquity*. Cambridge: Cambridge University Press.
Feke, J. 2012. Ptolemy's defence of theoretical philosophy. *Apeiron* 45(1): 61–90.
Feke, J., and A. Jones. 2010. Ptolemy. In *The Cambridge history of philosophy in late antiquity*, ed. L. Gerson, 197–209. Cambridge: Cambridge University Press.
Festugière, A.J. 1950. *La Révélation d'Hermès Trismégiste II: Le dieu cosmique*. Paris: Les Belles Lettres.
Gessner, S. 2010. Bons procédés entre érudits. Un mésolabe pour l'édition de l'Architecture de Vitruve (1567). *Revue de Synthèse* 131(4): 523–541.
Hadot, I. 2006. *Arts libéraux et philosophie dans la pensée antique: Contribution à l'histoire de l'éducation et de la culture dans l'Antiquité*. Paris: Vrin.
Jacob, C. (dir). 2007. *Lieux de savoir 1: Espaces et communautés*. Paris: Albin Michel.
Jones, A. 1994. The place of astronomy in Roman Egypt. *Apeiron* 27(4): 25–52.
Jones, A. 1999a. *Astronomical Papyri from Oxyrhyncus (P. Oxy. 4133–4300a)*, Memoirs of the American Philosophical Society 233. Philadelphia: American Philosophical Society.
Jones, A. 1999b. Uses and users of astronomical commentaries in antiquity. In (Most 1999), 147–172.
Kaster, R.A. 1988. *Guardians of language: The grammarian and society in late antiquity*. Berkeley: University of California Press.
Knorr, W.R. 1989. *Textual studies in ancient and medieval geometry*. Boston: Birkhaüser.
Knorr, W.R. 1996. The method of indivisibles in ancient geometry. In *Vita mathematica: Historical research and integration with teaching*, ed. R. Calinger, 67–86. Washington, DC: Mathematical Association of America.
Mansfeld, J. 1994. *Prolegomena. Questions to be settled before the study of an author, or a text*. Leiden: Brill.
Mansfeld, J. 1998. *Prolegomena mathematica. From Apollonius of perga to the late neoplatonists*. Leiden: Brill.
Marrou, H.I. 1965. *Histoire de l'éducation dans l'Antiquité*, 6th ed. Paris: Seuil.
Mercier, R. 2011. *Ptolemaiou Procheiroi Kanones. Ptolemy's Handy Tables. Vol 1b, transcription and commentary*, Publications de l'Institut Orientaliste de Louvain. Leuven: Peeters.
Mogenet, J. 1956. *L'introduction à l'Almageste,* Mémoires de l'Académie Royale de Belgique: LI fasc.2, Bruxelles.
Mogenet, J., and A. Tihon. 1985. *Le "Grand Commentaire" de Théon d'Alexandrie aux Tables Faciles de Ptolémée. Livre I*, Studi e Testi 315. Vatican City: Biblioteca Apostolica Vaticana.
Most, G. (ed.). 1999. *Commentaries: Kommentare*, Aporemata, Bd. 4. Göttingen: Vandenhoeck und Ruprecht.
Pingree, D. 1994. The teaching of the *Almagest* in late antiquity. *Apeiron* 27(4): 75–98.
Renaud, H. 1931. L'enseignement des sciences exactes et l'édition d'ouvrages scientifiques au Maroc avant l'occupation européenne. *Archeion* 13: 325–326.
Ritter, J., and B. Vitrac. 1998. Pensée grecque et Pensée 'orientale. In *Encyclopédie Philosophique*, vol. IV, ed. J.F. Mattéi, 1233–1250. Paris: Presses Universitaires de France. Text available online (Hal-SHS).
Rochberg, F. 2004. *The heavenly writing: Divination, horoscopy and astronomy in Mesopotamian culture*. Cambridge: Cambridge University Press.
Rome, A. 1931. *Commentaires de Pappus et de Théon d'Alexandrie sur l'Almageste. Tome I: Pappus d'Alexandrie, Commentaire sur les livres 5 et 6 de l'Almageste*. Text edited and annotated by A. Rome. Studi e Testi 54. Vatican City: Biblioteca Apostolica Vaticana.
Rome, A. 1936. *Commentaires de Pappus et de Théon d'Alexandrie sur l'Almageste. Tome II: Théon d'Alexandrie, Commentaire sur les livres 1 et 2 de l'Almageste*. Text edited and annotated by A. Rome. Studi e Testi 72. Vatican City: Biblioteca Apostolica Vaticana.

Rome, A. 1943. *Commentaires de Pappus et de Théon d'Alexandrie sur l'Almageste. Tome III: Théon d'Alexandrie, Commentaire sur les livres 3 et 4 de l'Almageste*. Text edited and annotated by A. Rome. Studi e Testi 106. Vatican City: Biblioteca Apostolica Vaticana.

Siety, A. 2001. *Mathématiques, ma chère terreur*. Paris: Calman-Levy.

Sluiter, I. 1999. Commentaries and the didactic tradition. In (Most 1999), 173–205.

Taub, L.C. 1993. *Ptolemy's universe, the natural philosophical and ethical foundations of Ptolemy's astronomy*. Chicago: Open Court.

Tihon, A. 1973. Les scholies des Tables Faciles de Ptolémée. *Bulletin de l'Institut Historique Belge de Rome* XLIII: 49–110.

Tihon, A. 1976. Notes sur l'astronomie grecque au Vème siècle de notre ère (Marinus de Naplouse, un commentaire au *Petit Commentaire* de Théon). *Janus* 63:167–184.

Tihon, A. 1976–1977. Le calcul de l'éclipse de soleil du 16 juin 364 p.C. et le 'Petit Commentaire' de Théon. *Bulletin de l'Institut Historique Belge de Rome* XLVI–VII: 35–79.

Tihon, A. 1978. *"Le 'Petit Commentaire' de Théon d'Alexandrie aux Tables Faciles de Ptolémée: Livre I". Studi e Testi 282*. Roma: Biblioteca Apostolica Vaticana.

Tihon, A. 1987. Le livre V retrouvé du *Commentaire à l'Almageste* de Théon d'Alexandrie. *Antiquité Classique* 56: 201–218.

Tihon, A. 1991. *Le 'Grand Commentaire' de Théon d'Alexandrie aux Tables Faciles de Ptolémée. Livres II et III*, Studi e Testi 340. Roma: Biblioteca Apostolica Vaticana.

Tihon, A. 1992. Propos sur l'édition de textes astronomiques grecs des IVè et Vè siècles de notre ère. In *Les problèmes posés par l'édition critique des textes anciens et médiévaux*, ed. J. Hamesse, 113–137. Louvain-la-Neuve: Brepols.

Tihon, A. 1999. *Le 'Grand Commentaire' de Théon d'Alexandrie aux Tables Faciles de Ptolémée. Livre IV*, Studi e Testi 390. Roma: Biblioteca Apostolica Vaticana.

Tihon, A. 2011. *Ptolemaiou Procheiroi Kanones. Les «Tables Faciles» de Ptolemée*. Volume 1a: Tables A1-A2. Publications de l'Institut Orientaliste de Louvain, 59a. Leuven: Peeters.

Vitrac, B. 2008. Promenade dans les préfaces des textes mathématiques Grecs anciens. In *Liber Amicorum Jean Dhombres*, ed. P. Radelet de Grave and C. Brichard, 518–556. Louvain-la-Neuve: Brepols. The complete text is available online (hal-SHS).

Part III
Comparative Approach

Chapter 6
Relationships Between French "Practical Arithmetics" and Teaching?

Stéphane Lamassé

6.1 Introduction

Guy Beaujouan distinguished practical arithmetics from university algorisms by two essential features: Practical arithmetics teach calculation written out with a pen and pencil and offer commercial applications.[1] For medieval historians, the expression "an arithmetic for merchants"[2] does not carry the meaning of a counter-top book, an aide-mémoire intended for everyday use in trade,[3] it is rather that of basic instruction.[4]

The geographical and chronological coincidence of the production of this type of work, the first known example being Leonardo of Pisa's *Liber Abaci* (1202), with what we know of the development of the world of commerce between the thirteenth and fifteenth centuries is quite surprising and demonstrates the existence of a close relationship between these two phenomena.

---

[1]This is the definition which the author (Beaujouan 1988, 73–88) uses to compare a group of 12 manuscripts that make up the corpus of practical arithmetics and are essentially those we will use. In this article we will not retain Bnf. lat. 7287, nor *Suma de la art de arismetica* by Francesc de Santcliment nor Bnf. lat. 7352. These manuscripts are used but will not be studied specifically as they open towards particular cultures (printing) or are clearly far from the others, like (Bibl. Nat. lat. 7287) which does not offer commercial development.

[2]See (Spiesser 2003a, 32–50).

[3]These are not textbooks of commercial practices in the way "*Pratica della mercatura*" is, they differ in this way from those presented by (Jeannin 2002).

[4]With regard to these arithmetics, Warren Van Egmond highlights that they make up the first level of mathematical literacy in the late Middle Ages (Egmond 2001).

S. Lamassé (✉)
Université Paris 1 EA 127, Modernité et Révolutions, Paris, France
e-mail: Stephane.Lamasse@univ-paris1.fr

A. Bernard and C. Proust (eds.), *Scientific Sources and Teaching Contexts Throughout History: Problems and Perspectives*, Boston Studies in the Philosophy and History of Science 301, DOI 10.1007/978-94-007-5122-4_6,

We know the children of Italian merchants found a basic education in the abacus schools which allowed them to forge a solid knowledge of mathematics,[5] under the aegis of a specific type of algorism professionals.[6] This is illustrated by the examples of Paolo dell'Abbaco[7] in the fourteenth century and of Benedetto da Firenze in the fifteenth century,[8] who both wrote treatises while teaching in these schools.

The content of these works reinforces this idea: we find mathematical rules taught using numerous references to commercial practice, with the rule of three playing a central role allowing the calculation of conversions of measurement, currencies, interest and rules for partnerships as well as a form of bartering known as barat.[9]

It is, however, difficult to accept that works of this type always received the same reception from merchants between the thirteenth and fifteenth century, or even that they all correspond to the same educational realities. While the number and diversity of texts continues to increase between these two dates, medieval society does not remain the same and nor do the merchants.[10] This is an historical fact that is rarely taken into account when opening these treatises for the first time and which requires an understanding of the actual production of these works.

Reading Franco-Occitan arithmetics raises to the question of differences in their production. Several features make them a good field for research. These arithmetics are, in fact, relatively few in number, ten or so, and their production is centered on the fifteenth century in a defined geographical area around urban centers such as Paris, Lyons, Marseilles, Nice and Pamiers. There are no known examples of this type of book[11] from previous centuries. This corpus, initially compiled by Guy

[5]On the teaching of arithmetic in Italy, there is an extensive bibliography. Only some references are here given, such as (Grendler 1989; Ulivi 2002a). Moreover, the disparity between the French-speaking and Italian-speaking worlds has already been highlighted (Benoit 1989a).

[6]Elisabetta Ulivi has been able to identify 76 abacus masters in the municipality of Florence between 1283 and 1500. She has also shown that the archives allow the masters' families and their networks of relations to be reconstructed, see (Ulivi 2002b).

[7]There is an extensive bibliography. For an introduction, see (Van Egmond 1977) and (Piochi 1984).

[8]He is the author of two works (Ulivi 2002b), which were quite successful, *Trattato di praticha d'arismetrica* in 1463 (of which 3 copies remain and which Elisabetta Ulivi describes as an encyclopedia) and *Trattato d'abacho* of which 18 copies exist –including two from the sixteenth century– and these are considered as conceived for teaching in these Bottegga. The investigation led by Elisabetta Ulivi on the environment surrounding the Florentine abacus schools has given us a fairly precise image of not only the teaching, but also the professional "milieu" of the teachers.

[9]On the presentation of these rules, see (Benoit 1989a).

[10]The introduction of (Van Egmond 1980) gives a typology of works that reflect this diversity.

[11]It is evident that they do not arrive on virgin ground, cultural life continued to be inhabited by the treatises produced in vernacular language from previous centuries. Treatises on algorism translated into vernacular language have existed since the thirteenth century. The works and collections of problems even circulated in Italian. Bartoli's dissertation kept in Avignon's municipal library was published in (Sesiano 1984a; Bozzolo and Ornato 1983, 116–118). Bozzolo and Ornato (1983, 366) offers a diagram reviewing medieval book production.

Beaujean and extended by Jean Cassinet and Alessandro Vitale-Brovarone,[12] has been the subject of numerous studies and is quite well known today.[13]

Nothing, or close to nothing is known about the teaching situations,[14] at best we can compare the Italian model and the heterogeneity of these books written in French. When men emerge from the silent darkness of the sources, it is very often from their work itself that we extract what we know.[15] The problem has frequently been addressed by those historians, for whom the very existence of these works emanates from a need to educate merchants.[16]

From a situation where information external to the documentation is, for the moment, incomplete if not non-existent, it is interesting to re-examine the relationship between "practical arithmetics" and teaching, based on a comparative study mainly focused on the treatises and paying attention to the greatest number of clues or traces.[17] The purpose of this article consists in returning to the rather classic historical approach of highlighting the diversity of production,[18] which will possibly allow us to refocus the hypotheses on the diversity of the readership and suggest new ones as to the underlying situation. As the bulk of the information comes from manuscripts, let us start with them and the clues they can give us by accepting the premise that this perspective may allow us to envisage groups of possible teaching situations that I shall call, for want of anything better at the moment, the implementation of a writing project.

Firstly we will examine the physical construction and the purpose of the works. For this we will pay attention to the place of the texts on practical arithmetic within them; to the way the authors themselves gave meaning to their treatise, by studying, where necessary, the introductory texts.

---

[12]This corpus was presented for the first time in (Beaujouan 1958, 84–87), continuing and adding to the work of Barthélemy de Romans in (Cassinet 1993) See note 30.

[13]For the composition of the corpus, see the table in Appendix 2.

[14]In this article we will take the meaning of this expression as given by Anne-Marie Chartier Chap. 2 in this volume.

[15]This is the case for men such as Barthélémy de Romans and Jean Adam, although we have some sources on the latter linked to missions given to him by the king. But for a man like Jean Certain, who produced a beautiful book in Marseilles, we still do not know anything more today. The approach induced from the sources themselves is the subject of new methodological and epistemological reflection among historians, a summary of which can be read in (Chastang 2008) and (Mackenzie 1991, 39–41).

[16]Henri Pirenne has shown that historical evidence attesting a certain merchant culture is quite abundant (Pirenne 1929). Paul Benoit and more recently Maryvonne Spiesser (2003b) have investigated the training of French merchants by studying the mathematical content, rules and problems in these works and, through comparison with the Italian model, have tried to recreate the teaching framework.

[17]Ginzburg (1980).

[18]This is one meaning of the typological work carried out by Guy Beaujouan, who investigated the manuscript traditions and suggests geographical differences between the works. This work remains the reference for classifications that can be made for these works (Beaujouan 1988).

Then we will center the analysis on the study of the paratext and the language in these treatises. We will thus address the methods of reading and knowledge transmission by the language used.

Finally, through the problems, we will discuss the implementation of a writing project. The analysis of this content leads to questioning the relationship these texts have with the mercantile world, and finally to re-examining the status of the practices they introduce.

6.2 Construction and Purpose of the Works

6.2.1 Interpreting the Differences in Facture of the Works

Outside the invaluable catalogues,[19] no quantitative or synthetic studies have, as yet, been carried out on scientific mathematics manuscripts, be they practical or not. The distinction between the works is essentially drawn from their content. Among the Italian manuscripts, Warren Van Egmond differentiates between those made up of collections of problems and those revealing a structured teaching method. Maryvonne Spiesser replaces this with a more detailed classification of the manuscripts, based on a corpus wider than the Franco-Occitan language, in order to position the *Compendy de la praticque des nombres* more precisely.[20]

Browsing through the body of French language texts, one is struck by the differences in their facture, and the fact that their structure can sometimes be quite complex. The treatise may be part of the book, but it can also be an element in a larger collection.[21] It is quickly possible to distinguish four groups by considering some summary descriptive elements of the support, such as the size, the number of folios, the ink, the type of support, the presence of copyists' marks, illuminations and paintings, and the layout of the work.

Three easily transportable works, bearing the marks of use and where arithmetic makes up the bulk of the book, form the first group. Two of them, the *Compendy de Pamiers,* first known manuscript (c. 1430) and BnF. fr. 2050 (c. 1460), have almost identical formats (210 mm × 140 mm, 205 mm × 145 mm). One can add

[19](Van Egmond 1980) is the most precise reference on the subject today. For a broader perspective on a European level see the International Catalogue of Mediaeval Scientific Manuscripts, http://jordanus.ign.uni-muenchen.de [consulted: 07/08/2011]. This on-line catalogue was created under the direction of Menso Folkerts.

[20]She offers a four-point classification: "complete" algorisms, collections of various problems, between collections of problems and "complete" algorisms and "encyclopedic" treatises (Spiesser 2003b, 62–63).

[21]The question of the composition of the works also quite naturally raises that of the compilation. It can be observed on the scale of the whole manuscript in the succession of the treatises, as well as the practical arithmetic text.

to this the *Petit cadrans* (early sixteenth century) which appears, even more than the two others, in a pocket format (140 mm × 50 mm). Apart from the *Compendy de Pamiers,* which also has some velum pages, all three were produced on paper supports. Judging by the presence of catchwords, the first two books appear to be the work of copyists. The books bear the marks of manipulation, but these additions cannot be dated with precision. In BnF. fr. 2050 there is a pen and ink line drawing of a profile. The manuscript produced in Pamiers presents a number of unique features; multiple operations as well as signatures at the beginning of the manuscript, which attest its utilization.[22] The structure of the *Petit cadrans* is marked by rather brutal changes of sources, which the copyist copies and seems to collect. Traces of compilation are found in each of the texts, but the homogeneity of the writing gives them the vigor of *ad hoc* works. There is no question as to the authors' willingness to produce a coherent, transportable text, even if we can reconstruct some of the processes used in creating the works. Thus BnF.fr. 2050 introduced a section on fractions identical to that in BnF. Lat. 7381, part of which ends with "*explicit de ruptis.*"

After this group, a second category emerges, made up of two much more luxurious works of different sizes. Bibl. St. Genev. 3143, written on paper, is comparable in size with the works in the first category (170 mm × 120 mm). The *Kadran aux marchans*, written on parchment, is bigger (272 mm × 200 mm). Both were produced with particular care, and feature ornamentations. The Bibliothèque Sainte Geneviève manuscripts have numerous ornamentations which are used for the representation of operations on the abacus.[23] The *Kadran aux marchans* presents two cadrans, one on the first sheet representing a cadran surrounded by four evangelical symbols and heraldic coat of arms of a still unknown family, the second, unfinished, was planned out on the first sheet. There we can see, thanks to a preparatory pencil drawing, two merchants measuring and weighing. These are not very voluminous manuscripts, having 75 and 84 folios respectively, and they have been copied very carefully.

Three quarto[24] compilations, on paper[25] and whose commercial arithmetic represents no more than one section of the book, can form a third group. One of them, manuscript 86 from Tournai is now destroyed, but its description in the catalogues is similar to the other two.[26] While these manuscripts may appear less well finished, they were written by a single hand and sometimes include paintings,

---

[22]"*hic liber ad me pertinet quy vocoo Criberiua Paullus*" (fol. 5v) and "*Cribreres*" (fol. 4v) can be read, and some comments like "*veryrat est bon compagnion*" (fol. 30r). To this we must add trial operations on the first and last pages.

[23]The 17 illuminations are all consultable on-line at the Liber floridus website: http://liberfloridus.cines.fr/textes/biblio_fr.html. [consulted 21/06/11]. Only the drawings have not been reproduced.

[24]Tournai 86: 210 mm × 70 mm; BnF. fr. 1339: 270 mm × 205 mm; Méd. Nantes 456: 286 mm × 215 mm.

[25]In BnF. fr. 1339 and Méd. Nantes 456 watermarks are identifiable but without being an exact, certain match, in (Briquet 1907).

[26]A description of this manuscript exists in (Faider and Van Sint 1950).

like BnF.fr. 1339.[27] They both contain a sequence of treatises that are not identical but do share a common core:

1. a "practical" arithmetic.
2. an arithmetic using counters that appears influenced by the *Algorismus linealis*[28];
3. a geometry.

This model of compilation is reminiscent of other texts of university origin. The anonymous writer of BnF nat. 1339 thus included a title for fractions evoking another treatise on the subject that was well-known at the time: the *Algorismus de minutiis,* of which there were numerous versions. The only difference to the other works is this anonymous author's marked concern to give a general introduction to the whole of his composition, thus reinforcing the unity of the work. These authors therefore include in their manuscripts a treatise presenting an approach close to that of the practical arithmetics, reflecting modes of appropriation diverse enough to constitute a "genre".

Only manuscript 456 in Nantes' municipal library ("Médiathèque de Nantes") carries any marks from a reader, as it bears later comments, complementary information and corrections.[29]

Finally there are two works, exceptional in both their size and content. These, the most important works in the corpus of fifteenth century practical arithmetics, are also those whose influence on their contemporaries is best known and were written by two mathematicians: Barthélemy de Romans and Nicolas Chuquet.[30] They are also the two thickest volumes with more than 300 carefully written, very readable, paper pages. These two works include the practical arithmetic treatises in a broader range of treatises and this changes our understanding of them. Each of these works

---

[27]Some geometrical figures have been cut out from the manuscript.

[28]Constructed on the model of the algorism inherited from Sacrobosco, the *Algorismus linealis* presents numeration and basic operations using a table with counters. This treatise can be found in BnF. fr. 1339 and Méd. Nantes 456, to which can be added the French part of Plut. 29 cod.43 at the Laurentian Library, containing an arithmetic using counters and a geometry. One can see reproductions of the manuscript and a description on the Laurentian Library website at http://teca.bmlonline.it/TecaRicerca/ (consulted 21/02/13). Manuscript BnF. fr. 1339 presents, in addition, a Greek algorism and a version of the *Traité des usages de l'astrolabe* by Jean Fusoris. For the treatise on fol. 80 v – 83 r, whose incipit is "Then come numbers that are made by certain figures that some call Greek algorisms" (*"S'ensuite apres nombre par certaines figures que aulcuns appellent algorisme grec . . . ")* see (King 2001; Sesiano 1985).

[29]The annotations born by Méd. de Nantes 456 date from the sixteenth century but it is not certain that they are from Fra Giovanni Giocondo's hand, see (Tura 1999). Adolfo Tura also gave an edition of the geometry known as Fusoris' with variants of the Méd. Nantes 456 and BNF. fr. 1339 texts in (Tura 2008, 119–171).

[30]Another excerpt from the work of Barthélemy de Romans is known in the practical arithmetic discovered by Alessandro Vitale-Brovarone (Vitale-Brovarone 1990, 29–32). This manuscript has the shelfmark Turin, Archivio di Stato, Archivio Biscaretti, Mazzo 29, n. 3. Furthermore we know that Barthélemy's work had a decisive influence on that of Nicolas Chuquet who had himself been partly motivated at the beginning of the sixteenth century by *L'arismeticque* then newly written by Etienne de La Roche.

thus puts the knowledge in these treatises into a broader perspective, introducing an approach which reflects the implementation of a mathematics project.[31]

*

What conclusions can we draw from the factual observations that differentiate these types of manuscripts?

The diversity of supports and investments reflects the importance of the presence of a practical arithmetic in each of these works. The importance given to this "genre" of treatise in the body of the various manuscripts is the clue that it had become permanently established as an element of mathematical knowledge and learning.[32]

The content of the works and the pedagogical dimension which can be attributed to their content cannot be analysed in abstraction of the status of the books as objects. To ignore this would not take account of the practical realities of the appropriation of knowledge related to the functional dimension of the object (heavy, large, valuable). On the contrary, taking these factors into account broadens the range of questions.

In the specific case of this corpus we see that the production of the works is not homogenous: there are rich, prestigious works; thick, heavy, scholarly constructions, more appropriate to libraries; compilations on the university model orientated towards practice; and works of practical knowledge that still bear marks of proven use, close to the notion we have of a textbook.

While such information, taken from the book as an object, can attest an intention, it remains difficult to identify where the intention came from: was it from the author, the person who commissioned the book or the copyist?

Inferring the teaching situation in which this information was used is no easier. So, for the last group this could have been individual tutoring as much as school teaching. The first two, on the contrary, assume a mode of appropriation more related to library attendance.

6.2.2 *Analyzing the Intention Through the Introductory Texts*

In addressing the reader directly, the authors themselves sometimes set out the objective of their text. If we put BnF. fr. 1339 to one side for the moment, all the authors have the desire to present their work, often to justify it and sometimes to introduce themselves. While, from the fifteenth century, the prologue became the

[31]The very precise codicological study by (Spiesser and Féry-Hue 2007) allows us to observe the relationships between the different treatises in mss. Cesena S.XXVI.6.

[32]The second, more theoretical treatise in Barthélemy de Roman's work is re-employed in manuscript Bnf. lat. 7381. These are fragments involved in another manuscript situation, indicating a diffusion of these works. One could also invoke a manuscript in Latin, Bibl. de. Dijon 447, which explores rules identical to those in the manuscripts in a different way: the rules of three, of association, and of false position.

place in a medieval book dedicated to justifying the object of the text,[33] only Jean Adam's manuscript leaves space for it, the other authors give elements that may offer an explanation to the creation of their work in short introductory remarks.[34]

These introductions and prologues, often reworking earlier texts, are frequently used by the authors for self aggrandizement. They include a number of platitudes protesting modesty and inferiority, which were common in the Middle Ages and whose purpose was to prepare the reception of the work.[35] We should, however, take an interest in this aspect, as Bernard Guenée was able to do with memory by showing that it was a particular *topos* in the work of historians. Thus are a number of substantives whose use could typify these introductions such as "number" (numbers), "art", "comprehension", "knowledge", "science" and "arithmetic".

The fact that these introductory remarks come in different lengths has an unavoidable weighting on their respective influence on the analysis. The amount of text they devote to this part, measured in the number of occurrences of words enables us to see the distribution of the manuscripts. This criterion reflects the diversity of the situations quite well. Thus, the text by Nicolas Chuquet is made up of 87 occurrences for 324 text folios, Jean Adam's text, written over his career has 668 for 75 folios.[36] Without going into a detailed study here, we can identify several problems.

The polysemy of the word "author" in the Middle Ages revolves around the idea of "authority" and that of "writer". Here I mean the person who actually composed the work. It does not always follow that he who wrote the work was credited with its authorship and half of the treatises do not have a declared paternity.

To the best of our knowledge, only two men are found in the situation of having written single works which are invoked in other texts. One of these authors,

[33]See the collective work (Baumgartner and Herf-Lancner 2002) and (Polo de Beaulieu 1999).

[34]A study of these texts to establish criteria for comparison might be helped by the list of variables compiled by Marie-Anne Polo de Beaulieu and Jacques Berlioz on Jean Gobi for comparing a corpus collection of *Exempla*. See (Polo de Beaulieu 1999) and (Berlioz and Polo de Beaulieu 2001). It was necessary to modify their list to these introductory texts, we have retained nine points: length of the passages (number of occurrences); designation of the work; presentation of the arithmetic; reason for the production of the treatise, planned use for the work; designation of the author; religious authority; announcement of the plan. This set of criteria allows us to examine the differences between these arithmetics through a limited number of indications. The context of this article does not allow development of these elements, nor discussion of the value and the quality of these indicators. They were used, however, to carry out the present work.

[35]We also find them in narrative texts, see (Bourgain 2001; Guenée 1983) Bernard Guenée recalls that Ernst Robert Curtius was the first to demonstrate the importance of a catalogue of these *topoi* whose evolution and changes may accompany changes in thinking. See also (Hamesse 2000, ix–xiii).

[36]The following is a breakdown of the number of occurrences (forms) by manuscript: Bnf. lat. 7352: 160(87); *Compendy de Pamiers*: 161(124); Bnf. fr. 2050: 104(63); Bnf. fr. 1339: 341(143); *Traicté de la praticque*: 199(107); *Compendy de la praticque* 374(165); *Triparty en la science des nombres*: 84(57); Méd. Nantes 456: 187(101); *Kadran aux marchans*: 571(292); BSG. 3143 978(445); *Compendio de lo abaco*: 168(113); *Petit cadrans*: 92(63).

Barthélemy de Romans, even acquired the status of an authority alongside Aristotle, Plato and Alexandre de Villedieu in a long list drawn up by Jean Adam. Some of his choices and procedures are the subject of criticism by Jean Adam and Nicolas Chuquet. Barthélemy de Romans was a theologian and the treatises currently attributed to him can be found in the Cesena manuscript. This manuscript gives a different introduction for each treatise;[37] there are thus three of them and this diversity emphasizes the singularity of each element in the manuscript, by identifying the addressee and the scope of the treatise. He calls his first treatise the "Treatise on the practice of algorism" (*traicté de la praticque d'algorisme*), and considers algorism as a synonym for arithmetic.[38] It is the most similar to the rest of the corpus of practical arithmetics. Barthélemy de Romans emphasizes that "this science, which affects accounting, is useful and convenient for all kinds of people"[39] and he lists money-changing, geometry and astronomy. Arithmetic therefore contributes to general training and as it can lessen the "great tiredness and headache"[40] of doing accounts: it is thus essential learning. This idea that the learning of arithmetic is likely to facilitate the learning of other disciplines is found in most of the texts.

As Maryvonne Spiesser has shown apropos the Cesena S-XXVI-6 manuscript, Barthélemy de Romans explicitly positions the second treatise, which he calls the *Compendy de la praticque des nombres*, as a supplement to the first. This gives the feeling that together the treatises form a kind of progression to his third work, the *Spéculative des nombres*.[41] Barthélemy de Romans dedicates his second treatise to Guillaume de Claro[42] and also "[for] the understanding of those who have, or who have had this Compendy and have owned the previous work."[43] The dedication serves a twofold purpose. Guillaume de Claro was a wealthy citizen of the town,

---

[37]It should be emplasized that it is not certain that all these treatises are from Barthelemy (Spiesser M. and Féry-Hue, 2007).

[38]This identity is not uniformly accepted when you consider that Jacques Legrand in his *Archiloge Sophie* (c. 1415) distinguishes quite clearly between algorism and arithmetic.

[39]"*cest science qui touche fait de comptes est elle utile et bien seant a toutes manieres de gents*".

[40]"*grand fatigation et rompement de teste*".

[41]He wrote in *Compendy de la praticque:* "to give greater clarity and understanding to those who have or would have this compendy [meaning, the practical treatise] and would have the current work") fol. 149^{r}.

[42]*Compendy de la praticque des nombres*: "I have made the collection of all the rules of this compendy for the honor, respect and the singular love of the noble Guille de Claro, of Aix en Provence." *("pour honneur et reverence et singulier amour de noble homme Guille de Claro, habitant de la cite d'Aix en Provence, ay fait la suyvant recollection de toutes les regles de celluy compendy").* Some traces of this man can be found in the archives (Spiesser and Féry-Hue 2007). He also appears in a "confirmation of the statutes and ordinances, on the subject of auditors of accounts in the charter and the regulations for recourse to their services." ("*confirmation des statuts et ordonnances, sur le faict des auditeurs des comptes tutelaires, & reglement sur le recours d'iceux*") given by King René in 'The collection of antiquities and privileges in the town of Bourges and several other capital towns of the kingdom', Paris, 1621, p.440.

[43]"a l'entendement de ceulx qui ont, ou qui auroient celluy compendy et auroient le precedent ouvrage"

who was able to wield influence through his social rank. He may also have been an immigrant from Florence. Could this be one of the reasons for writing this book which offers a mathematical explanation of the Italian and French practical arithmetic textbooks, which his contemporaries would have been familiar with at the time?

The most abstract treatise in this manuscript, the last, is justified by a request that could be qualified as pedagogical:

> It is because the reverend master, the source for this book, was in Carcassonne teaching the science of algorism or *arismetic*, that he was asked by some students to produce a concise treatise which would give them a clear understanding of proportional numbers, which would encompass all understanding. He wrote this short treatise which gives knowledge appropriate to proportions, because it is necessary both for a clear understanding of the calculations which form the third general part of the *Compendy*, and which have been used in the practice of numbers.[44]

It is not surprising that a Dominican friar was found in an urban teaching situation. Carcassonne is one of the Order's centers and it was one of the Order's functions to teach in the city. It is unlikely, given the content of the text, that the teaching took place in a small school: it could have been the Order's *studium,* the *escoliers* could have been clerics, and we can envisage that it was also open to young laymen.

The second writer who could be described as an author is Nicolas Chuquet. He only reveals three features of his identity, and then only at the end of his work: this "Parisian" had obtained the university degree of "bachelor in medicine" and he wrote a book in Lyons to which he gave the title *Le triparty de Nicolas en la science des nombres*. Hervé L'Huillier noted his presence in the city of Lyons's register of taxes in 1480 as "Master Nicholas, Writer" (*Maistre Nicolas, escripvain)* and in 1488 as *Nicolas Chuequet, algoriste*. It is possible that this change in designation is an indicator of a degree of professionalization made possible by the growth in economic and cultural activity and the presence of a large Italian community in the city.[45] He was probably the teacher to whom Etienne de la Roche refers in his text of 1520 as one of the "experts and masters of this art."[46] Nicolas Chuquet gives neither explanation nor justification of his work.

In the arithmetic treatises using counters, which are less well known than the others, one written in the vernacular incorporates commercial arithmetic.[47] This

---

[44]"*Car le maistre reverend conditeur du livre estant au lieu de Carcassonne enseignant la science d'algorisme ou arismetique fut par aulcuns des escoliers pry de leur faire aulcun brief traictié qui leur donnast clere congnoissance des nombres proporcionalz, esquelz gisent et sont toutes raisons, fist ce petit traicti lequel donne competant cognoissance de leurs proporcions, tant comme sont necessaires a avoir clere congnoissance des raison qui est la tierce partie generale du compendy quil avoit fait aultres foiz de la pratique des nombres*". fol. 269^{v}.

[45]L'Huillier (1976).

[46]"*maistres expertz en cest art.*"

[47]The work follows a progression similar to practical arithmetics by showing calculations with counters; it develops the rule of three and the rule of association, which it adapts to fiscal problems.

is the work by Jean Adam.[48] Jean Adam was a man who made a career as an accountant in the service of princes. The princes' partiality for these specialists seems to have accompanied the development of accountancy which occurred during this century.[49] Jean Adam was possibly secretary to Charles de France in 1467 and became a Counselor and an Examiner of Accounts in 1471.[50] We know of the 4 years he spent in the service of Louis XI. While he was writing his book in Paris in 1475, he was a notary and secretary to the King and had been so since October 5, 1474. He dedicated his work to the upwardly mobile Nicole Tilhart then the King's *clerc des comptes*, under whose protection he placed himself.[51] On March 26, 1476 he took the office of Receiver General in the new and insecure province of Roussillon (1462–1493).[52] He left that job in May 1478. Is he the same man mentioned in St. Pourçain on October 7, 1494[53]? We do not know. Like the other works written at the time for men of some social importance, Jean Adam's work starts with a prefatory letter (*epistre de l'acteur*), in which he thanks the dedicatee, Nicole Tilhart. This could be surprising for book on calculation techniques using an abacus intended for account auditors. Like many men of his time and station, Tilhart liked books and, in the fifteenth century, there was a taste for books on science.[54] This case is a good example: it is a book for specialists written by a man who must also have been one, it is a way for Jean Adam to show both his deep understanding of the question and the extent of his knowledge and experience.

This work is truly a quality work, with a form of high culture: thus Jean Adam does not hesitate to litter the text with carefully chosen passages in Latin which he always translates a few lines further on. These notes set the value of Jean Adam's works.[55]

[48]This work attracted the attention of Thorndike, see (Thorndike 1959).

[49]One could continue this line of thought. Indeed, Jean Adam's book is littered with pejorative observations on those who do wrong, those who cheat, an aspect which reminds (Mattéoni 2007).

[50]Lapeyre and Scheurer (1978, 3–4).

[51]Nicole Tilhart was appointed by letter on May 29 1470 to the office of *clerc extraordinaire de la chambre des comptes*. In 1476 he was commissioned to the distribution on ordinary and extraordinary finances of the Dauphiné, Valentinois and the Diois. He was promoted to General of Langue d'Oïl in 1477 (Lassalmonie 2002, 356–357).

[52]The difficulties of the administration of this province were highlighted recently in (Lassalmonie 2002, 359, 513).

[53]A goldsmith called Jean Adam appeared in "Découverte d'un trésor du XVIe siècle à Montélimar", *Bulletin d'archéologie et de statistique de la Drôme*, 1887, 16 p.

[54]There are no studies on the liking of these men for books on science, but there are traces of purchases. Thus manuscript Bnf. fr. 2021 (dating from the thirteenth century and containing an ecclesiastical date calculator, a geometry and an algorism) bears on page 154, "This is the book of Pierre Cautet, procurer in parliament who bought it at the royal palace in Paris in one thousand iiii and fifty". This man is mentioned among the lawyers to the court of treasury, see (Dupont-Ferrier 1936).

[55]The references he cites are classic, he re-uses the etymologies of Isidore de Seville and Boethius. In 1520, Etienne de La Roche used exactly the same citation.

Beyond those obvious aspects Adam also reveals his command of the subject through his writing project, which consists in the application of methods and practices to "counters by following as closely as possible in order and form, Galician *gect* and arithmetic according to the style of counters used in the chamber of accounts of the very Christian king in Paris."[56]

This book transmits knowledge to, and disseminates it in another cultural milieu, and some of this knowledge comes from arithmetic. He also emphasizes his work's intended readership, "this present treatise has been written only for people in the accounts of finance and goods to know the casting of counters according to the rules of the arismetic art."[57] Thus, it was a shrewd, skilled man who had already demonstrated his quality that Louis XI sent to the province of Roussillon.

Manuscript 2904, kept at the Bibliothèque de l'Arsenal, is much more explicit in its introduction.[58] The author, Jean Certain, presents himself as having recently moved to Marseilles, the city where he completed his text in 1485, and says that he came from Bilbao. This incomplete information is not supplemented by any other documentation, apart from a small book which is an abridged copy of the first. Nevertheless, he gives his work a programmatic title. Jean Certain entitled his work the *Kadran aux marchans* and immediately stressed his willingness to "do nothing but that which is pleasant and enjoyable."[59] Jean Certain's book does not have a particular addressee, and we are yet to identify the family we could be led to by the shield on the first leaf. In addition, the presence of *tu* in the prologue could simply be a device to establish a link with the reader. Like Jean Adam before him, he adds a large number of oratorical precautions addressed to his anonymous dedicatee. One of these consists in working on the text that before them: "Pray my coveted friend that you would carefully re-examine, consider and understand it."[60] three verbs that indicate a progression that begins with a "pencil in hand" re-reading, then reinforced by the verb 'to consider' which is less practical, and which to 'understand', meaning, to appropriate the text. As a result, when he then asks the reader to make corrections where necessary, this cannot be done without thoroughly

[56]"*merreaulx en ensuivent les plus pres que poussible est l'ordre et la forme du gect et calcul gallican sellon le stille de gecter en la chambre desdit comptes du Roy tres Chretien a Paris*", fol. 3^r^.

[57]"ce dit present traicté n'est fait que pour gens de finances de comptes et de marchandise pour savoir gecter sellon les regles de l'art d'arismetique" (fol. 4^r^). He notes that he will compose with "God's help, another treatise that deals with the progression for extracting square and cube roots (...) that cannot be practiced with tokens. This is intended for land surveyors and astronomers, so will not be mentioned here." *("l'aide de Dieu, ung aultre traicté a part c'est assavoir de progression et extraction de radices et quarreures cubic (...) qui bonnement ne se peuvent praticquer par gectoners et aussi se sera pour les geometriencs et estronomiens et ne sera icy faicte aultre mencion").*

[58]It has the highest number of occurrences (578) along with BSG. 3143. See the comparative presentation Table 1 in the annex. On the presentation of the *Kadran aux marchans,* see (Benoit 1989b).

[59]"faire aucune chose qui soit aggreable et plaisante".

[60]"*te prie mon desire amy que le vueil les revisiter, considerer et entendre*".

reading the text.[61] Then, after some confessional understatements on his capacity in understanding, the author presents his work as a compilation[62] whose interest lies in the purpose which he assigns to it, that of offering an overview of arithmetical practices in the world of commerce and as a reference tool:

> This is why I want to call [this book] the *Kadran aux marchans*, for in the same way as the cadran is a help and guide on the road for all types of people to know the extent of time and of days, this treatise will be a guide, teaching and enlightening all merchants to know how to calculate well, and so take and give fairly when selling and buying, each according to his fair and just right.[63]

It is, therefore, an instructional text presenting the state of the practices, but the purpose of the training must be for the merchant to know the rules to avoid being misled (*decevoir*). This word must be understood in the double meaning of 'error' and 'deceit'. There is some ethics of trade at work here.

A few pages further on while addressing the section on currencies he says:

> And because the knowledge of this part is, or should be, the Masters General of currencies, money-changers and receivers, and other people in finance.[64]

This presentation is entirely original, just as much of the content is. Thus, in the first part of his book in the presentation of algorithms for the calculation of integers, where all the other authors present one or two algorithms for multiplication, Jean Certain presents five, and he operates in a similar way for division.[65]

The *Compendio dell lo abaco,* printed in 1492 – which is possibly a re-working of a 1460[66] text – contains little information on its editor, Francès Pellos, but expressly

---

[61]The request for correction of mistakes and errors is a cliché which is found in a large number of works, such as Jean Adam's: "[I] beg most humbly that those who discover any omissions please correct them and excuse the errors" ("*supplie tres humblement a tous que le verront qui leur plese croiser les obmissions et beguinement corriger et supplier les faultes*"). It is also found also in the texts of various types, see (Bourgain 2001).

[62]*Kadran aux marchans*, fol. 1^r, "[I] started to set about collecting several other [texts]" ("*me suis mys à faire et rassembler de pluseurs autres*").

[63]*Ibid.* "*Et pourtant le veulx appeller le kadran aux marchans car tout ainsi que le kadran est guide conducteur et chemin a toutes manieres de gens pour congnoistre la limitacion du temps et du jour, ainsi ce present traicté sera guyde, enseignement et declaracion a tous marchans de bien savoir compter pour justement prandre et donner en vendant et achatant a chascun son loyal droit.*"

[64]"*Et pour ce que la congnoissance de ceste partie est ou doit estre aulx generaulx maistres de monnoyes, changeurs et recevers ou autres gens de finances.*" *Kadran aux marchans*, fol. 67^v.

[65]*Kadran aux marchans*, fol. 10^r-22^v, and for a description of the algorithms we find correspondences in (Van Egmond 2001).

[66]This idea is advanced by (Sesiano 1984b) who identified a problem dating from 1460. It is perfectly possible that the work printed in 1492 compiles earlier collections. One of the two master printers, Nicolo Benedeti appears to have published several mathematical works, as well as participating, much later, in Juan de Ortega's publication in Barcelona. Information on the printers can be read in (Polain 1970).

names its addressee as the town of Nice.[67] Around 60 years earlier, the anonymous[68] *Compendy de l'art d'algorisme,* put itself under the protection of Saint Antonin "captain governor and defender of the university and the municipality of Pamiers."[69] The municipality and the university are two facets of the town of Pamiers' identity and being under this dual authority was necessary when carrying out educational activity,[70] whether employed by the municipality or not. The learning purpose assigned to this book is learning to do calculations.[71]

Education was also an urban phenomenon and it is known that, even while there was no general education, cities in southern France did devote funds[72] to it. Does this mean that the works were intended to be used for teaching in urban schools, or was it to gain the protection of municipalities?

BnF. fr. 2050 and Méd. Nantes 456, follow the tradition of Sacrobosco's *Algorismus* and both open with declarations that the purpose of mathematics is to enter into the understanding of "natural things", and continue:

> The author's intention is to deal in this text with what he was able to understand in the art of arithmetic that was commonly known as cipher or algorism.[73]

Highlighting the idea of compilation is a cliché in these types of work.[74] The Latin introductions to the algorisms are passed on here. And in the case of the BnF. Fr. 2050, we should not put too much faith in its rhetorical provisions which are all copied; the author shows his capabilities and arithmetical skill later in the same text.[75] Manuscript BnF. Fr. 1339 follows this approach, but is much more original

[67] *Compendio de lo abaco*, fol. 4^{r}: "this treatise will be as concise as is possible for me, for the citizens of Nice are shrewd and have knowledge of all things, and especially the said art" ("*losquals tracteray sub brevibus tant coma a mi sera possible, per los citadins de la Citatat de Nisa son subtile et speculatieus en ogni causa et specialment de las dichas arts*").

[68] François Baby (1982) proposed Jean Toloza as author of *Compendy de l'art d'algorisme*, this person was appointed to be teacher by the municipality in 1423.

[69] "*capitain governador et defendor de la universitat et communa de la ciutat de Pamias*".

[70] At the end of the thirteenth century, Boniface VIII endowed Pamiers a *studium generale*. While the university has never existed, the trustees of the town have always paid great attention to the management of their school. There are still concerns in 1439 over whether the papal bull allowed the teaching of all the sciences. (Lahondès 1883; Baby 1982).

[71] "captain governor and defender of the university and the municipality of the city of Pamiers" ("*capitain governador et defendedor de la universitat et communa dela ciutat de Pamias.*" fol. 16^{r}.

[72] Verger (1997).

[73] "*l'intencion de l'acteur est de en treter cy apres celon ce que il en a peu comprandre en l'art d'arismetique qui continuellement ce nomme chiffre ou agolisme.*" BnF. fr. 2050. fol. 1r.

[74] A very similar sentence is found in Méd. Nantes 456 produced in the last quarter of the century. "This art will be addressed later according to what the author was able to understand and learn for both written whole numbers and fractions and using an abacus" ("*En quel art sera traictié cy apres selon que l'acteur en a peu comprendre et savoir tant de numbre entier que rompu et tant par la plume que par le giet*").

[75] For example, he performs a compound interest calculation in which he manipulates a number of 16 figures.

in its presentation; it makes reference to the traditional authorities, but in fact hardly more than in other works: God, the Holy Spirit, Plato, Boethius and St. Augustine. He then starts the text with a presentation of the creation of the world.

*

The purpose the authors, be they identifiable or not, give to their work, is shown most concretely in these introductory texts. Considering these avenues in the analysis allows us to consider the context of the transmission of knowledge and of learning.

The teaching situations are largely understood from the introductions expressly given by the authors. Only three of these treatises appear to attest pedagogical activity, and it is even the *raison d'être* given by Barthélemy de Romans for at least one of them, which leads to re-reading his entire work and gives it one meaning and one direction: from the practical towards abstraction. This is also what can be deduced from the texts dedicated to a town: Pamiers or Nice. Once again, it is still necessary to be cautious over which part is rhetoric and which part is promoting the text for political ends.

Two others position themselves as vectors for the transmission of knowledge: the *Kadran aux marchans* which appears more like a work of learning through the application of rules and problems and provides the keys to commercial practice, negotiation and money in the broadest sense; and Jean Adam who puts his knowledge to the service of competence in the area of accountancy, his objective being to transfer knowledge, to transmit it in a developing sector.

That leaves those who again take up the tradition of algorisms by declaring no other object than teaching how to "calculate", to handle numbers. These elements are found as basic points in all the arithmetics.

6.3 Simplify, Guide and Memorize

The study of a medieval text cannot be carried out without studying its paratext, just as its image cannot be abstracted from it. Here we will try to observe the traces of appropriations to which the text could have been subject, as well as the elements put in place by the writer to guide the reader. In this regard the study of the language used in these treatises can be used to examine the transmission of knowledge.[76]

6.3.1 Assistance in Reading

The texts in these treatises are hierarchically structured into several sections and each section into parts, then rules and finally problems. The structure is visible

[76] Analysis of a treatise by Bernard de Gordon by Claude Thomasset can serve as a model, see (Thomasset 1994; Polo de Beaulieu 1994).

through the text and typographical elements such as initials or color. In most cases these segments are introduced by sayings, in the *Compendy de Pamiers*, for example, where we find the titles "Here is the second part of this compendy that speaks of fractions",[77] and for sections of lower rank "The third chapter which teaches multiplication"[78] and even "How you should add".[79] Even in situations where the compilations distinguish clearly between the treatises, as is the case for the anonymous Bibl. nat. fr. 1339, one finds a similar structure. In this work, one notable difference comes from the way it presents the problems, which are introduced by a title, sometimes enhanced with colors, as are the answers. These graphical elements reinforce a dialectic rhetoric that one finds explicitly in the *Kadran aux marchans.*[80] The numbers are often the subject of colorful framing, which allows them to be found easily in the text or in graphics, algorithmic manipulations are frequently put in frames, which are also used for tables for memorization.

Reading is therefore guided by indications on its order, but the texts also bear other elements suggesting that they are likely to be subject to non-linear consultation. One can observe, as Frankwalt Möhren has done for the *Pratike de geometrie*,[81] that the paragraphs have repeated introductory formulas such as "if you wish" (*se tu veus*), and the use of "Idem" (*Item* The familiarity, the use of *tu*, when it is employed, corresponds to a quite well established rhetoric in medieval teaching. There are other methods of orientation in the paratext such as notes often introduced by an index or decoration. Thus these treatises are designed to be used and practiced and bring the author and reader closer together. Apart from these aspects, there are few, if any comments and traces of pictures are rare.

*

This concern for clarity, for structuring and the introduction of elements for non-linear consultation reflect in the first place relationship with books and reading.

These observations accentuate the idea of works produced for personal consultation, where orientation is a necessity but they do not allow us to infer a real

---

[77]"*Siec se la segonda partita daquesta compendi que parla de nombre ro*" fol. 50^r.

[78]"*Lo tertz capitol qui ensenha de multiplicar*" fol. 24^r.

[79]"*Com deves ajustar*" fol. 19. In the case of *Traicté de la praticque* and *Kadran aux marchans* the quality of the copy accentuates this structure.

[80]During the development of these problems the *Kadran aux marchans* qualifies the latter as "request" and proposes a living relationship centered around a question/response dialectic, evoking dispute in a problem. *Kadran aux marchans*, fol. 40^r: "And these questions are difficult to understand, as it may be that the applicant wishes for a solution using the rule of three, and the defendant, or the respondent answers in another way, but it is enough to answer" ("*Et tieulx questions sont difficiles à entendre car par aventure peult estre que le demandeur la demandra par la reigle de 3 et le deffendeur ou le respondeur respondra en autre maniere mais il soufist de responder*").

[81]Möhren (2006, 98–99).

pedagogical situation. There is no information on the type of reading that could have been carried out with these books, there is no *scholium*.[82]

Are there books for the teacher or for autonomous study? It is impossible to make the distinction. Helped by these clues the reader will at best differentiate between the structure of Bib. Nat. fr. 1339 endowed with a well supported text, requiring little outside intervention, and that of the *Compendy de Pamiers*, denser, more difficult to read but also highly structured and furnished with out-of-text support. These distinctions are difficult to interpret.

6.3.2 Language Oriented Towards the Practice of Knowledge

The choice of lexicon used in this corpus does not only carry the cultural branding of the authors' training. The anonymous works from northern France certainly present a more specialized vocabulary than those from the south; they use a division of numbers into "digit", "article" and "composite" following a university algorismic heritage. It is also true that the lexicon in BnF. fr. 2050 is marked by the use of a few words of Italian origin, which seems to attest an influence. But there are common characteristics that go beyond this cultural aspect, such as the small number of nouns used, to the benefit of mainly practice-oriented verbs.[83] The word 'practice' can be understood in two ways, for those verbs give information on social practices, but they can also refer to the teaching of arithmetic through the practice of algorithms. One can understand the title *Traicté de la praticque* in the same way. This is a language orientated towards pedagogical preoccupations which, as shown by Maryvonne Spiesser for Barthélemy de Romans, is based on teaching and the explanation of techniques and not on broader arguments.[84]

This explanation of techniques uses reading the discourse out loud, graphical representation and prompting the memory. This, of course, is reminiscent of the comments by authors advising the memorization of certain points of a text, as in manuscript BnF. Fr. 2050 which offers the use of the phrase "by heart" several times, or indeed Jean Certain who emphasizes a certain multiplication *par dande* is the most efficient provided one knows the multiplication tables by heart.[85] But practice also implies a repetition that structures the memory. This repetition is not only the "little song" evoked by Maryvonne Spiesser, but is also linked to the representation of an operation or even an algorithm. That there is a constant interchange in these

---

[82]Only Méd. Nantes 456 bears comments from a later reader aiming at attaching the text to other elements of a more humanistic mathematics culture.

[83]Benoit (1985).

[84]Spiesser (2003b, 188).

[85]In this algorithm for multiplication, almost no intermediate result is noted. And it is difficult to implement when the numbers become quite large.

works between text and graphical representation, a part of the former serving to enlighten the latter, is shown by the study of the most common verbs in the corpus.[86]

The rule of three is practically reduced to two phrases, one formalizing the statement: *se tant vault tant que vaud tant*[87]; the other the method of resolution: *multiplier et partir*.[88] The first statement relies on orality it is requested to be said and at the same times it must be written down on paper.[89] For the rule of double false position[90] the interchange between the text and the graphical layout makes up a significant part of the training in the rule and leads to formalizing the problem with the objective of resolution.

*

Following the avenue of the language used has broader implications than the few points raised here. It is interesting to observe how lexical and syntactical analysis leads to a study of manipulation, and representation, and to a very concrete practice of mathematics. It should be possible to compare these observations with other mathematical texts expressed in other languages, such as Latin.

In any event, the organization and writing reveal a concern for learning, for which the book is the support. This passage by the written word entails the writer considering teaching in a deferred way and from the side of the reader. The analysis of the language can restitute, in a way, how the knowledge constructs itself in the reader, through practice.

6.4 Mathematical Problems: Carrying Out the Project

6.4.1 Problem Sequences and the Construction of the Works

Mathematical problems are the flesh and blood of these books and far from being catalogues of recipes, they reflect the realization of the projects of their various authors and embody the idea that they have of the knowledge that they teach and of its transmission.

The study of series of problems allows pedagogical concerns to be identified. Thus, for example, teaching integers, fractions and compound numbers, is for the authors a learning issue, the importance of which is revealed by certain blocks of problems. The *Compendy de Pamiers* is a good example, as it re-uses the same

---

[86]Lamassé (2012).

[87]That is: "if this is worth this, what is the value of that?".

[88]That is: "multiply and divide".

[89]*Ibid.*

[90]In modern terms, this is a rule of linear interpolation.

series of numbers several times in the sequence of problems.[91] This illustrates how these works put effective teaching methods to use, focusing on memorization and the exploration of algorithms: the construction of the sequence is specific to each author.

The mathematical problem also reflects a real "common culture", in the sense that some can appear commonplaces, obligatory points of passage. This can manifest itself in statements whose form is common and others that are re-used word for word. Faithful copying makes it possible to identify the transmission of knowledge between certain works, almost constituting a "network": this is what is brought out in Appendix 1. Without repeating the detail of the analysis here,[92] there are quite clearly, through varying intensities of influence, six texts sharing a number of identical problems attesting a closer relationship. The *Compendy de Pamiers* (1430) played a central role in the dissemination of knowledge which may have set the pattern, subsequently Barthélemy de Romans (1470)[93] seems to have taken a position more than that of a mere go-between: he restructures the ensemble.

The construction of mathematical texts here does not escape the medieval text construction which uses tradition as one of the conditions for the validation of what it presents.[94] Repeating nearly identical problems can be explained by the writing project. As Pierre Chastang noted recently: "All text is constructed in an intertextual relationship which places the present work in a network of earlier writings. It is through this difference, where re-use and disparity combine through the versions, that the meaning shapes itself".[95] In his third appendix, Nicolas Chuquet repeats many elements from the *Traicté de la praticque*, notably the rules of barat, in their entirety. This is surprising in so far as this rule whose practical resonances are well known could have been the subject of greater originality.

Repeating rule of three and rule of association problems, Chuquet explicitly states the relationship between the mathematical problem and its possible use in the commercial world and in so doing, positions the reading of his work.[96] Nicolas

---

[91]This approach is particularly well developed in his section on the rule of three, where he presents a series under a general rule and then applications with currency, with the following numerical expressions:

4	7	12
$4+\frac{1}{2}$	7	12
$4+\frac{1}{2}$	$7+\frac{2}{3}$	12
$4+\frac{1}{2}$	$7+\frac{2}{3}$	$13+\frac{3}{4}$

Bnf. fr. 1339 and the *Compendio de lo abaco* operate in a similar way with other values.

[92]Lamassé (2007).

[93]It is essentially the content of the manuscript of Cesena, i.e. the *Traicté de la praticque*, the *Compendy en la science des nombres* and a third part containing geometry. The *Speculative* that belongs to the same book does not appear to have had the same success.

[94]Chastang (2008, 269).

[95]*Ibid.*

[96]Paul Benoit was the first to note that Nicolas Chuquet raised the problem of the relationship of an interest problem with the commercial reality to which it was supposed to correspond. Refusing the

Chuquet does not just relocate a block of problems, he integrates them into his approach, alters the statements and changes the pedagogy applied.[97] Thus in the first treatise of the Cesena manuscript the rule of association is separate to a group of problems presenting statements which could be associated with situations of business partnership, involving merchants and workmanship. The first follows the rule of three, a far cry from the second which includes more recreational problems.[98] In the *Triparty en la science des nombres,* on the contrary, the two are put one after the other.[99] This modification was accompanied by an alteration to the problem statements, through which Nicolas Chuquet explores all the algorithmic parameters by envisaging situations first of profit, then of loss, for the enterprise.[100] The *Traicté de la praticque* prefers to center the organization of the sequence of problems differently, around the merchant – workmanship relationship and breach of contract.[101]

A substantive dispute separates the two authors on these problems: Nicolas Chuquet disqualifies the usual resolution procedure as it does not consider compound interest[102] while Barthélemy rightly distances himself from this possibility.[103] Nevertheless, Nicolas Chuquet pursues his problems by repeating the procedures "according to the style and opinion of some" (*selon le stile et opinion d'aulcun*), maybe because he deals with this aspect elsewhere. Examining the organization of the problems allows us to observe the different purpose assigned to each of these works with a keener eye.

---

solution to this problem provided by other authors, he writes: "thus such calculations(/problems) are without value and I think that there are no partnerships of this kind between merchants " (ainsi telles raisons sont nulles et croy que entre marchand nulles telles compagnies se font)". Benoit (1988).

[97]For a more precise study on these copies, see (Lamassé 2007).

[98]In folios $41^r - 43^r$ et $116^v - 122^r$ respectively.

[99]In folios $279^r - 285^r$ et $285^r - 288^v$.

[100]The contribution of Alexei Volkov Chap. 10 in this volume proposes a similar reflection by the analysis of lists of problems obtained by the variation of one parameter.

[101]Indeed, he reserves a section to interest problems.

[102]*Triparty en la science des nombres*, fol. 280^v: "For some the calculation is finished. For such calculations to be valid, one must assume that it is the principal or the capital that earns, and not what [the capital and the interest] earns. However, this is not really so, as the earnings and what the earnings earn for the merchant can be used and increase proportionally like the principal day by day, month by month and year by year" ("*Et est faicte la raison selon le stile et opinion d'aulcun. Et a ce que telz comptes puissent estre valables il convient presupposer que le principal ou le chatel tant seulement gangne et non le gaing. Et pour tant qu'il n'est pas ainsi car le gaing et la gaing du gaing faiz en marchandise peuvent profiter et gangner proporcionellement comme le principal de jour en jour de moys en moys et de an en an par quoy il peult entrevenir plus grant gain.*").

[103]*Traicté de la praticque*, fol. 117^r: "Note that in this calculation, as for those that follow, it must be understood that the only earnings come from the principal, and not from what the principal earns, because otherwise these calculations could not be done". ("*Et note que en ceste raison et aussi es autres cy apres ensuivant l'on doit entendre que le principal gaigne et non pas le gaing car autrement telles raisons ne se pourroient pas bien faire.*").

6.4.2 *The Relationship to Reality, a Pedagogical Anchorage*

These treatises are able to refer more specifically to commercial practice through the mathematical concerns they present. The rules of three, of association, and of barat are presented around notions of trade and value, which are put into practice through situations calculating prices, interest and the fair distribution of profits.[104] More often than not, the methods offered for resolution can be applied perfectly to commercial situations; moreover this is the same as is found in contemporaneous Italian works and in regards to the problems on simple partnership, sometimes similar to what we know as notarial deeds.[105] We can see the adequacy between these works and commercial realities extremely well in the basic problems in which the authors propose calculations with "complex" numbers,[106] as well as in the notes and brief rules given throughout the texts.

The use of commercial vocabulary was not necessarily intended only for merchants and there is no need for reference to a commercial context to present algorithmic practices which could be useful in trade situations. Thus in the books which we think of as close to the world of money-changers, teaching the rule of three, is done without any link to a concrete reality.[107] *A contrario*, BnF. fr. 2050 offers a compound interest calculation exercise in its section on false position, a rule for which there are not usually commercial examples.[108]

The way in which the problems are organized can give indications on the usage the writers made of the realities of the commercial world. Let us take the *Kadran aux marchans* again, the peculiarities of whose construction we know. The analysis of the structuring of the problems in this work can lead us to question the nature of the transmission it envisages for this commercial knowledge. In the section on the rule of three, Jean Certain distinguishes, as had been done before in the *Compendy de Pamiers*, the usage of the rule for currencies, for weights and for measurements, a tripartite of merchants' concerns, then he finishes with a group of problems which he entitles "issues which can be discussed in the rule of three"[109] and in which he proposes inverse or chained rules of three problems.[110] Jean Certain justifies this method of writing by explaining that he hesitated between grouping together all the situations which are not directly useful at the end of the book so as not to

---

[104]For example in the *Kadran aux marchans*: "fair and just right" ("*loyal droit*").

[105]One can compare the partnership with the contractual problems presented in (Wolff 1954).

[106]This name is given to numbers expressed with monetary subdivisions. Jean Certain gives an example of these arithmetical complexities in subtraction exercises, *Kadran aux marchans*, fol. 6-9^{r}.

[107]These manuscripts are presented in (Bompaire 2006). All these money changer manuscripts present problem sets that are quite general in their presentation, as evidenced in mss. Bnf. fr. 1514.

[108]This is about a bank placement by a man in favor of his daughter who has just been born, BnF. fr. 2050, fol. 93^{v}.

[109]"*les questions deputables de la reigle de troys*".

[110]Jens Høyrup has tried to write a history of this type of problem (Høyrup 2000, 64–67).

disturb linear reading.[111] The rule of partnership is treated in the same way, the author starts with an overview of the rule, then adds "they make several types of partnership with covenants and pacts which are difficult to judge",[112] that is to say, to interpret, and finishes by "Other questions on partnership that are more for debate than use".[113] Clearly, the construction of sets of problems within the sections gives a documentary aspect which he ascribes to his work, consisting in covering the information in the best possible way and in making distinctions in the knowledge diffused, either warning as to the innocuousness of certain uses or to emphasize the complexities therein.

The *Compendy de Pamiers* adopts a radically different approach by beginning the sections with some "exemples generals", meaning examples without explicit reference to commercial practice, and later offering only one adaptation to commercial reality. It is similar in this approach to manuscript Bibl. Nat. Fr. 1339, manuscript Med. Nantes 456 and to the work by Francès Pellos.

The argument of the problem statements' resonance with social reality, such as this can be discovered by the historian, is not necessarily sure evidence of a proven teaching situation. It appears that these fifteenth century men liked not only to play with possible reality, but also with abstraction. The world of commerce does not have the same status in all these works, it could simply be pedagogical dressing, but may also be a reality to be explored, as in the *Kadran aux marchans,* which may have lead the authors to set problems that go beyond the mathematical content ascribed by the "genre" of practical arithmetics.[114]

*

It seems to me that we can group the texts by the layout of the problems. Jean Certain, faithful to his writing project, offers a didactic presentation, separating the difficult, useful to know cases from those that are not.

---

[111] Which he justifies in this way: "In the rule of three there are some issues that are more for discussion than for immediate use. To know how to answer these questions I have given you several examples, though not as many as I would like wished, because I have so much desire to keep to the point that I have left them for the end of the book, even if reason demands that they are organized in order [of the rule]. Nevertheless, as I have read that happiness always comes first and is necessary (fol.39r) although I do not put a useful example in the order of the rule through oversight or ignorance, I do put it at the end with the others." ("*En la reigle de troys se font pluseurs questions qui sont plus desputables que proufitables. Et pour savoir respondre à tieulx questions vous en mectray si apres pluseurs et non mye toutes celles que je vouldroye bien car j'ay si grant desir de parfournir les choses principalles que je laisseray partie des accidentalles pour la fin de ce livre, bien que raison feroit que elles fussent par ordre. Neantmoins j'ay ouy dire que le bonnure seroit tousiours premier au plus necessaires et aussi (fol. 39* ^{r}*) si je laisse quelque chose neccessaire dire pour leur ordre pour oubly ou non savance je auray matiere de les acompaigner avecques les autres alla fin*").

[112] "*ilz se font pluseurs manieres de compaignies avecques convenances ou paches expres qui seroient difficilles à juger*", fol. 45^{r}.

[113] "*[A]utre raisons de compaignies y a qui sont plus disputables que prouffitables*", fol. 46^{v}.

[114] Nicolas Chuquet's remark on the discrepancy between real commerce and the solution proposed for calculating interest points to the shortcomings of arithmetics that circulated at this time.

Nicolas Chuquet's project is broader, it is a question of presenting the mathematical knowledge of his time, and for each element in his *Triparty*, he uses problems to explore the algorithms while wondering about the uses which are made of them.

The *Traicté de la praticque* is representative of a large number of works, similar to Nicolas Chuquet in the usage of problems, it displays a very accomplished pedagogy, paying attention to all the elements of the problem and notably to the manipulation of numbers, which remains the epicenter of the difficulties posed by these problems.

6.5 Conclusion

The body of works which appears is fairly complex, as there are almost as many writing projects as there are texts. Some of them reflect the circulation and discussion of the knowledge included in these arithmetics in the Franco-Occitan world, others, independent of the first, demonstrate a broad interest in this genre in the fifteenth century. The teaching of longhand calculation in this form of practical arithmetics became important enough to become an integral part of works that may appear to be far removed from the practice of trade, as is the case, for example, of manuscript Bibl. Nat. fr 1339.

Each of these texts is part of what Yves Gentilhomme calls a deferred communication.[115] While it is necessary to accept several types and levels of reading for these arithmetics, considering the notion of a "writing project" distinguishes each work. There is then the implication that the writer will make a certain number of choices in relation to the hypothetical reader, in order to make himself understood. Some of these may be pedagogical in character (memorization, number practice, writing method). The aim of the transmission of knowledge intrinsic to a writing project partly effaces the teaching situations which may have predated it and it is difficult to envisage the practices to which the books were subject (reading out loud or to oneself).

All the authors pay great attention to the explanations of rules and procedures. They make use of simple language and a graphical form of thinking facilitating access to quite complex algorithms without providing mathematical proofs. The writing is based on the triptych of written/oral, graphical representation and memorization. The transmission of knowledge through problems involves practice, an endeavor to understand through replication. In this respect, the importance of presenting the procedures for resolution is a great support to the reader; the few implicit references can lead us to envisage either guided or nearly autonomous learning. These books do not presuppose prior knowledge other than knowing how to read. If they are the books for masters, they are not necessarily masters in scholarly mathematics.

[115] Genthilhomme (2001).

Through their structure, and the position they give to practical arithmetic, these books reveal different reading projects and orientations. We have to differentiate the treatises displaying basic teaching leaving a fundamental part to commerce, like BnF. fr. 2050 and the *Compendy de Pamiers*, which we could qualify as normal,[116] from those taking a more mathematical direction. The latter is evident in the construction used by Barthélemy de Romans as well as in the appropriation that Nicolas Chuquet makes of it. In these two works there is a shift in the structuring of mathematical knowledge, one of the points of which is practical arithmetic. We know, moreover, that both were teachers, but we do not know if they taught what they wrote, and as such, have formalized it. Finally, there are the treatises which move away from the content of the normal treatises to offer a new proposition. These books are sometimes beautiful, like the *Kadran aux marchans* and the work by Jean Adam, which shows the interest wealthy men had in the subject. In this case they are works for libraries exhibiting mercantile knowledge or using the rules ascribed to it. In this sense, being as close as possible to actual mercantile reality is important for these texts. Hence the abundance of rules focusing on monetary exchange, the calculations of distribution between trading partners and the part devoted to metals.

There is a link to the commercial world in the broad sense, but the use of examples does not constitute proof that their intended readership was the children of merchants. The commercial link manifests itself in two ways: On one hand, the willingness to correspond to reality by identifying procedures and good practice, and on the other, a formalization of reality which allows the mathematical tools to be internalized.[117] For these authors, the commercial world is therefore both a practical reality and serves as a vehicle for intellectual endeavor. It remains difficult for us to appreciate this reality as it is not always as explicit as is written by Nicolas Chuquet and can appear incidental. However, research into the such incidents is not without worth and may even be the only window we have onto contents and underlying situations, which were evident at the time.[118]

[116]In a sense quite close to that used by Kuhn in the expression "normal science".

[117]The latter are particularly well shown in (Labarthe 2005). In a section on false position, BnF. fr. 2050 teaches the calculation of compound interest taking as an example the opening of an account on the birth of a daughter. Without doubt its author shows his command of arithmetic by this, which contributes to his prestige, but it also reveals practices that do not use the usual formalization.

[118]In a section on false position, BnF. fr. 2050 teaches the calculation of compound interest taking as an example the opening of an account on the birth of a daughter. Certainly it shows by the same his command of arithmetic, which contributes to his prestige, but it also reveals practices that do not use the usual formalization.

Appendices

Appendix 1

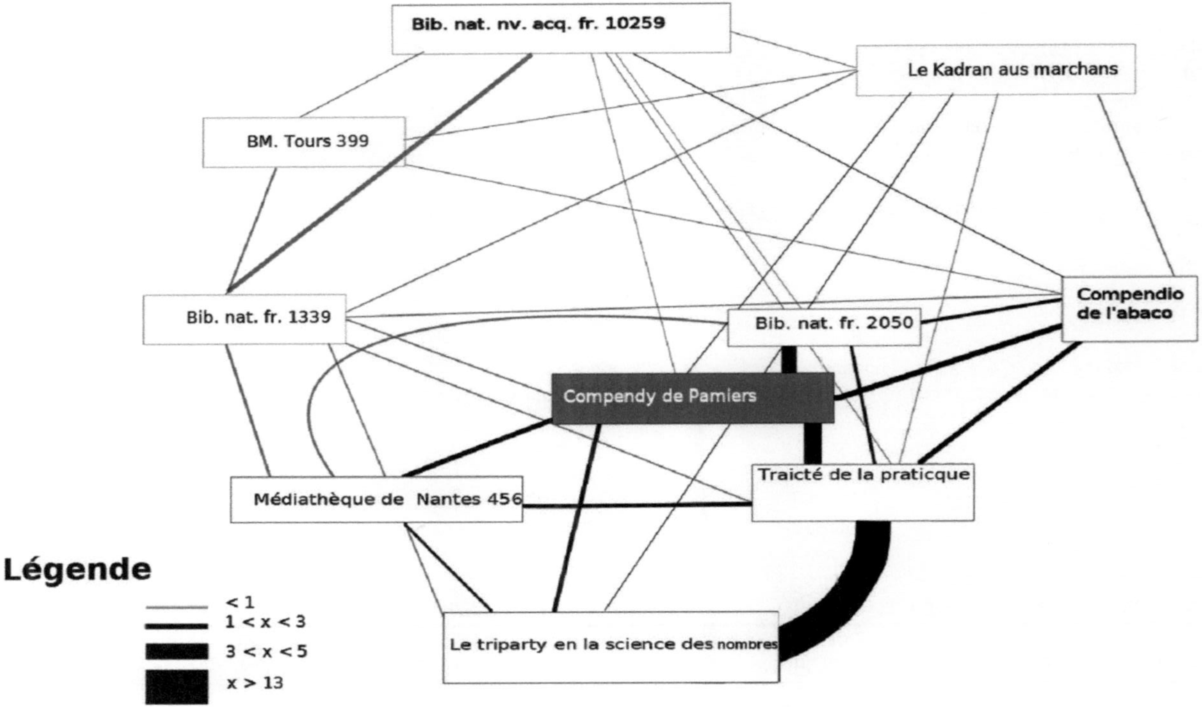

Relationship densities for problems shared by the texts (The value "x" is a link indicator explained in Lamassé 2007)

Cote	Bibl. Nat. Nv. Acq. Fr. 4140	Bibl. Nat. Fr. 2050	Bib. Nat. fr. 1339	Bibl. St. Genv. 3143	Cesena S-XXVI-6	Med. Nantes 456	Bibl. nat. Fr. 1346	Bibl. Nat. Ars. 2904	Bibl. nat. Res P-V- 351	Bibl. nat. Nouv. Acq. Fr. 10259
Title	*Compendy de l'art d'algorisme* for clarity we use *Compendy de Pamiers*				(1) *Traicté de la praticque*, (2) Compendy de la praticque des nombres		Triparty en la science des nombres	Kadran aux marchans	Compendio dell lo abaco	*Petit cadrans*
Author	Anonyme	Anonyme	Anonyme	Jean Adam	Mathieu Préhoude and Barthélemy de Romans	Anonyme	Nicolas Chuquet	Jean Certain	Francès Pellos	Anonyme, Jean Certain(?)
Dates	First third of the XV century	Around 1460	Around 1460	1475	1476	1488	1484	1485	1492	Start of the XVI century
Place of production	Pamiers	?	Parisian region	Paris	?	Normandy?	Lyons	Marseilles	Nice	?
Geometry	No	No	Yes	No	Yes	Yes	Yes	No	Yes	Start
Dedicatee	Ville de Pamiers			Nicolle Tilhart	Guillaume de Claro			inconnu	The people of Nice	
Passage size	161	104	341	978	(1) 198, (2) 374	187	84	570	978	168
Manuscript number used in the text	[1]	[2]	[3]	[4]	[5]	[6]	[7]	[8]	[9]	[10]

Appendix 2: Brief Profile of the Corpus

Brief description and comparison of the manuscripts

Bibliotheque Malestiana **S – XXVI – 6**, 354 folios[118]	Bibliothèque nationale **Français 1346**, 324 folios
Barthélemy de Romans	Nicolas Chuquet
fol. 7–140 v: Anonyme. *Traicte de la praticque d'algorisme*	fol. 1–147 r: Le Triparty en la science des nombres
fol. 149–268 v: Barthélemy de Romans, *Le Compendy de la praticque des nombres*	fol. 148 r – 324: [**Appendices**]
fol. 269–290 v: Barthélemy de Romans, *La speculative des nombres*	1. fol. 148 r – 205 v: [First rules]
fol. 291–294: [Sequitur de proportionibus]	2. fol. 206–210 v: [Problems]
fol. 300–330: Traicte de la praticque de geometrie et de la composition du quadrant	3. fol. 211–262 v: [Geometrical treaty]
	4. fol. 264–324: [Practical arithmetic]

References

Baby, Fr. 1982. Pamiers ville occitane (1350–1550). In *Histoire de Pamiers*, 115, 231. Paris: Promodis.

Baumgartner, E., and L. Herf-Lancner. 2002. *Seuils de l'œuvre dans le texte médiéval*. Paris: Presses de ma Sorbonne nouvelles.

Beaujouan, G. 1958. Les arithmétiques françaises des XIVe et XVe siècles. In *Actes du VIIIe congrès international d'histoire des sciences: Florence 3–9 septembre 1956*, 84–87. Paris: Hermann.

Beaujouan, G. 1988. The place of Nicolas Chuquet in a typology of fifteenth-century French arithmetics. In *Mathematics from manuscript to print 1300–1600*, ed. Hay Cynthia, 73–88. Oxford: Clarendon.

Benoit, P. 1985. Recherches sur le vocabulaire des opérations élémentaires dans les arithmétiques en langue française de la fin du Moyen Age. *Documents pour l'histoire du vocabulaire scientifique* 7: 77–95.

Benoit, P. 1988. The commercial arithmetic of Nicolas Chuquet. In *Mathematics from manuscript to print 1300–1600*, ed. Cynthia Hay, 96–116. Oxford: Clarendon.

Benoit, P. 1989a. Calcul, algèbre et merchandise. In *Eléments d'histoire des sciences*, ed. Michel Serres, 197–221. Paris: Bordas.

Benoit, P. 1989b. Marchands et mathématiques: le cas français. In *Le marchand au Moyen Age: XIXe Congrès de la Société des historiens médiévistes de l'enseignement supérieur public*, Reims, June 1988, 195–210. Paris: SHMES.

Berlioz, J., and Polo de Beaulieu. 2001. Les prologues des recueils d'Exempla. In *Les prologues médiévaux, Textes et études du Moyen Age*, ed. Ecole française de Rome, 276–321. Turnhout: Brepols.

[118] A complete description is found in (Spiesser and Féry-Hue 2007).

Bompaire, M. 2006. Compétences et pratiques de calcul dans les livres de changeurs français (XIVe-XVe siècles). In *Écrire, compter, mesurer. Vers une histoire des rationalités pratiques*, ed. N. Coquery, F. Weber, and F. Menant, 143–163. Paris: Éditions rue d'Ulm.

Bourgain, P. 2001. Les prologues des textes narratifs. In *Les prologues médiévaux, Textes et études du Moyen Age*, ed. Ecole française de Rome, 245–273. Turnhout: Brepols.

Bozzolo, Carla, and Ezio Ornato. 1983. *Pour une histoire du livre manuscrit au Moyen Âge, trois essais de codicologie quantitative*. Paris: Éditions du CNRS.

Briquet, Charles-Moïse. 1907. *Les filigranes, dictionnaire historique des marques de papier dès leur apparition vers 1282 jusqu'en 1600*. 1923. Leipzig: K. W. Hiersemann.

Cassinet, J. 1993. Le manuscrit XXVI de Cesena, important maillon occitan de transmission de l'algorisme au XVe siècle. *Bolletino di Storia delle Scienze Matematiche* 13: 251–285.

Chastang, P. 2008. L'archéologie du texte médiéval: Autour de travaux récents sur l'écrit au Moyen Âge. *Annales EHESS* 2: 245–269.

de Lahondès, J. 1883. Les écoles dans une petite ville avant la Renaissance? *Mémoires de la Société archéologique du midi de la France* 12: 392–403.

Dupont-Ferrier, G. 1936. Les avocats à la Cour du Trésor de 1401 à 1515. *Bibliothèque de l'école des chartes* 97(1): 5–81.

Faider, P., and Jan P. Van Sint. 1950. *Catalogue des manuscrits conservés à Tournai (Bibliothèque de la Ville et du séminaire)*. Gembloux: J. Duculot.

Genthilhomme, Y. 2001. Termes et textes mathématiques: réflexions linguistiques non standard. *Cahiers de Lexicologie* 76: 57–89.

Ginzburg, C. 1980. Signes, traces, pistes. Racines d'un paradigme de l'indice. *Le débat* 6: 3–44.

Grendler, P.F. 1989. *Schooling in renaissance Italy. Literacy and learning, 1300–1600*. Baltimore/London: Johns Hopkins university press.

Guenée, B. 1983. Histoire, mémoire, écriture: contribution à une étude des lieux communs dans les préfaces des œuvres historiques. *Comptes rendus des séances de l'Académie des Inscriptions et des Belles Lettres* 1983: 441–456.

Hamesse, Jacqueline. 2000. *Les prologues médiévaux: Actes du colloque international organised by l'Academia Belgica and l'Ecole française de Rome with the FIDEM concours (Rome, 26–28 March 1998)*, ed. Jacqueline Hamesse. Textes et études du Moyen Âge, 15. Turnhout: Brepols.

Høyrup, J. 2000. *Jacopo da Firenze's Tractatus Algorismi and early Italian abbacus culture*. Bâle/Boston/Berlin: Birkhäuser.

Jeannin, P. 2002. Les manuels de pratique commerciale imprimés pour les marchands français. In *Marchands d'Europe pratiques et savoirs à l'époque moderne ?* Paris: éd. ENS Ulm.

King, D. 2001. *The ciphers of the monks, a forgotten number-notation of the middle ages*. Stuttgart: Franz Steiner.

L'Huillier, H. 1976. Éléments nouveaux pour la biographie de Nicolas Chuquet. *Revue d'Histoire des Sciences* 19: 47–50.

Labarthe, M.-H. 2005. Les règles de compagnie, dans les premières arithmétiques imprimées des Espagnes: de la règle marchande à l'outil mathématique. *Revue d'histoire des mathématiques* 11: 257–313.

Lamassé, St. 2007. *Les problèmes dans les arithmétiques commerciales en langues française et occitane de la fin du Moyen Âge*. Ph.D. dissertation, Université de Paris 1 Panthéon-Sorbonne.

Lamassé, Stéphane. 2012. Les traités d'arithmétique médiévale et la constitution d'une langue de spécialité. In *Sciences et langues au Moyen Age*, ed. Joëlle Ducos, 65–104. Heidelberg: Universitätsverlag Winter.

Lapeyre, A., and R. Scheurer. 1978. *Les notaires et secrétaires du roi sous les règnes de Louis XI, Charles VIII et Louis XIII*. Paris: BNF.

Lassalmonie, J.-Fr. 2002. *La boîte à l'enchanteur, politique financière de Louis XI*. Paris: Comité pour l'histoire économique et financière de la France.

Mackenzie, D.F. 1991. *La bibliographie et la sociologie des textes*. English trans. M. Amfreville. Paris: Edition du cercle de la Librairie.

Mattéoni, O. 2007. Vérité des comptes et paroles du juge. Réflexions sur la procédure de contrôle des comptabilités en France à la fin du Moyen Âge. *Revue Historique* 641: 31–69.

Möhren, Fr. 2006. Les débuts de l'écriture française de la géométrie au XIII siècle. In *L'écriture du texte scientifique au Moyen Âge*, ed. Claude Thomasset, 93–116. Paris: PUPS.

Piochi, B. 1984. Il Trattato di Paolo dell'Abbaco. *Annali dell'Istituto e museo di storia della scienza di Firenze* 9: 21–40.

Pirenne, H. 1929. L'instruction des marchands au Moyen Âge. *Annales d'histoire économique et sociale* 1: 13–28.

Polain, Louis. 1970. *Catalogue général des incunables des bibliothèques publiques de France: Pacatus – Perrinus, Johannes*, vol. 16. Nendeln: Kraus-Thomson organization.

Polo de Beaulieu, M.-A. 1994. Didactisme ou persuasion, Les recueils d'Exempla au Moyen Age. In *Éducation, apprentissages, initiation au Moyen Âge*, 397–410. Montpellier: Les Cahiers du CRISIMA.

Polo de Beaulieu, M.A. 1999. *Éducation, prédication et cultures au XIVe siècle: essais sur Jean Gobi le jeune*. Lyon: PUL.

Sesiano, J. 1984a. Les problèmes mathématiques du Memoriale de F. Bartoli. *Physis* 26: 129–150.

Sesiano, J. 1984b. Une arithmétique médiévale en langue provençale. *Centaurus* 27: 26–75.

Sesiano, J. 1985. Un système artificiel de numération au moyen Âge. In *Mathemata: Festschrift für Helmuth Gericke*, ed. Menso Folkerts, 165–196. Stuttgart: F. Steiner.

Spiesser, M. 2003a. Les manuels d'arithmétiques pour les marchands dans les manuels du XVe siècle. *Bulletin de l'APMEP* 444: 32–50.

Spiesser, M. 2003b. *Une arithmétique commerciale du XVe siècle, le Compendy de la praticque des nombres de Barthélemy de Romans*. Brepols: Turnhout.

Spiesser, M., and Féry-Hue Fr. 2007. Le recueil mathématique de Cesena, Biblioteca Malatestiana, S.XXVI.6, au cœur du réseau franco-occitan des mathématiques pratiques du XVe siècle. *Archives Internationales d'Histoire des Sciences* 57: 47–88.

Thomasset, Cl. 1994. La "question dispute" comme élément d'une stratégie didactique en langue vulgaire. In *Éducation, apprentissages, initiation au Moyen Âge*, 513–527. Paris: Les Cahiers du CRISIMA.

Thorndike, L. 1959. The arithmetic of Jehan Adam. In *Science and thought in the fifteenth century*, 150–160 and 302–307. New York: Columbia University Press.

Tura, A. 1999. Codici di matematica di fra Giocondo. *Bibliothèque d'Humanisme et Renaissance* 61: 701–711.

Tura, A. 2008. *Fra Giocondo & les textes français de géométrie pratique*. Genève: Droz.

Ulivi, E. 2002a. Maestri e scuole d'abaco a Firenze: la 'Bottegga di Santa Trinita. *Bolletino di Storia delle Scienze Matematiche* 22: 43–91.

Ulivi, E. 2002b. Benedetto da Firenze (1429–1479), un maestro d'abaco del XV secolo. Con documenti inediti e con un 'Appendice su abacisti e suole d'abaco a Firenze nei secoli XIII-XVI. *Bollettino di storia delle scienze matematiche* 22(1): 243.

Van Egmond, W. 1977. New light on Paolo dell'Abbaco. *Annali dell'Istituto e museo di storia della scienza di Firenze* 2: 3–21.

Van Egmond, W. 1980. *Practical mathematics in Italian renaissance: A catalog of Italian abbacus manuscripts and printed books to 1600*. Florence: Istituto e Museo di Storia della Scienza.

Van Egmond, W. 2001. Abacus, algorism, abbacus: Methods of reckoning in the merchant cultures of Mediterranean. In *Commerce et mathématiques du Moyen Âge à la Renaissance, autour de la Méditerranée: Actes du Colloque international du centre international d'Histoire des Sciences Occitanes*, Beaumont-de-Lomagne, 13–16 mai 1999, ed. Maryvonne Spiesser, 21–54. Toulouse: CIHSO.

Verger, J. 1997. *Les gens de savoir dans l'Europe de la fin du Moyen Age*. Paris: PUF.

Vitale-Brovarone, A. 1990. Considérations sur la production de textes mathématiques en France et sur leur diffusion manuscrite et imprimée. *Le Moyen Français* 25: 23–53.

Wolff, Ph. 1954. *Commerce et marchands de Toulouse, v.1350–v.1450*. Paris: Plon.

Chapter 7
On the Transmission of Mathematical Knowledge in Versified Form in China

Andrea Bréard

7.1 Introduction

Versified mathematical algorithms and problems make their appearance in China around the end of the Song dynasty (eleventh century) and are widely spread among the sources of the Ming (1368–1644) and the Qing (1644–1911) dynasties.[1] Certain rhymes were accompanied by a reference to a well known tune according to which they were supposed to be chanted. Mathematical problems are even classified according to these tunes in the most widely read Ming dynasty mathematical source, the *Unified Lineage of Mathematical Methods* (*Suanfa tongzong* 算法統宗, 1592) as well as other texts.

For mathematical verse, there are no explicit contemporary accounts concerning the social and educational context of their utilization, or the related reading practices. The exercise of memorization of the Confucian classics, which figured among the texts to be learnt by heart for participation in the provincial and imperial examinations in China until the early twentieth century, nevertheless is well known. Against this traditional educational background, one might conclude that memorization was used to acquire different sorts of knowledge, including for example memorization verse that gives the basic rules and prescriptions for a mathematical procedure.

In this article, I will attempt a classification of the content and functionality of the different types of mathematical verse that can be found in the studied corpus. My contribution is based on available primary source material, including sources of the encyclopedic genre, from the Yuan (1271–1368) and Ming dynasties, which

[1] A list of Chinese dynasties and their respective dates is given in the appendix.

A. Bréard (✉)
Laboratoire Paul Painlevé, Université Lille I, Villeneuve d'Ascq Cedex, France
e-mail: andrea.breard@math.univ-lille1.fr

A. Bernard and C. Proust (eds.), *Scientific Sources and Teaching Contexts Throughout History: Problems and Perspectives*, Boston Studies in the Philosophy and History of Science 301, DOI 10.1007/978-94-007-5122-4_7,

contain mathematical verse and jingles. Instead of interpreting the abundance of versified material in terms of popularization of mathematical learning within a context of commercial book publishing and a higher social mobility, I have tried to place the studied corpus into a larger educational context. I shall argue that mathematical verses show certain parallels with mnemonic exercises linked to educational primers in China and were formulated for pedagogical purposes aiming not only at mathematical literacy but also at acquiring reading and writing skills more generally.

In the following section, I will give a general outlook on the traditional curriculum in China, which emphasized the ability to memorize and recite classical material (much of it in metrical form) and to match rhyming couplets. Orality thereby plays an important part in the process of reciting and *in fine* comprehending the texts. The third section suggests a typology and gives examples of the kinds of mathematical verses that can be found in Yuan and Ming dynasty texts, in either a variety of wordings of equivalent rhymes or in a stabilized form of concise jingles. Without qualifying this material a priori as forms of didactic poetry, this typology allows to identify the categories of verse eventually intended for memorization and to differentiate their modes of transmission and pedagogical or recreational functions. Finally, based on the typology of the selected case studies, I will draw some tentative conclusions about the role of memorization and orality in the transmission of mathematical verse in an educational context.

7.2 Learning and Reading the Classics

Chinese education traditionally centered upon memorization and reproduction of Confucian texts. Rote memorisation was fundamental to acquiring the beginnings of literacy and continued to the most advanced stage of study.[2] "The child's ability to memorize was thus highly prized among literati and in popular culture."[3] The goal being to be ultimately selected for civil service through a system of state examinations.

Basic education, such as learning to read, write, and compose essays, was provided in private or government schools since early times. Children from 8 or 9 years of age on were trained in community schools (*shexue* 社學), "the primer vehicle in Ming society for the dissemination of basic literary and writing skills."[4] But literacy was acquainted even earlier in their homes, where they learned to recognize a certain amount of characters by reciting primers written in rhymed verses. By

[2]Rote learning or learning by rote is learning things mechanically by repeating them without thinking about them or trying to understand them.

[3]Elman (2001, 268).

[4]Lee (2000, 623).

the Ming, the *Three-Character Classic* (*Sanzijing* 三字經),[5] the *Hundred family names* (*Baijiaxing* 百家姓) and the *Thousand-Characters Essay* (*Qianziwen* 千字文), which simultaneously inculcated fundamental moral principles had become the most widely used textbooks.[6] There was little if any explication and some scholars complained that local schools focused too much on "reciting the texts and matching the lines" to the detriment of moral teaching:

> Today people simply teach students to recite and match rhyming lines. If the students recite and match well, then they are thought to be intelligent, but the teachers know not of the nature and sensibilities of their students nor of their character, nor of how they serve their relations and their peers.[7]

Ben Elman's book on education and the examination system goes into some detail about the part memorisation played in acquisition of classical literacy,[8] which was a three-stage learning process[9]:

> The ability to write elegant examination essays was the crowning achievement for educated men (and women). This creative learning process began with rote memorization during childhood, continued with youthful reading, and concluded with mature writing. As a graded sequence of learning, its foundation was long hours of oral memorization spent by children aged three to eight *sui* 歲.[10]

Premodern Chinese literati, such as the Ming loyalist turned educator Lu Shiyi 陸世儀 (1611–1672) believed, that committing to memory, learning by heart and reciting out loud were not useful when reading the *Histories*, but indispensable for memorizing and understanding the *Five Classics*, the *Four Books* or other Confucian classics. Otherwise, he says, "their (philosophical) doctrine and argumentation cannot emerge."[11]

Lu Shiyi also supported the idea, "that the memory (*chi-hsing* 記性) was strongest at an early age, while understanding (*wu-hsing* 悟性) was a gradual achievement that was derived from mastering the literary language and its moral and historical content. Lu thought students were educable at age five or six, but mature learning could commence only when they reached fifteen *sui*."[12] He gave a selected list of books, proposed for a 10-year period from 15 to 25 years of age, for

---

[5] Also called *Trimetrical Classic*, see Giles (1900, iv): "a hornbook for boys".

[6] See Lee (2000, 439–440, 452–453, 459–461). See also Zhang (1962, 25–27) on the role of these three primers in Ming education.

[7] Citation by Gui Yougang (1507–1571) according to McLaren (2005, 167–168).

[8] Full literacy is sometimes called *vocality*, meaning the ability to express oneself in writing.

[9] See Elman (2001, 261–270).

[10] In Chinese age reckoning newborns start at 1 year old, and according to the lunar calendar the person's age is incremented each passing of a New Year.

[11] Cited from Lu Shiyi's *Notes on Reflections and Disputations* according to Zhou (1895, 38A), a compilation of citations concerning the reading of books:

> 記誦之功讀史不必用若五經四書太極西銘之類必不可不成誦不成誦則義理不出也陸桴亭思辨錄.

[12] Elman (2001, 262–263).

a broad knowledge of traditional learning, written down in "enough books to make the ox carrying them sweat."[13]

In order to acquire elementary education and later master the Classics, the usual techniques for memorization were chanting and reading out loud. Vocalization, either to oneself or to an audience, was the most favored reading practice and was facilitated by versification. As McLaren points out, "much early educational material was composed in verse, in part to make vocalization and memorization easier. Even before formal education began, parents would teach children to recite classical poems. During the late Ming and Qing periods, potted histories written in parallel prose or verse were in common circulation."[14] Even today, Chinese schoolchildren still read and recite the above-mentioned *Three-character classic*, a popular three-syllable-a-line primer. It starts with an enumeration of the first powers of 10, before moving on to the fundamental elements in Chinese natural philosophy:

知某數	Learn to count,
識某文	and learn to read.
一而十	Units and tens,
十而百	tens and hundreds,
百而千	hundreds and thousands,
千而萬	thousands and tens of thousands.
三才者	The Three Forces
天地人	are Heaven, Earth, and Man.
三光者	The three Luminaries
日月星	are sun, the moon, and the stars...

As for more specialized education, students educated to be artisans, doctors or merchants were taught to read and chant concise verses relevant to their specific trades.[15] Farmers learned jingles with a limited number of characters about climate, the calendar and the seasons.[16] Basic mathematical knowledge, needed by merchants for commercial transactions, was also taught through easy verses and rhymes on various aspects of arithmetic. Under the Ming, publications incorporating versified material flourished,[17] but individual mathematical works and

---

[13]The Chinese saying "enough books to make the ox carrying them sweat" (*han niu chong dong* 汗牛充棟) refers to an immense number of books. See Zhou (1895, 38A–40B):

> 書籍之多千倍於古學非博不可然**汗牛充棟**將如之何偶思得一讀書法節十年訟讀自十五歲致二十五....

[14]McLaren (2005, 167–168).

[15]Fishmongers for example used the *Seven-words-a-line song of fish names* (*Yuming qi yan ge* 魚名七言歌. See Wang (1994, 55).

[16]The thirteenth-century book on divination *Yuanhai ziping* (海子評) by Xu Sheng 徐升, for example, gives a seven- and a five-syllable-a-line verse for respectively memorizing the sequence of the 12 earthly branches and the 10 heavenly stems, as well as fortune-telling verses. See Wang (1994, 50–51).

[17]In Bréard (2010) I do relate this phenomenon to the development of printing and the increasing social mobility under the Ming dynasty. See also Honda (1994).

encyclopedias containing short and easy verses can be found from as early as the thirteenth and fourteenth century.

It is possible that the absence of mathematical verse in the surviving sources before the Song dynasty was related to a social differentiation between a high tradition of mathematical writing in discursive algorithmic formulation, and a 'popular' tradition, which emphasized easy, oral memorization and was destined to a less literate public. Both traditions might have existed in parallel, or one possibly evolved out of the other, but since only highly valued canonical texts have survived before the era of commercial printing, earlier 'popular' traditions got lost. I shall discuss these issues, of which some can in parallel be observed in the realm of medicine, in the beginning of the following section. I will then attempt a classification of mathematical verses that have survived the processes of selection and preservation in private and imperial libraries. Such typology has no explicit equivalent in the Chinese sources, but it allows us to draw some conclusions upon the diverse functions of versification in a mathematical context and to characterize the didactic intentions that are at work behind the different metrical forms.

7.3 Versification in Mathematical Texts

From the Ming onward, scholarly knowledge, not sanctioned by any formal authority, became accessible to a larger audience because of the commercial printing of an ever increasing number of books. For mathematics, as for medicine and other fields of knowledge, much printed material was no longer ratified by the State. In parallel, no imperial or specialized institution provided any normalized teaching program for mathematics, despite the crystallization of a mathematical scholarly, mainly algebraic, tradition during the Song-Yuan period (eleventh–fourteenth century). Earlier, during the Sui dynasty (581–681), a mathematical academy had been established to train future examination candidates,[18] but the school was removed from the landscape of formal education in the mid-eleventh century.[19]

It is then that a large number of introductory mathematical texts began to appear. It even appears, that the Ming Directorate of National Youth (*Guozi jian* 國子監) compiled a special mathematical textbook for use in schools. But this work, *The Mathematical Methods* (*Suanfa* 算法), did not survive after the Ming era. Judging from the texts that are preserved,[20] these have several points in common. First, they contain versified mathematical problems or algorithms.[21] Second, either through excerpts, explicit reference, system of classification or choice of mathematical

[18]The now still extant *Ten Books of Mathematical Classics* (*Suanjing shi shu* 算經十書) were especially compiled to serve as textbooks in a specific sequence. See Siu and Volkov (1999).

[19]See Lee (2000, 517–525).

[20]See Li (1954–1955) for a bibliography of extant and lost Ming dynasty texts.

[21]Cheng (1990, 991) even refers to one book which explicitly indicates versified content in the title, the now lost *Sanyuan hua ling ge* 三元化令歌.

problems, they regularly referred to a similar body of mathematical knowledge, in particular the canonical *Nine Chapters on Mathematical Procedures* (*Jiu zhang suan shu* 九章算術). Third, there was obvious simplification of content and language[22] compared to the classics. Finally, most authors of these texts claimed, often in the preface, that they were intended for their students. For example, the Song dynasty author Yang Hui 楊輝 conceived his *Mathematical treasure of variations upon the multiplication and division* (*Cheng chu tong bian suan bao* 乘除通變算寶, 1275) especially for the instruction of beginners and even proposed a detailed curriculum for studying mathematics.

Although in Ming sources verses and rhymes for easy memorization were numerous, they could not include everything one should learn. Authors often stressed the importance of knowing by heart verses which were related to division and multiplication. The programmatic verse on 'Methods of Practicing/Studying Mathematics' in the opening of the first scroll of Cheng Dawei's *Unified Lineage of Mathematical Methods* ingeniously included an enumeration of the ten numerals in its to-learn list[23]:

習算之法

一要先熟讀九數　二要誦歸除歌法
三要知加減定位　四要知量度衡畝
五要知諸分母子　六要知長闊堆積
七要知盈朒互隱　八要知正負行例
九要知勾股弦數　十要知開方各色

1. First, one has to recite fluently the nine numbers[24];
2. One has to know by heart the verses for the division methods;
3. One has to know how to determine the positions in addition and subtraction;
4. One has to know measures of content, length, width and surface;
5. One has to know anything concerning the numerator and the denominator in fractions;
6. One has to know length, width, heaps and volumes[25];

---

[22]The texts were flawed by a mixture of simplified and variant writings of characters.

[23]Cheng (1990, *juan* 1: 1A-1B). The same versified 'Outline of mathematical methods' (*Suanfa tigang* 算法提綱) can be found earlier in Wu Jing's book (Wu fl. 1450).

[24]The expression 'nine numbers' probably refers to the multiplication table up to nine times nine. See for example the Outline in Zhu (1839, 1127), where this multiplication table is referred to as the 'Method of the Nine Numbers' (九數法).

[25]The Chinese term for heaps refers to an accumulation of discrete objects like grains, in a certain geometric shape, whereas 'volumes' refers to a continuous geometric solid.

7. One has to know excess, deficit, mutual elimination[26];
8. One has to know positive and negative numbers, rows and columns[27];
9. One has to know the numbers for the three sides of rectangular triangles;
10. One has to know all the colors in the extraction of square roots.[28]

The multiplication table and rhymes for division had a special status within the things to learn: they are mentioned in the first two lines of the verse as those that one has to 'master fluently' and 'know by heart', whereas more advanced mathematical methods need to be 'known', or 'understood'. In the section of his book, where Cheng gives the division verses, he underlines again the fact, that these, as well as verses for multiplication, need to be thoroughly memorized:

> The above method [for division] and the 'combined numbers of nine nine'[29] can easily be confused. But when memorizing, we only need to distinguish how the lines are constructed: when the greater number comes first and the lesser number afterwards, then it is a line from the 'nine returns' (*jiu gui* 九歸).[30] Like in 'eight six seventy-four' being a phrase for division, 'six eight fourty-eight' being a phrase for multiplication[31] where the former way of constructing the lines is connected to the latter. It is all these kinds of verses for memorization that every student shall thoroughly commit to his memory.
>
> 右法與九九合數相混但記句法惟辨多數在先少數在次即九歸之句如八六七十四是歸六八四十八是因之類已上句法併後各樣歌訣皆學者所當熟記.[32]

7.3.1 *Typology*

So far, I have loosely used terms like jingles, rhymes, songs, tunes, etc. when speaking about versified mathematical content. The following typology will make explicit how to distinguish these terms from each other and to which concepts they correspond in the Chinese texts. Although this typology is not explicitly formulated in the studied corpus, I simply wish to propose a scheme that clarifies how each kind of verse might play a certain role in an educational context, and how, once committed to memory, some eventually help in performing a calculation through certain mnemonic devices.

---

[26]The terminology here refers to the method of double false position.

[27]This is necessary to solve systems of linear equations.

[28]Liu Hui's commentary (dated 263) to the algorithm for square root extraction in the *Nine Chapters* makes use of colors to refer to the geometric entities in a square corresponding to the terms in the algorithm. See Chemla and Guo (2004, 322–329). For a further discussion of the algorithm of square root extraction in its versified form see Sect. 7.3.1.6.

[29]Denoting a set of multiplication jingles for $x \cdot y$, where $x \leq y$ and $x, y \leq 9$.

[30]Technical term for division by one-digit numbers $x \div y$, where $x, y \leq 9$ and $x \geq y$.

[31]Mathematically meaning, that in division of 60 by 8, one obtains 7 and a rest of 4, when multiplying 6 by 8 one obtains 48. The smaller characters here reflect the fact that this phrase is typeset as a commentary in the original text.

[32]See Cheng (1990, *juan* 1: 9B).

Four types of content of mathematical verses can be distinguished:

1. Those that are stating a mathematical problem;
2. Verses which are related to their solution algorithm;
3. Verses that articulate a larger vision of mathematical practice or depict a narration linked to mathematical activities, instruments or practitioners;
4. And finally, verses for rote memorization (*koujue* 口訣) of the basic arithmetical operations, which I refer to as 'jingles'.

Verses of the first three kinds are either written in regular verse (called *ge* 歌, *gejue* 歌訣 or *zong jue* 總訣, 'master rhyme') or composed to be chanted according to a certain well-known tune. A single tune may have many variant forms both in terms of basic structure, line lengths, rhymes, intonation, mood, etc. From a literary point of view, the fourth kind of verse is distinctly different. A small and sometimes irregular number of characters form a line, each one made up mostly of number words (see Sect. 7.3.1.10). Examples for each type are discussed in the following subsections.

7.3.1.1 Versified Mathematical Problems

Scrolls 13–16 of Cheng Dawei's *Unified Lineage of Mathematical Methods* systematically distinguish mathematical problems formulated in rhymes (*ge* 歌), and problems formulated in verses for a specific tune name. Figure 7.1 shows how Cheng classified the number of 'Difficult problems' of the four scrolls. First, it is divided into nine types of mathematical topic according to the canonical *Nine Chapters*. Then, within these categories, Cheng counts each of the number of problems written in rhymes and the number of problems following a certain tune:

Scroll Thirteen:

– Rectangular Fields Seven problems

Rhymes Five *Fengqi wu* 鳳棲梧 (Phoenix perching on a Chinese Parasol Tree) One *Shuang daolian* 雙搗練 (Pounding Silk with a pair [of sticks]) One.

– Millet and rice Thirteen problems

Rhymes Eight *Xijiang yue* 西江月 (Moon of the Western River) Three *Meiqi qing* 梅氣清 (Plum Fragrance) One *Shuixian zi* 水仙子 (Ondine) One.

Scroll Fourteen:

– Proportional distribution Twenty-nine problems

Rhymes Twenty-four *Xijiang yue* 西江月 (Moon of the Western River) Three *Zhegu tian* 鷓鴣天 (A Partridge in the Sky) One *Shuixian zi* 水仙子 (Ondine) One

Scroll Fifteen: ...[33]

Cheng made a clear distinction between rhymes and problems following tunes not only in the index to scrolls 13–16, but also typographically. Rhymes are written

[33]Cheng (1990, 800).

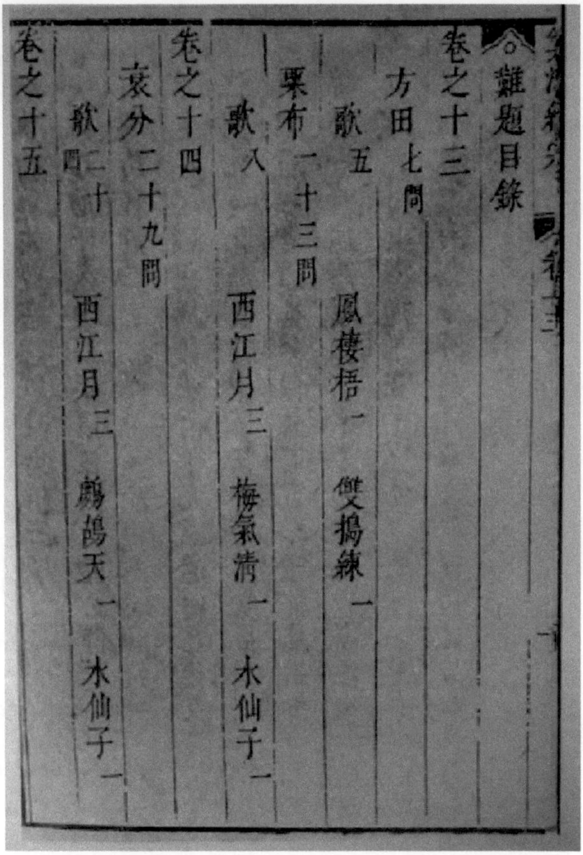
難題目錄
卷之十三
方田七問
歌五 鳳棲梧一 雙搗練一
粟布一十三問
歌八 西江月三 梅氣清一 水仙子一
卷之十四
衰分二十九問
歌四 西江月三 鷓鴣天一 水仙子一
卷之十五

Fig. 7.1 Index to difficult problems

out in lines, separated by spaces, with a constant number of syllables per line, whereas problems following tunes are similar to discursive passages of the book, sometimes punctuated.[34] This may have facilitated recitation by helping the less literate reader recognize the pattern of the tune according to which the problem should be chanted.

Cheng Dawei's index in Fig. 7.1 was not the sole occurrence of such a scheme of classification. Earlier, Wang Wensu 王文素 collected "more than three hundred problems in rhymes and regulated verse during the summer and appended them"[35] for his *Precious mirror of old and new mathematical learning, newly collected and thoroughly proofed* (*Xinji tongzheng gujin suanxue baojian* 新集通証古今算學寶鑒, preface dated 1524). Like Cheng, he listed mathematical problems in the

[34]Such is the case in the 1716 edition reprinted in Cheng (1990), but not in Cheng (1882).

[35]Translated from the preface to Wang (2008, xxiii).

table of contents following the tunes according to which they were to be recited.[36] Unfortunately, the only extant manuscript copy transmits only the first 41 scrolls, but not the last section of 12 scrolls entirely devoted to mathematical rhyme and regulated verse.

7.3.1.2 Example 1 of a Versified Problem: Piles of Countable Objects

I shall first illustrate the different ways of versification related to distinct reading practices by two subsequent examples drawn from scroll 15 of Cheng's book.[37] They both concern situations, where discrete objects are accumulated in the shape of a truncated pyramid with a rectangular and a triangular base, respectively.

西江月[38]
今有酒**瓶**一垛。要積一百六十。下長多廣整七枝。廣少上
長三隻。堆積**槽**坊**圓**內。上下長廣**難**知。煩公仔細用心幾[。]
借問各該有幾

Moon of the Western River
Let us suppose that we have a heap of wine jars.
We want to pile up one hundred and sixty.
The length at the bottom is precisely seven bottles more than the [upper] breadth.
The [lower] breadth is three bottles less than the upper length.
The pile shall fit within the boundaries of the storage space.
Its upper and lower breadth and length are hard to know.
Let me bother the honorable youth to apply their mind to this,
and venture to ask how much is each of these [unknowns] ?

---

[36]See the Table of contents (Wang 1524, xxxiii–xxxv). The tunes were *Shuixian zi* 水仙子 (Ondine), *Xijiang yue* 西江月 (Moon of the Western River), *Luomei feng* 落梅風 (Wind from Falling Plums), *Shanpo yang* 山坡羊 (Sheep on a Precipice), *Chao tianzi* 朝天子 (Son of Heaven), *Qing jiang yin* 清江引 (Clear River, a Prelude), *Zui taiping* 醉太平 (Drunk in a Peaceful Time), *Hong xiuxie* 紅繡鞋 (The Red Embroidered Shoes), *Wu ye'r* 梧葉兒 (Leaves of Chinese Parasol Tree), *Yan er luo* 雁兒落 (Falling Swan), *Zhegu tian* 鷓鴣天 (A Partridge in the Sky), *Pu tian le* 普天樂 (The World being Overjoyed).

[37]Cheng (1990, 888–890).

[38]A number of variant and simplified character forms rarely used in formal writing before the Ming are included in the printed version I have consulted. I have adopted their nowadays standard form in my transcription.

The problem here is enunciated according to the Tang dynasty tune *Moon of the Western River* (*Xijiang yue* 西江月),[39] which follows a 50-syllable pattern with an intricate set of formal rules. In addition to the tonal constraints,[40] these forms required parallelism between certain couplets. The lines in these couplets had to contain contrasting content, with the characters in each line usually in the same part of speech.

Several features of this verse support the hypothesis that these texts were destined for a less literate public. Not only are the verses punctuated, but there are also different characters, and not only homophone variants of the same character, which were used to convey the same meaning.[41] Such diversity was not common in the mathematical classics, where terminology showed a high degree of homogeneity. All this eventually helped in learning to read and write, and thus promoted the acquistion of a higher degree of literacy, while studying mathematics.

The problem immediately following the solution to the one translated above states a similar problem in regulated verse with seven syllables in each of the four lines:

歌
紅桃一垛積難知
共該六百八十枝
三角垛來尖上一
每面底子幾何為

A rhyme
Hard it is to know the accumulation in a pile of red peaches.
Altogether there should be six hundred and eighty pieces.
Coming to the top of a triangular pile we have one.
How many form each side at its base?

[39]The name of the tune stems from a Tang dynasty poem by Li Bai 李白, *The Ruin of the Gu-Su Palace* 《蘇台覽古》「只今惟有西江月，曾照吳王宮裡人」.

[40]The patterns required a certain sequence of level/even tones (*ping* 平) and oblique/non-level/uneven tones (*ze* 仄). The tune here follows the pattern:

(中)仄(中)平平仄，(中)平(中)仄平平。
(中)平(中)仄仄平平，(中)仄平平(中)仄。
(中)仄(中)平平仄，(中)平(中)仄平平。
(中)平(中)仄仄平平，(中)仄平平(中)仄。

where the middle tone (*zhong* 中) in brackets signifies an optional choice between even and uneven tone.

[41]This is the case for example for the two homophone measure words for bottles: *zhi* is either printed as 隻 or as 枝. There are also two different expressions used for a heap (or pile): a variant form of *duo* 垛 and *dui* 堆.

These versified mathematical problems – examples with specific numerical values – provide a model for problems of the same kind. Yet, they do not indicate the solution procedure. In an educational context, one might view them as canonical examples, which provided a framework for composing further mathematical problems. This would put Cheng's *Unified Lineage* very much on the same level as contemporary letter writing guides or other how-to encyclopedias that provide models for literary composition. But according to the explanation of what Cheng understands to be a 'difficult problem' in these four scrolls, where all the problems are stated in versified form, one might also view them as riddles, as amusement literature for leisurely distraction of the mathematically versed, who are already familiar with the methods of the *Nine Chapters*:

> Concerning the difficult problems: from 1404 on the Honorable Liu Shilong 劉仕隆 from Linjiang [in the province of Jilin][42] together with all the Noblemen of the Grand Secretariat prepared and collated the *Great Compendium*.[43] When he retired from his official duties, he compiled the 'difficult problems' and appended them to the *Perspicuous mathematical methods from the Nine Chapters* (*Jiu zhang tongming suanfa* 九章通明算法). Also, in the *Analogical categories to the Nine Chapters* (*Jiuzhang bilei* 九章比類) by Wu Xinmin [Wu Jing] from Qiantang and in the *Mathematical methods from all schools* (*Zhujia suanfa* 諸家算法), the collected poems, verses and jingles/slogans/verbal instructions (*Shici gekuo kouhao* 詩詞歌括口號) are named 'difficult problems'. 'Difficult' means difficult. But they only seem difficult, and in reality they are not difficult. It is only their wording that is tricky. Once exposed by a mathematical master, one realizes that it is not true that nobody knows how to manipulate them, and that the difficult methods that one does not know all do not go beyond the *Nine Chapters*. The 'difficult problems' outside the *Nine Chapters* only consist in establishing the method. [...] Now, we list separately the explanations and discussions of the solution methods according to the *Nine Chapters*, and append a collection of miscellaneous methods at the end of the *Unified Lineage*.

7.3.1.3 Example 2 of a Versified Problem: Quadratic Equations

Versified mathematical problems do appear in sources earlier than Ming dynasty. In the *Jade Mirror of Four Unknowns* (*Siyuan yujian* 四元玉鑑, 1303), Zhu Shijie grouped them together in a single chapter, entitled 'Some ask in versified form' (*Huo wen ge tuan* 或問歌彖). No typographical indication, like spaces between lines or punctuation, was given for how to read the problem texts in this chapter.[44]

[42]Author of a now lost book from 1424, entitled *Entirely Understandable Mathematical Methods from the Nine Chapters* (*Jiu zhang tongming suanfa* 九章通明算法). See Li (1954–1955).

[43]Reference to the *Great Compendium of the Yongle Reign-Period (1403–1422)* (*Yongle dadian* 永樂大典).

[44]At least such is the case in the commented 1836 edition by Luo Shilin, which is the earliest printed edition preserved. There is an earlier manuscript version, dated 1819 in the National Library of China (Ancient Rare Books Section N° 16000), and another manuscript in the National Palace Museum in Taipei, which was copied during the Jiaqing reign (1796–1820), but I have not had a chance to revisit these for this article. See Bréard (1999, Chap. 4, n. 19).

The versified mathematical problems from 'Some ask in versified form' certainly showed features common to the explicitly versified problems in later texts and distinct from the classical tradition, particularly in the way the question about the unknowns was asked. Since punctuation did not give a regular rhyme in either problem in this particular chapter, thus, I assume that their formulation followed a familiar tune pattern. The first problem, illustrated in Fig. 7.2 reads[45]:

或問今有方池一所。每面丈四方停。葭生西岸
長其形。出水三十寸整。東岸蒲生一種。水上一
尺無零。葭蒲稍按水齊平。借問三般怎定。
答曰水深一丈二尺　蒲長一丈三尺
葭長一丈五尺

Someone asked: Let us suppose that there is a square pool.
Each of its sides is [1] *zhang*, 4 [*chi*], and just stops there.
A reed is growing on the West edge, long is its shape.
It extends precisely 30 *cun* above the water.
One rush is growing on the East edge,
it extends 1 *chi* above the water and no more!
If the tops of the two are tied together they will meet on a level with the water.
Now I venture to ask, how we can decide the three questions.
The answer is: Depth of the water 1 *zhang* 2 *chi*; length of the rush 1 *zhang* 3 *chi*; length of the reed 1 *zhang* 5*chi*.

The problem is solved through consideration of right-angled triangles in three ways, each leading to a quadratic equation in one of the three unknowns. It appeared again, in a simplified version with only one plant and two unknowns (see Fig. 7.3), in Cheng Dawei's *Unified Lineage of Mathematical Methods* (1592).[46] As in the previous example cited from Cheng's book, it was supposed to be chanted according to the melody *Moon of the Western River*, with punctuation indicating the various lines:

[45]Translated from the edition in Zhu and Luo (1937, 429). The puncutation is mine, and results in a pattern, that can be chanted according to the tune *Moon of the Western River* mentioned in Footnote 40. Zhu and Guo (2006, vol. II: 329), gives a translation, which is only followed partially here.

[46]See Cheng (1990, *juan* 16: 935).

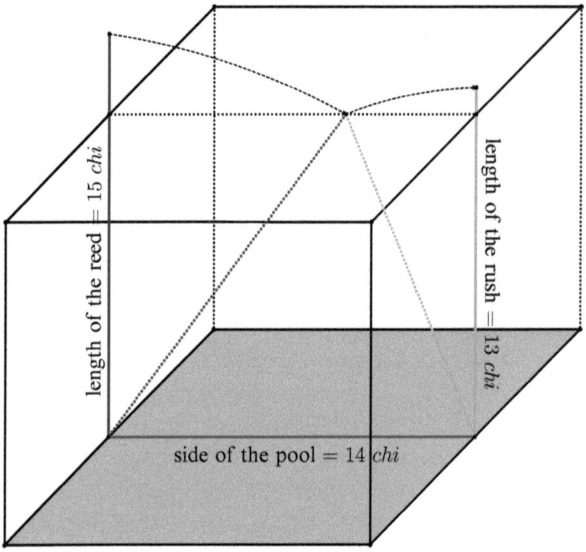

Fig. 7.2 Illustration of the problem of two plants in a square pond in Zhu Shijie's *Jade Mirror of Four Unknowns* (1303)

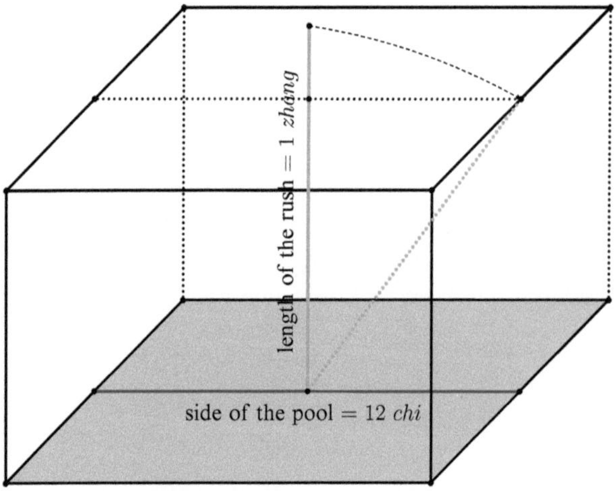

Fig. 7.3 Illustration of the problem of one plant in a square pond from Cheng Dawei's *Unified Lineage of Mathematical Methods* (1592)

今有方池一所。每邊丈二無移。中心蒲長一根肥。出水過
於二尺。斜引蒲稍至岸。適然與岸方齊。請君明算更能推[。]
蒲長水深各幾 右調西江月
答曰蒲長一丈 水深八尺

Let us suppose that there is a square pool.
Each side is [1] *zhang* 2 [*chi*] and no more.
In its center grows one rush having a root and a tail.
It extends 2 *chi* above the water.
When bending the rush across [the pool] to the edge,
it will reach on a level with the edges of the square [pool].
I beg you, Sir, who understands calculations, to solve this.
How much is each, the length of the rush and the depth of the water?
The above is tuned according to the *Moon of the Western River*.
The answer is: length of the rush 1 *zhang*, depth of the water 8 *chi*.

The situation here is such, that the entire length of the rush corresponds to the distance between the center at the bottom of the pool and the middle of one of its sides. In this form, the problem is similar to one in the *Nine Chapters on Mathematical Procedures* from Han dynasty, and differs only by its numerical values and the name of the plant.[47]

There are many more examples of versified problems that one can trace back to the *Nine Chapters*. But were classical problems rewritten and transmitted in a different textual form to facilitate learning? Or did oral transmission foster the versification of mathematical problems, which later authors then collected and put into print?[48] I will come back to such questions related to orality in the conclusion. Rather than interpreting similarities with the classics in terms of textual developments related to the transmission processes of written material, I suggest that transformations of mathematical problems and algorithms were a kind of mathematical activity with a didactic perspective. In the following section, a

---

[47] See Problem 9.6 in Chemla and Guo (2004, 711).

[48] Cf. the situation in medicine as described by Leung (2003, 132):

> Yang highly praised the work [the *Yijing xiaoxue* 醫經小學 (Primary Study on Medicine), an important medical introductory textbook written in 1388], as its rhymes and verses written to facilitate memorization by beginners were based on classics.

diachronic approach to the prominent example of the Chinese Remainder Theorem shows how a fourth-century solution algorithm was rewritten in different terms, calling upon knowledge from fields other than pure mathematics.

7.3.1.4 Versified Algorithms

A second type of verse content found in mathematical texts were algorithmic solution procedures formulated in rhymes. Generally, the verses I found only alluded to the algorithms, which prescribed sequences of operations that should be performed in order to solve an entire class of mathematical problems.

They allowed merely for the recollection or reconstruction of the operations from memory, but did not stand by themselves as an entire algorithm. The same phenomenon can be observed in medical primers.[49]

7.3.1.5 Example 1 of Versified Algorithms: 'The Chinese Remainder Theorem'

Although not precisely a theorem, but rather an algorithm, this term refers to a solution procedure found as early as in *Master Sun's Mathematical Manual* (*Sunzi Suanjing* 孫子算經) compiled approximately in the fourth century. The statement of the problem goes as follows[50]:

今有物不知其數三三數之賸二五五數之賸
三七七數之賸二問物幾何
答曰二十三
術曰三三數之賸二置一百四十五五數
之賸三置六十三七七數之賸二置三十
并之得二百三十三以二百一十減之即
得凡三三數之賸一則置七十五五數之
賸一則置二十一七七數之賸一則置十
五一百六以上以一百五減之即得

Now there are an unknown number of things.
If we count by threes, there is a remainder 2;
if we count by fives, there is a remainder 3;
if we count by sevens, there is a remainder 2.
Find the number of things.

[49]Cf. Leung (2003, 141):

> Although the verse form could not express the exact composition of the various recipes, it provided the principles of the combination of various drugs in recipes for different purposes.

[50]Here problem 26 of *juan* 3. I only give the translation of the problem here from Lam and Ang (2004, 219). The Chinese text follows the edition in Qian (1963, vol. 2: 318).

To solve this problem,[51] Sun's text gives two algorithms formulated discursively as in the *Nine Chapters*. They offer three "magic" numbers which, combined in a proper way, yield the least positive solution:

$$140 + 63 + 30 - 210 = 23.$$
$$2 \cdot 70 + 3 \cdot 21 + 2 \cdot 15 - 105 \cdot 2 = 23.$$

The second formulation of Sun's solution procedure can be interpreted as valid for a more general class of problems, and not only the given problem with three specific numerical values for the remainders r_1, r_2, and r_3. It links these variables to the respective factors 70, 21, and 15 and the number 105, allowing to solve problems of the following kind:

$$x \equiv r_1 \pmod 3$$
$$x \equiv r_2 \pmod 5$$
$$x \equiv r_3 \pmod 7.$$

with the second algorithm:

$$r_1 \cdot 70 + r_2 \cdot 21 + r_3 \cdot 15 \equiv x \pmod{105}.$$

This type of mathematical problem and its solution procedure has circulated widely in China. It was mentioned in mathematical and literary sources. Yang Hui, in his *Selection of Strange Mathematical Methods in Continuation of Antiquity* (*Xugu Zhaiqi Suanfa* 續古摘奇算法, 1275) commented that "this is commonly referred to as 'the secretly selected army of the King of Qin' or 'the procedure of shooting repeatedly'."[52] Zhou Mi 周密 (1232–1308) called it 'the calculation from the devil's valley' (*Guigu suan* 鬼谷算), or the 'calculation of cloisoned walls' (*Geqiang suan* 隔牆算), while Cheng Dawei mentioned that it was also named 'the secretly selected army of Han Xin' (*Han Xin dian bing* 韓信點兵).

Versified solutions were equally in circulation. Zhou Mi, in his *Random Jottings from the Hall for Pursuit of Elegance* (*Zhiyatang zachao* 志雅堂雜鈔), after

[51]In modern terminology, the problem can be transcribed as a system of simultaneous linear congruence equations:

$$x \equiv 2 \pmod 3$$
$$x \equiv 3 \pmod 5$$
$$x \equiv 2 \pmod 7$$

[52]In Yang (1275, *juan* 上: 1100).

explaining in detail the solution procedure to the more general class of problems, gave the following riddle-like rhyme[53]:

有一詩隱括云
三歲孩兒七十稀
五留廿一事尤奇
七度上元重相會
寒食清明便可知

There is a poem, which I include secretly, that says:
Seventy children three years old, that's rare!
Five remaining twenty-one, that is an even more unusual affair!
After seven degrees, one meets again the half-moon.
From the winter solstice to the middle of the third month, that is how one can know [the answer].

This verse did not explicitly give the steps to calculate, nor did it explicitly indicate all the numerical constants required to solve the problem. A certain amount of basic astronomical knowledge was necessary to figure out that it takes 15 days for a passage of 7° (*du*) along the ecliptic, and 105 days from the beginning of the year to the middle of the third moon in the Chinese calendar. Having resolved these questions, the verse then only allowed to recall the fact that the number 70 is linked to the remainder r_1 of (mod 3), 21 to the remainder r_2 of (mod 5), the remainder r_3 to (mod 7) and it finally gave a hint to the number 105. What to do with these numbers was not stated in the verse, and therefore the student needed to recall the algorithm from his memory with the help of the verse.

The *Unified Lineage of Mathematical Methods* (*Suanfa tongzong* 算法統宗, 1592) gave the same problem as Master Sun and was preceded by a similar rhyme as the one found in Zhou Mi's *Jottings*. It was slightly more straightforward for only the constant 15 needed to be deduced as the number of days in half a month, the other constants were given[54]:

---

[53]My own translation follows rather literally the edition in Zhou (1969, *juan xia* 下 *Yin-yang suanshu* 陰陽算術: 2a–3a (92–93)). Volkov (2002, 402) based on (Libbrecht 1973, 286) both give more literary translations.

[54]Translated from Cheng (1990, *juan* 5: 430). Libbrecht (1973, 291–292) translates the passage as follows:

> Three septuagenarians in the same family is exceptional.
> Twenty-one branches of plum-blossom from 5 trees.
> seven brides in ideal union precisely the middle of the month.
> Subtract 105 and you get it.

Volkov (2002, 401) says the text "may approximately be rendered as follows:

> Three men walking together, [then] **seventy** are rare/dispersed (?);
> Five trees, **21** branches of plum-blossoms;
> **Seventh** [month's] gathering is in the **middle of the month** [=**15**th day];
> Subtract **105** and then [you] can know [the answer]."

物不知總孫子歌曰又云韓信點兵也
三人同行七十稀　五樹梅花廿一枝
七子團圓正半月　除百令五便得知
今有物不知數只云三數剩二箇五數剩三箇七數剩二箇問共若干
答曰共二十三箇

'Tis rare to find one man
Of seventy out of three,
There are twenty one branches
On five plum blossom trees.
When seven disciples reunite
It is in the middle of the month,
Discarding one hundred and five
You have the problem done.

The inclusion of variant rhymed and unrhymed versions of the 'Chinese Remainder Theorem' in diverse sources suggests that this was a widely circulating 'algorithm', transcending social, temporal, disciplinary and regional barriers. Oral transmission in the way of a popular riddle or in an educational context where a student's mathematical and astronomical knowledge was equally challenged, was probably the source for varying wordings and metaphors in the verses. These were not stand-alone instructions for mathematical operations, but had to be complemented by further instruction, either through an oral explanation by a teacher or a reading of *Master Sun's Classic* or Yang Hui's manual. But even the non-versified versions in these two texts were not entirely explicit about how to solve the remainder problem. Nevertheless, they were an exceptional case in the history of the modalities of writing algorithms in the Chinese classical tradition, where the entire list of operations for solution of a class of problems was generally stated. In the following, I will give another example of a versified general algorithm, where the Classics (and their early commentaries) do indeed provide all the steps necessary for extraction of the square root of a natural number.

7.3.1.6 Example 2 of Versified Algorithms: Square Root Extraction

Scroll six of Cheng Dawei's *Unified Lineage* was a rich collection of rhymes, glossaries, jingles, problems, diagrams and commentaries devoted to techniques of root extraction,[55] and in extension of the algorithm to the solution of quadratic equations. Constructed with a pedagogical focus, it catered in particular to beginners, as the author stated in the preface to the chapter:

> Existing commentaries on square and cube root extraction give you a total headache. For those who begin learning [mathematics] they are difficult. I now append commentaries and explanations to make things simpler.[56]

What followed, was a first rhyme on square root extraction, which, as in the previous example, was not a stand-alone algorithm, but required further elements for the student or practitioner of mathematics in order to perform the correct operations. Cheng first included the following verse[57]:

開平方法認商歌
一百一十定無疑 一千三十有令餘
九千九九不離十 一萬總(終)為一百推
得商方除倍作廉 次商名隅併廉除
餘數續商隅又倍 只依此法取空虛

Rhyme for knowing the quotient in the method of 'opening a flat square'
One hundred – ten, for sure, there is no doubt!
A thousand – thirty something and a remainder.
Nine thousand nine nine [i.e. 9,999] does not go beyond [multiples] of ten;
In the case of ten thousand, one has to advance to one hundred.
The quotient obtained, one eliminates its square, by doubling it one makes the [divisor of the] edge.
The next quotient will name/determine the [divisor of the] corner, one has to eliminate it together with the [divisor of the] edge.
For elimination of numbers in the following quotients, one doubles furthermore the [divisor of the] corner.
It is only according to this method, that one can reach the void!

What was rather unique for rhymes in contemporary mathematical writing (as far as I can judge from the available sources), was that Cheng added a glossary following immediately the above rhyme. Glossaries were a common feature in word or phrase books and helped to enhance a student's literacy, in this case: mathematical

---

55 The goal of these techniques, referred to as 'the opening of the square', is to eliminate successively surfaces of a square, having at the outset a surface equivalent to the number one wants to extract the square root of. The algorithm comes to an end, when one has exhausted the entire surface of the square. For a detailed discussion of the algorithm for square root extraction in the *Nine Chapters* and Liu Hui's commentary to it, see Chemla (1994).

56 Cheng (1990, *juan* 6: 1A (445)).

57 Voir (Cheng 1990, *juan* 6: 1A–1B (445–446)).

literacy. Citing technical expressions and the first four lines of the rhyme, Cheng explained their meaning in geometrical terms, followed by a set of concise verse for memorization of square numbers[58]:

> The explanation says: A 'flat square' is the surface one obtains by multiplying by itself the side of a square. To 'open' is to search for the side of a square. 'One hundred – ten, for sure, there is no doubt!' means that if the surface is one hundred [square] *bu* one can divide to obtain ten *bu* as the side of the square, and there is nothing to doubt [i.e. the result is precise]. 'A thousand – thirty something and a remainder' means 'Nine thousand nine nine does not go beyond tens' means… These phrases are memorization verses for division of the initial quotient. Furthermore, in the following are all [the verses] for quotients and [their corresponding] surfaces.[59]

○開平方初商定首位訣是自乘之數
商一步 積一步 商一十步 積一百步
商二步 積四步 商二十步 積四百步
商三步 積九步 商三十步 積九百步
…
商九步 積八十一步 商九十步 積八千一百步

Memorization verses for determining the highest position of the initial quotient in square root extraction. These are numbers multiplied by themselves.
Quotient one *bu*, surface one *bu*; quotient ten *bu*, surface one hundred *bu*.
Quotient two *bu*, surface four *bu*; quotient twenty *bu*, surface four hundred *bu*.
Quotient three *bu*, surface nine *bu*; quotient thirty *bu*, surface nine hundred *bu*. …
Quotient nine *bu*, surface eighty-one *bu*; quotient ninety *bu*, surface eight thousand one hundred *bu*.

7.3.1.7 Verses with More General Mathematical Content

Besides the more technical verses we have seen so far, there is a particular category of rhymes which goes beyond the mathematical context and conveys a narrative, or a 'moral' lesson. Their narrative style and rich vocabulary might be an indication that these rhymes were not simply meant to be memorized but enhanced students' reading and pronunciation abilities. Above, we have already seen one such example from the very beginning of the *Unified Lineage*, where Cheng Dawei put in verse how one should proceed in studying mathematics. Two more examples from the

---

[58] I will discuss such kind of jingles with further examples in Sect. 7.3.1.10.

[59] This sentence is printed in smaller characters. This usually marks a commentary or a later addition to the original edition.

same book, which also appear in similar wording in many contemporary daily-use encyclopedias, aimed at convincing the reader to trust new mathematical methods, replacing the instrument of the abacus with written calculations. These rhymes also reminded the student to apply his heart-mind to the understanding of the principles, so that he could mobilize his knowledge in analogical situations.

7.3.1.8 Example 1 of General Content: *Brocade in a Single Brush Stroke*

一筆錦歌

巧算一筆錦為奇　不用算盤數可知
垛積合總乘除法　各行寫數莫差池
但看直行未後數　逐位合數似走之
照式用心明其理　厘毫絲忽不須疑

The brocade in a single brush stroke rhyme[60]
Just how extraordinary are calculations done by the brocade in a single brush stroke!
Without the use of the abacus one can know the numbers.
For adding up the total of accumulated piles, for methods of multiplication and division;
in each column one inscribes a number, none has to go beyond its pond.
One only has to look along the horizontal line if there are no more numbers following;
position by position one unites the numbers, it seems that they are pursued.
By following the model pattern and applying one's mind, one can understand its internal logic;
the units of *li, hao, si*, and *hu* suddenly don't need to be doubted!

[60] See Cheng (1990, *juan* 17: 6a–b). The corresponding layout of the numbers involved in this kind of written calculation is shown in Fig. 7.4 from Cheng (1990, *juan* 17: 5a–b).

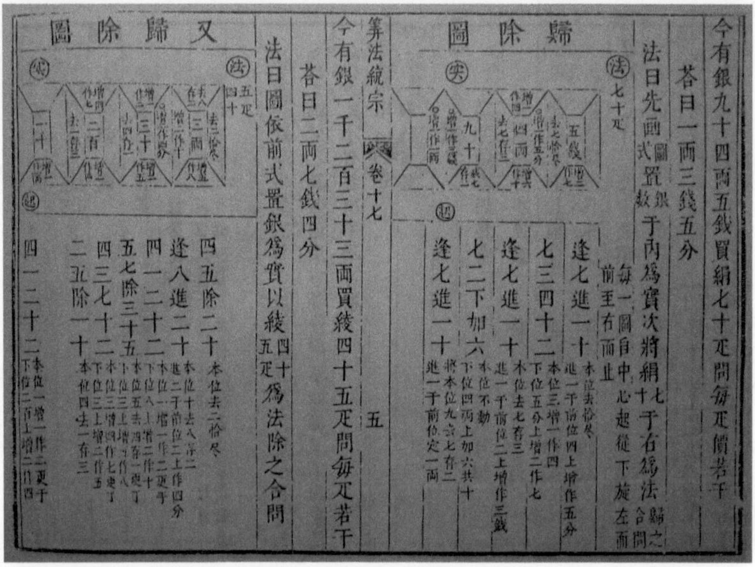

Fig. 7.4 Layout of numbers in the case of written division

7.3.1.9 Example 2 of General Content: *Written Calculation*

寫算即鋪地錦

歌

寫算鋪地錦為奇　不用算盤數可知
法實相呼小九數　各行寫數莫差池
記零十進於前位　逐位數上亦如之
照式畫圖代乘法　厘毫絲忽不須疑

Rhyme on written calculation[61]
Amongst written calculation, the *Pudijin*[62] is extraordinary !
Without the use of the abacus one can know the numbers.

[61]Translated from Cheng (1990, *juan* 17: 3b (p. 946)).

[62]Multiplication method of Indian origin. It appears in China as early as in 1450 in Wu Jing's *Great compendium of analogical categories to the Nine Chapters* (*Jiu zhang suanfa bilei daquan* 九章算法比類大全). See Martzloff (1997, 92)) and Wu (fl. 1450).

Divisor and dividend mutually call upon each other by the nine unit numbers;
in each column one inscribes a number, none has to go beyond its pond.
Inscribing a zero when a ten appears, we advance by one position;
position by position augmenting the numbers, is equally like this.
Following the pattern/model, one draws a diagram, which replaces the [former] method for multiplication;
the units of *li, hao, si* and *hu* suddenly don't need to be doubted!

Comparing the two verses, there is a striking similarity in their wording, although both related to different techniques of written calculation: the first to a method of division, the second to multiplication. The 'moral' lesson nevertheless was the same, and was even expressed with the same words in both rhymes. But there were many more verses praising the *Pudijin* multiplication method. Another cluster of rhymes with similar wordings could be found in daily-use encyclopedias from the beginning of the early seventeenth century, all of them printed commercially in Jianyang, Fujian province.[63] Although they contain lines identical to Cheng Dawei's *Unified Lineage*, they were clearly distinct from the latter source.

7.3.1.10 Jingles

Sets of concise lines, containing few characters beyond the ten numerals, and not necessarily representing entire grammatically correct phrases represent the kind of mathematical verse that I refer to as 'jingles'. Their spread and survival until today seems intimately linked to the usage of the abacus (*suanpan* 算盤). But jingles such as the 'nine nine' multiplication verse, or the 'nine returns' discussed below, were actually available both in earlier manuscript or printed sources when mathematicians calculated with counting rods. Recipes for the abacus, an instrument widely used among scholars and merchants under the Ming dynasty, were exclusively written in concise jingles, and did not give any further explanation about the operations to be followed mechanically.

Xu Xinlu 徐心魯 indicated in the complete title of his *Methods for Calculating on an Abacus* (*Panzhu suanfa* 盤珠算法)[64] that the *Secret Jingles* contained in his book were *handed down through family instruction* (*jiazhuan mijue* 家傳秘訣). It is doubtful, that the title referred to any precise reality in the teaching of arithmetic through 'secret verses for memorization'. A vast number of commercial imprints from the Song dynasty on carried such expressions in their titles in order to attract curious book-buyers and readers, making thembelieve that they would

[63] See the appendix to Bréard (2010), which gives the variant rhymes from four Jianyang imprints.
[64] See Xu (1573).

Fig. 7.5 Women teaching the Six Arts (1599)

acquire knowledge hitherto reserved to a limited social circle. Family instruction, on the other hand, played an important role in early childhood education, as illustrated in Fig. 7.5.[65]

7.3.1.11 Example: *Jiugui*: The Nine Returns

The jingles for a certain type of division by a single-digit number have a strikingly long term history: they survived the instrumental changes from calculation rods to the abacus, and even in written calculation, a particular layout of the numbers is developed that allowed for the application of the same jingles. Their textual stability over several centuries might underline the importance of rote memorization of these rules and their fundamental role in mastering basic arithmetic operations. Jingles are thus in stark contrast with other types of mathematical verse which were characterized by an unstable text and textual history.

[65]Illustration from Yu (1599).

According to Yang Hui the ¡¡nine returns¿¿ were already in use in the *Guide to arithmetical methods* (*Zhinan suanfa* 指南算法, ca. 1078–1189). Zhu Shijie in his 1299 manual points out, that such verses belonged to a minor or irregular tradition, but were easier to learn:

> Ancient methods make much use of 'division/elimination with the quotient' (*shangchu* 商除).[66] Because it is difficult for beginners, it has been replaced by this method [the *jiugui* division/elimination method] which is not an orthodox technique (*fei zheng shu* 非正術).[67]

The ¡¡nine returns¿¿ (*Jiugui* 九歸) referred to a set of jingles corresponding to the number and position resulting from division of one position by a number n from 2 to 9. For division by n one needed to memorize n lines, altogether there were thus $2 + 3 + \ldots + 9 = 44$ jingles. Division by three, for example, requires to know the following three lines[68]:

三一三十一	Three-one: thirty-one.
三二六十二	Three-two: sixty-two.
逢三進一十	Coming upon three, advance ten.

Dividing one position by the number seven was summarized by the following seven jingles[69]:

七一下加三	Seven-one: add three to the next.
七二下加六	Seven-two: add six to the next.
七三四十二	Seven-three: forty-two.
七四五十五	Seven-four: fifty-five.
七五七十一	Seven-five: seventy-one.
七六八十四	Seven-six: eighty-four.
逢七進一十	Coming upon seven, advance ten.

The entire set of rules relied on the fact, that the numbers of the dividend and the quotient of the division would appear in the same line (or on the same abacus), the former shifted by one position to the left of the latter. This is, for example, the reason why dividing a certain decimal position of the dividend containing the number five, and being divided by seven, made 'seventy-one': seven on the decimal position to the left and the remainder one remaining on the position where originally was the five, since $50 = 7 \cdot 7 + 1$.

[66]Division by estimation of the quotient, originally using three lines on the calculation surface in counting rod calculation.

[67]Translated from Zhu (1839, 1127).

[68]See Cheng (1990, *juan* 2: 8b (78)).

[69]*Ibid. juan* 2: 9a (79).

Yang Hui, in his *Mathematical treasure of variations on multiplication and division* (*Chengchu tongbian suanbao* 乘除通變算寶) from 1275, noted that by understanding the principles of the method, one could significantly accelerate the process of rote memorization of the jingles[70]:

> For the study of the ¡¡nine returns¿¿ (*jiugui* 九歸), if one does not spend five to seven days on the memorization (*ji* 記) and the reading (*nian* 念) of the method, one will not be fluent. But nowadays, if one looks in detail at the commentaries to the *jiugui* problems and procedures in the *Detailed Explanations to the Nine Chapters of Mathematical Methods*,[71] one can easily fill the gaps where the intention [was not explicit] and thus memorize how to read the method and use the method in a single day!

7.4 Conclusion: Mathematical Verse in an Educational Context

The growth of mathematical introductory textbooks during the Ming-Qing period not only reflected the development of mathematics as an important tool for merchants, but it was also the result of a booming printing industry and an increased literacy beyond the scholarly elite. Instead of interpreting this trend as a popularization of mathematical knowledge, I have tried to link sources containing mathematical verse to a larger context of learning to read and write. As Elman points out,

> Repetition as a habit of learning based on reciting and copying was the key to developing the memory as a pedagogic tool to produce shared linguistic conventions by education.[72]

Re-placing versified mathematics into the larger context of acquiring literacy, including mathematical literacy or numeracy, can help elucidate some of the features of the types of verse I have identified in the present study. Reciting mathematical poetry on various aspects of mathematics could enrich a child's vocabulary and capacity to recognize characters. The category of jingles discussed in Sect. 7.3.1.10 nevertheless were an exception to this, since even the poorly literate could decipher the lines with knowledge of only the number words and a few more characters. Verses – with the exception of jingles – were adapted to the reading capacities of a larger audience on the basis of the available mathematical classics. But they were flawed by a mixture of simplified and variant writings of characters, and duplication of the same word in a single rhyme rarely occured. This implies that word recognition might have been built even into mathematical writings: less literate readers could read the texts if they had already heard, recited or chanted the verses or lyrics.[73]

---

[70]In Yang (1842, *juan* 1: 2a), translated into French in Friedsam (2003, 60).

[71]A book by the same author, Yang Hui, the *Xiangjie jiuzhang suanfa* 詳解九章算法, written in 1261.

[72]Elman (2001, 268).

[73]See also Lowry (2005, 93):

Generally, reading practices in China were and are still related to vocalization. As shown in Sect. 7.3.1.1, one type of mathematical verse also related to practices of chanting according to a certain tune and with a certain sequence of intonations. These tunes then became actors' categories for classification of mathematical problems. But chapters containing such versified mathematical writings were either not easily transmitted in textual form – as reflected in their neglected preservation and historiography – or they appeared with an unstable text in diverse Ming dynasty sources.[74]

My hypothesis is that a different process of learning was attached to jingles as opposed to other kinds of rhymes. Jingles gave sets of elementary arithmetic rules necessary in simple commercial transactions. The flourishing of these easy to memorize verses under a stable textual form over nearly a millennium until today, bears witness to the normative role of rote memorization in the transmission of mathematical knowledge in oral and written form. As the opening rhyme in the *Unified Lineage* (cited page 160) as well as Yang Hui's curriculum (translated on page 181) suggest, only jingles were the kind of verse one had to know by heart. They were the most elementary in mastering the basic arithmetic operations. Their stable textual version and survival beyond several changes in the calculation devices they were applied to might be related to the fact that these verses were thoroughly committed to the memory of a large audience. Unlike these elementary arithmetical jingles, which had a stable text through centuries, versified mathematical problems or algorithms were subject to change in wordings. Familiar tunes certainly helped in recalling and reciting mathematical rhymes whenever necessary, even if they were not memorized thoroughly. In particular, they served as mnemonic devices that were acquired as an aid in remembering how to perform an algorithm.

Versified mathematical problems were probably also part of oral question and answer sessions between students and a teacher, as was suggested in the illustration reproduced in Fig. 7.6.[75] A master and two pupils are depicted, one with an abacus and the other with what seems to be paper and a brush for written calculation. It was placed at the beginning of the last four scrolls on 'Difficult Problems' in the *Unified Lineage*, which contained only versified mathematical problems. Furthermore, the preface to this section of the book underlines that it was not only the solution procedure, but also the wording of the mathematical problems that was challenging. Therefore a master needs to give further explanations to the students. This supports not only my argument that texts analyzed here might have taught the knowledge of

The bowdlerized written forms suggest that, though perhaps not the most discerning, the reading public for such materials had sufficient familiarity with the verses (as well as recognizing a number of characters) to make sense of variant forms used by printers or copyists.

[74]Whether variations were bound to regional dialects as to maintain the rhyme patterns, still needs to be explored.

[75]Illustration from Cheng (1882), originally published in 1592.

Fig. 7.6 'Teacher and student discussing difficult [problems],' (*Shi sheng wen nan*) 師生問難 in Cheng Dawei's *Unified Lineage of Mathematical Methods*

words, although this was not their primary purpose, but also that orality played a crucial role in the process of transmission of mathematical knowledge in versified form.

Appendix: Chinese Dynasties After the First Unification

- Qin Dynasty 221 BCE – 206 BCE
- Han Dynasty 206 BCE – 220 AD
- Three Kingdoms – Period of Disunity 220 AD – 265 AD
- Jin Dynasty 265 AD – 420 AD
- Southern and Northern Dynasties 386 AD – 589 AD
- Sui Dynasty 581 AD – 618 AD
- Tang Dynasty 618 AD – 906 AD
- Five Dynasties Period 907 AD – 960 AD
- Song Dynasty 960 AD – 1279 AD
- Yuan Dynasty 1271 AD – 1368 AD
- Ming Dynasty 1368 AD – 1644 AD
- Manchu or Qing Dynasty 1644 AD – 1911 AD

References

Bréard, Andrea. 1999. *Re-Kreation eines mathematischen Konzeptes im chinesischen Diskurs: Reihen vom 1. bis zum 19. Jahrhundert*. Boethius, vol. 42. Stuttgart: Steiner Verlag.

Bréard, Andrea. 2010. Knowledge and practice of mathematics in late-ming daily-life encyclopedias. In *Looking at it from Asia. The processes that shaped the sources of history of science.* Boston studies of philosophy of science, ed. Florence Bretelle-Establet et al. Dordrecht: Kluwer. pp. 305–329.

Chemla, Karine. 1994. Similarities between Chinese and Arabic mathematical documents (I): Root extraction. *Arabic Sciences and Philosophy* 4: 207–266.

Chemla, Karine, and Guo, Shuchun. 2004. *Les Neuf Chapitres sur les procédures mathématiques*. Paris: Dunod.

Cheng, Dawei 程大位 (1533–1606). 1882. *Zengbu Suanfa tongzong daquan* 增補算法統宗大全 *(Complemented collection of the Unified Lineage of Mathematical Methods)*. 泰山堂 Taishantang edn.

Cheng, Dawei 程大位 (1533–1606). 1990. *Suanfa tongzong jiaoshi* 算法統宗校釋 (Annotated edition of the Unified Lineage of Mathematical Methods). Hefei: Anhui jiaoyu chubanshe 安徽教育出版社. Reprint of the 1716 edition.

Elman, Benjamin A. 2001. *A cultural history of civil examinations in late imperial China*, Reprint edn. Taibei: SMC Publishing Co. Originally published Berkeley: University of California Press, 2000.

Friedsam, Manfred. 2003. L'enseignement des mathématiques sous les Song et Yuan. In *Éducation et instruction en Chine, vol. II. Les formations spécialisées*, ed. C. Despeux, and C. Nguyen Tri, 49–68. Bibliothèque de l'INALCO, vol. 5. Paris – Louvain: Editions Peeters.

Giles, Herbert A. (trans.) 1900. *Elementary Chinese. San Tzu Ching* (三字經). Shanghai: Messrs. Kelly & Walsh. Reprinted in Peking, China 1940.

Guo, Shuchun 郭書春 et al. (eds). 1993. *Zhongguo kexue jishu dianji tonghui. Shuxue juan* 中國科學技術典籍通彙. 數學卷, 5 vols. Zhengzhou: Henan jiaoyu chubanshe 河南教育出版社.

Honda, Seiichi 本田精一. 1994. 宋元明代における児童算数教育("Children's mathematical education during the Song, Yuan and Ming dynasties"). *Kyūshū Daigaku Tōyōshi ronshū* 九州大東洋史論集 *(Oriental studies)*, 37–72.

Lam, Lay Yong, and Tian Se Ang. 2004. *Fleeting footsteps: Tracing the conception of arithmetic and algebra in ancient China*, Revised edn. Singapore: World Scientific.

Lee, H.C. Thomas. 2000. *Education in traditional China. A history*. Handbook of oriental studies, vol. 13. Leiden/Boston/Köln: Brill.

Leung, Angela Ki Che. 2003. Medical instruction and popularization in Ming-Qing China. *Late Imperial China* 24(1): 130–152.

Li, Yan 李嚴. 1954–1955. Mingdai de suanxue shuzhi 明代的算學書志 ("Bibliography of mathematical works of the Ming Dynasty"). In *Zhong suan shi luncong* 中算史論叢. (Collected essays on Chinese mathematics), vol. 2. 86–102. Beijing: Kexue chubanshe. Reprinted from *Tushuguan xue jikan* 圖書館學季刊 (Library science quarterly) 1.4 (1926), pp. 667–682.

Libbrecht, Ulrich. 1973. *Chinese mathematics in the thirteenth century. The Shu-shu chiu-chang of Ch'in Chiu-shao*. Cambridge: MIT.

Lowry, Kathryn A. 2005. *The Tapestry of popular songs in sixteenth- and seventeenth-century China: Reading, imitation, and desire*. Sinica Leidensia series, vol. 67. Leiden/New York: Brill.

Martzloff, Jean-Claude. 1997. *A history of Chinese mathematics*. New York: Springer.

McLaren, Anne E. 2005. Constructing new reading publics. In *Printing & book culture in late imperial China*, ed. Cynthia J. Brokaw and Kai-Wing Chow, 152–183. Berkeley/Los Angeles: University of California Press.

Qian, Baocong 錢寶琮. 1963. *Suanjing shi shu* 算經十書 (Ten Books of Mathematical Classics). Beijing: Zhonghua shuju 中華書局.

Sakade, Yoshinobu 出祥伸 Ogawa, Yōichi 小川陽一, and Sakai, Tadao 酒井忠夫[監修] (eds). 2000. *Chūgoku nichiyō ruisho shūsei* 中國日用類書集成. Tokyo: Kyūko shoin 汲古書院.

Siu, Man-Keung, and Alexei Volkov. 1999. Official curriculum in traditional Chinese mathematics: How did candidates pass the examinations? *Historia Scientiarum* 9(1): 87–99.

Volkov, Alexei. 2002. On the origins of the *Toan phap dai thanh* (great compendium of mathematical methods). In *From China to Paris: 2000 years transmission of mathematical ideas*. Boethius, vol. 46, ed. Yvonne Dold-Samplonius, 369–410. Stuttgrat: Franz Steiner Verlag.

Wang, Wensu 王文素. 1524. *Xinji tongzheng gujin suanxue baojian* 新集通証古今算學寶鑑 (Mathematical mirror of old and new mathematical learning, newly collected and corrected). Manuscript copy of 41 *juan* from Beijing Library. Reprint in (Guo et al. 1993, vol. 2: 337–971).

Wang, Ermin 王佴敏. 1994. *Zhongguo chuantong jisong zhi xue yu shiyun koujue* 中國傳統記誦之學與詩韻口訣 ("Chinese traditional recitation methods and arts of rhyming"). *Zhongyang yanjiuyuan Jindai shi yanjiusuo jikan* 中央研究院近代史央研究所集刊 (Journal of the Institute of Modern Chinese History at the Academia Sinica), 23: 31–64.

Wang, Wensu 王文素. 2008. *Suanxue baojian jiaozhu* 算學寶鑒校注 (Critical edition of the Precious mirror of mathematical learning). Beijing: Kexue chubanshe 科學出板社.

Wu, Jing 吳敬. fl. 1450. *Jiu zhang suanfa bilei daquan* 九章算法比類大全 (Great compendium of analogical categories to the Nine Chapters of Mathematical Methods). Reprinted in (Guo et al. 1993, vol. 2: 1–333).

Xu, Xinlu 徐心魯. 1573. *Panzhu suanfa* 盤珠算法 (Methods for Calculating on an Abacus). Xiong tainan 熊台南 edn. Complete title: 新刻訂正家傳秘訣盤珠算法士民利用 (*Newly engraved and corrected edition of the* Method of Calculating on an Abacus, *with verses for memorization transmitted secretly through family, for the use of scholars and commoners*. Preserved in Japan, Naikau bunko,子 56 函 5 號. Reprinted in (Guo et al. 1993, vol. 2: 1143–1164).

Yang, Hui 楊輝. 1275. *Xugu zhaiqi suanfa* 續古摘奇算法 (Selection of Strange Mathematical Methods in Continuation of Antiquity). Manuscript edn. Reprinted in (Guo et al. 1993, vol. 1: 1095–1117).

Yang, Hui 楊輝. 1842. Chengchu tongbian suanbao 乘除通變算寶 *("Mathematical treasure of variations on multiplication and division")*. In *Yang Hui suanfa* 楊輝算法, 宜稼堂叢書 yijiatang congshu edn.

Yu, Xiangdou 余象斗 (ed). 1599. *Santai wanyong zhengzong* 三台萬用正宗 (complete title: Santai's orthodox instructions for myriad uses for the convenient perusal of all the people in the world, newly engraved *Xinke tianxia simin bianlan Santai wanyong zhengzong* 新刻天下四民便覽三台萬用正宗). Housed in Tōkyō Daigaku Tōyō Bunka Kenkyūjo Niida Bunko 東京大學東洋文化研究所仁井田文庫 (Institute of Oriental Culture, University of Tokyo, Niida Collection). Reproduced in (Sakade et al. 2000, vols. 3–5).

Zhang, Zhigong 張志公. 1962. *Chuantong yuwen jiaoyu chutan* 傳統語文教育初探 (Preliminary inquiry into traditional language education). Shanghai: Shanghai jiaoyu chubanshe 上海教育出版社.

Zhu, Shijie 朱世傑. 1839. *Suanxue qimeng* 算蒙 (Introduction to Mathematical Learning). 揚州刻本 Yangzhou edn. Reprinted in (Guo et al. 1993, vol. 1: 1123–1200).

Zhou, Yongnian 周永年. 1895. *Xianzheng dushu jue* 先正讀書訣 (Recipes of Effective Reading by Ancient Wisdom). Guiyang: Yan Xiu 嚴修.

Zhou, Mi 周密. 1969. *Zhiyatang zachao* 志雅堂雜鈔 (Random Jottings from the Hall for Pursuit of Elegance). Biji xubian 筆記續編; 31. Taibei: Guangwen shuju 廣文書局.

Zhu, Shijie 朱世傑 and Guo, Shuchun 郭書春 (trans.) and Chen, Zaixin 陳在新 (Engl. trans.). 2006. *Siyuan yujian* 四元玉鑑 (Jade Mirror of Four Unknowns). Shenyang: Liaoning jiaoyu chubanshe 遼寧教育出版社.

Zhu, Shijie 朱世傑 and Luo, Shilin 羅士琳 (comm.). 1937. *Siyuan yujian xicao* 四元玉鑑 (Detailed Explanations of the Jade Mirror of Four Unknowns). Wanyou wenku 萬有文庫, vol. 2–700. Shanghai: Shangwu yinshuguan 商務印書館. Originally published 1836.

Chapter 8
Mathematical Progress or Mathematical Teaching? Bilingualism and Printing in European Renaissance Mathematics

Giovanna C. Cifoletti

In mathematical books of sixteenth-century Europe the question of teaching is present throughout. The very need for mathematical books is presented as the need of explaining mathematical contents effectively and of training the readers in mathematical techniques. The book should make understanding and learning mathematics more accessible: while for most books their printing was justified by being spiritually and morally edifying, for mathematical books the explicit purpose was to make mathematics clearer to follow and easier to absorb by a larger group of people. Mathematical books were intended to make the task of teaching lighter and even superfluous, providing the basis for self-teaching. These are the statements outlining the explicit principles of the actors. These statements open some questions as to the actual use of these books. The more basic question is which kind of teaching, preceptorial, private or public, was available at that time for these subjects, the following is in what teaching situation the books were present and used: here I shall work on the interplay between bilingualism and printing.

The examples will be some arithmetical and algebraic books which seem to give more direct information concerning the teaching situations. In particular I shall look at Gemma Frisius' *Arithmeticae practicae methodus facilis* (1540)[1] Oronce Finé's *La Théorique des Ciels, mouvements, et termes practiques des septs planètes* (1528), *Les Canons et documents très amples, touchants l'usage et practique des communs Almanachz que l'on nomme ephemerides. Briefve et isagogique introduction sur la judiciaire astrologie* (1543), and the posthumous *La composition et usage du*

[1] Frisius (1540).

G.C. Cifoletti (✉)
Directrice d'études à l'Ecole des Hautes Etudes en Sciences Sociales,
Centre Alexandre Koyré, 27, rue Damesme 75013, Paris
e-mail: cifolet@gmail.com

A. Bernard and C. Proust (eds.), *Scientific Sources and Teaching Contexts Throughout History: Problems and Perspectives*, Boston Studies in the Philosophy and History of Science 301, DOI 10.1007/978-94-007-5122-4_8,

quarré géométrique, par lequel on peut mesurer fidèlement toutes les longueurs, hauteurs et profunditez (1556). By Jacques Peletier du Mans we shall examine his *Aritmetique* (1549),[2] *L'Algebre* (1554)[3] and *De Occulta parte numerorum quae algebra et almucabala vulgo dicitur* (1560),[4] *De Usu geometriae* (1572) and *De l'usage de la Géométrie* (1573), finally Pedro Nunes' *Libro de Algebra* (1567).[5] Notice, however, that these books will be taken into consideration here in different ways: as examples of textbooks in public or in private teaching, as sources for facts concerning teaching and learning, as reading of sixteenth-century theories about mathematical teaching and learning. I hope this could be taken as a contribution to refine our future questions about the connection between scientific books and scientific teaching in early modern Europe.

8.1 Gemma Frisius' *Arithmeticae Practicae Methodus Facilis*. The Integration of Abacus Mathematics with *Algorismus* at the Colleges

Gemma Frisius (1508–1555) had an impact on the history of arithmetic and algebra because of his work on arithmetic: *Arithmeticae practicae methodus facilis*.[6] This book is in Latin and is explicitly addressed to college teaching. Textbooks for college teaching were traditionally published in Latin, so that, in general, the translation into Latin indicates, in France, the passage of a book to college teaching.

It has been the most popular Latin arithmetic of the time, with about 74 editions at the end of the sixteenth-century. At first, its format made it belong to the genre of the *algorismus,* though adapted to the taste of the day. *Algorismus,* or university practical arithmetic, was the medieval genre explaining the use of indo-Arabic numerals, mostly meant to teach elementary arithmetic (with indo-arabic numerals) as a basis for astronomy and cosmography. It was in Latin, the humanistic neo-Latin, namely a style that revived the models of classical Latin and addressed to college students as a prerequisite to university's teaching. It contained traditionally a definition of numbers and digits, the four operations with integers, the four operations with decimal fractions, the four operations with sexagesimal fractions. The last section was meant to instruct the students to use astronomical tables. This was crucial because future physicians could then determine horoscopes, whereas others could know simply measure time or the longitude. But Gemma Frisius casts the topic of arithmetic and its four operations in an entirely different framework.

---

[2]Peletier (1549).

[3]Peletier (1554).

[4]Peletier (1560).

[5]Nunez (1567).

[6]Frisius (1540) in this connection, see (Van Ortroy 1920; Echalier Serra 2011). On Gemma Frisius and his bibliography, see (Hallyn 2008).

At folio 5 we read:

> There are four sorts of arithmetic and by means of them all the rules and almost all questions are solved. We call species certain forms of operation by means of numbers: as in Dialectic the forms of the arguments are included in four species, i.e. syllogism, induction, enthymeme and example. The first of these is addition.[7]

We recognize here the reference to the teaching situation of the colleges, and even not to any kind of college, but to humanistic colleges[8]: one of the main features of humanistic education was the rhetorical reform of logic[9] which, precisely, stressed the role of *example* and *enthymeme*, which are the rhetorical versions of the logical notions induction and syllogism: the first indicates the instance, the case, whereas the latter indicates probable syllogisms. The book will be about the four operations, in analogy with the four sorts of reasoning, applied to different sorts of numbers. This presentation will be crowned by the rules based on the rule of three, presented as the theory of proportions applied to commerce.

This short book (69 folios in 8°) accomplished in a concise way what had been Pacioli's program: integration of abacus schools' arithmetic with university's *algorismus.*[10] Its language and structure makes it fit into the genre of *algorismi*, but it goes farther than contemporary books of the genre in the treatment of the rules of commerce and in the implicit use of algebra. In this sense it is a proper introduction of the mathematics of business in the context of university arithmetic.[11]

After the four operations on integers and geometric progressions, Frisius introduces "the rule of proportions, or of the three numbers". This is a very interesting mixture of abacus schools' mathematics of commercial rules with the theory of proportions. The third part, according to the genre of the *algorismi*, would have normally been about astronomical or sexagesimal fractions, the two adjectives being used as synonymous. Instead, it is about the rules of commerce: "De Regulis vulgaribus". Gemma Frisius deals with some variants of the rule of company, of the rule of alloys and then at length with the rule of false position. There are three sections corresponding to the three rules of the Thing or of algebra, which give

---

[7]Quatuor omnino sunt Arithmetices species, per quas omnes regulae quaestionesque omnes fere perficiuntur: vocamus autem species certas operandi per numeros formas, quemadmodum in dialectice argumentorum formae, quatuor comprehenduntur speciebus, syllogismo scilicet, inductione, enthimemate et exemplo. Prima harum est Additio. (Frisius 1540, f. 3v). All translations are mine.

[8]By this I mean the colleges developed following the pedagogical prescriptions present in new schools of late fifteenth and early sixteenth centuries Italy and Northern Europe. See (Grafton and Jardine 1987).

[9]I mean the transformation of logic which culminated with Petrus Ramus's *Dialectique.* See (Cifoletti 2004, 2006c).

[10]For these schools and for *algorismi* and *abacus* traditions, see the Chaps. 6 and 9 by S. Lamassé et E. Caianiello.

[11]The importance of this innovative feature is stressed in (Zemon Davis 1960). As to the difference between "abacus" and "abbacus", I take it as a variant without meaning, given the instability of Italian spelling: I therefore follow Caianiello not accepting Van Egmond convention.

three cases which in our terminology would be ascribed to equations up to the third degree, followed by the extraction of square and cubic roots.

Notice however that Gemma Frisius' book does not contain any algebraic notation or symbol for the unknown quantity, therefore we cannot talk about equations. Gemma Frisius explicitly says that it would be much easier to deal with these questions by algebra, but given that this is a whole topic of its own and that he hopes to write a treatise about it, he omits algebra for now. Yet, he calls 'rules of algebra' some applications of the rule of false. The rule of false was a procedure at least as old as Babylonian mathematics. The rule says: in a problem expressible in terms of one thing, let us insert an arbitrary value (the false position). Then the expression of the problem in that value will be false, but proportional to the expression of the problem in the value sought, so that we can apply the rule of three again.

The example given by Frisius here is the classical problem of three men, which in our notation would be a system of three equations, in which we use the method of substitution.

Having the algebraic technique in mind allows Gemma Frisius to introduce some algebraic arguments without an algebraic notation, that is, equations with unknowns. In fact, he worked on examples in which the application of the rule amounts to the extraction of the square root. This simple technique could make the readers of Frisius' arithmetic open to the use of solution formulas for algebraic equations. In fact, Michael Stifel in 1544 will take the same move. In folio 31 Frisius explains why he prefers to arrange all the matter by focusing on the *regula falsi*, without dealing with algebra proper. He declares that he is aware of the fact that algebra would provide a more certain and by far easier procedure, because nobody has seen a mathematical art more excellent and elegant. However, many have written about it[12] and he also plans to write an algebraic treatise himself.

Gemma Frisius' book continues with the section on the rule of algebra in fractions, which is relatively unusual. The fourth part is again on proportions, but here Frisius deals with mean proportionals. Gemma Frisius proposes a constant translation between the theory of proportions and commercial rules. What is most striking is that in the 1547 edition Frisius adds an absolute novelty: at this point he introduces a section on usury. It is noteworthy that this section was added by Gemma only in the 1547 Antwerp edition,[13] which was the fourth edition: this was after the publication of Peletier's Parisian edition. Frisius develops both simple and compound interest.

With respect to his notable omission, astronomical fractions, Frisius explicits his view point in the second edition, where he reintroduces astronomical fractions in an appendix. He argues that it is not difficult to deal with them and relegates the topic to an appendix: a significant change in the structure of the genre, but also in the conception of the book. Furthermore, it is not what one would expect now from an

---

[12]He only mentions Christoff Rudolff's *Die Coss,* published in Strasbourg in 1525.

[13]Frisius (1547).

astronomer of renown. Yet, the book was hugely successful, so apparently it did not miss its target. We have perhaps an indication of a reason for this in the words by which Frisius introduces his appendix:

> I don't see any important difficulty in physical or astronomical fractions, so that in order to make easier to young people the road to excellent disciplines, for which especially we meant to help the reader, I shall note some points which can appear difficult.[14]

Frisius suggests that of course astronomy and physics are the interesting applications of arithmetic, but once practical arithmetic in his new sense is absorbed, together with the rule of the Thing, sexagesimal fractions will be easily manageable.

This book is important in the history of mathematics not only because of its content, but also because it is an extraordinary bestselling textbook. Furthermore, the various editions included many corrections and additions. Special attention should be given, first of all, to the additions during Frisius life. So, if we take into consideration the editions until 1555, additions consist of references or of entire sections concerning extra topics. The references are to classical sources, such as Euclid's *Elements* and Ptolemy's *Almagest*. The main topic added is usury, included by Frisius himself in 1547. As to the later authors, the more substantial additions have been introduced by Peletier. Jacques Peletier du Mans (1517–1582) published an annotated edition of Frisius's text already in 1545.[15] His additions are *Compendium de fractionibus astronomicis*, devoted to sexagesimal fractions, traditionally included in the *algorismi*, and *De Cognoscendis per memoriam calendis*.

This change introduced by Peletier indicates that the book was used in the universities, where the new topics introduced by Frisius, as well as his style of presentation, were appreciated, but also where astronomical fractions were taught, so that a written presentation was needed. The other additions seem to be of interest for a larger audience, dealing with the basics for calendars and almanacs: *De cognoscendis per memoriam Calendis, Idibus, Nonis, Aureo numero, et loco Solis et Lunae in Zodiaco*. It contains the basic tools for Latin calendars and horoscopes. Peletier added also a short section on algebra: in fact he gives the proofs for the extraction of square and cubic roots, or rather transformed into demonstrations the intuitive arguments Gemma Frisius had provided for the algebra section, but again without an algebraic notation. This edition of Frisius' book was subsequently reprinted several times, and was in fact one of the great publishing successes of its time. The editor and translator of the Italian edition, the professor of rhetoric Orazio Toscanella, was particularly interested in reproducing Peletier's contribution on astronomical fractions, which he praised profusely in his introduction.[16] Pierre Forcadel, who became *lecteur royal ès mathématiques* in Paris, provided a French

---

[14]Frisius (1547, fol. 64r).

[15]Frisius (1545). This first edition published by Richard was followed by a 1563 edition published by Cavellat and a 1572 edition published by Marnef.

[16]See (Frisius 1567).

translation of Frisius's arithmetic: *L'Arithmétique de Gemma Phrison.*[17] It is important to notice that both French editors were former students of a *lecteur royal ès mathématiques*, Oronce Finé.[18]

In conclusion, Gemma Frisius managed to produce a compact textbook for colleges of Northern Europe. In terms of the mathematical content, it was sensitive to the new humanistic pedagogical reform, combining abacus mathematics with the arithmetic of *algorismi*, and doing this in a "short and clear" way, according to the rhetorical reform. Also, it was in Latin because it was addressed to colleges, but his language followed the rules for humanistic neo-Latin. His initial choice of inclusion of commercial rules was successful, and was soon confirmed by the inclusion of sensitive topics such as usury. However, Frisius' decision to understate the role of astronomical calculations was soon corrected: in following editions and Peletier's addition in this matter was highly appreciated.

8.2 Bilingualism and the French Vernacular Scientific Book Project: The Pedagogical Programs of Finé and Peletier

8.2.1 Finé

We have seen, so far, an example of innovation in genre coming in Latin scientific literature. In fact, two new genre of text appeared in this period, which can be seen as intermediate with respect to previous genres: some, like Frisius's *Arithmetica*, included "vernacular" material in a university textbook in Latin, others, like some new vernacular texts, gave a classical, literate shape to vernacular material. The latter genre took explicitly some distance from university teaching while shaping these books. However, if the book got some success, it was not unusual that the author himself translated these books into Latin in order to introduce the scientific content, but also the new presentation, in university teaching.

But things could also go the other way, so that university professor felt committed to publish in the vernacular, so again the same author published his book first in Latin and then in the vernacular. In fact many university professors were strongly engaged, at the time, in the promotion of scientific literature in the vernacular. First, this was connected to the general tendency in humanistic culture, the so called "*umanesimo volgare*". The larger program was to create a new encyclopedia in the vernacular. For mathematical sciences, this was also connected to the non-academic origin of the disciplines, often transmitted from Arabic through abacus schools.

---

[17]Published in Anvers by Jean Withage, in 1584.

[18]Oronce Finé (1494–1555) was a contemporary and fellow cosmographer of Gemma Frisius, teaching mathematics at the Collège Royal from 1530 to 1555.

There is also a more general aspect which should be taken into account: in fact, what we see is that both the authors and the readership of sixteenth-century books are mostly bilingual and often could have access to both versions of the books. The choice was mostly connected to a different *reading context*, so that paradoxically different readerships could consist partially of the same people. With this in mind, looking at the choice of language is still a strong indication of the teaching context.

Oronce Finé was the first in France to launch the program completed by the next generation of mathematicians, namely to publish in French a new genre of mathematical book, in a new style, the French vernacular scientific book.

This new genre took the form of craftman's manuals rather than that of textbooks connected to teaching practice, and was aligned with the tradition of arts and crafts in the vernacular. Manuals could coincide with textbooks, but were also conceived of as a distinct genre: by and large, manuals are connected to abacus schools and *Rechenmeister*'s shops, whereas textbooks are related to colleges. The French vernacular scientific book project is of interest to us precisely because the authors started publishing these books in the vernacular, as a record of the teaching which was going on in the vernacular teaching situations, whereas textbooks took as models the Latin medieval texts. Oronce Finé published three works in French. He picked his topics both from the Latin tradition among university disciplines and from the vernacular tradition outside the university. The purpose was for him, as for many later sixteenth-century French authors, to launch a new literature which followed new criteria, according to the taste of the day.

To the first category belong two books, published in 1528, was *La Théorique des Ciels, mouvements, et termes practiques des septs planètes*. It was a version of Peurbach's *Theorica Planetarum*, who had innovated in the genre[19] inaugurated by the medieval text attributed to Gerardus of Cremona, the Latin *Theorica planetarum*. It should be noted that Finé published a Latin translation of his own book, *Nova Universi Orbis Descriptio* in 1532, just after his appointment at the Collège Royal.[20]

To go back to *La Théorique des Ciels,* we should notice that it was an up-to-date astronomical treatise, and that it offers a good example of how a book could travel between Latin and the vernacular. First, the university book by Peurbach appears in Vienna. Then, Finé gives it a new form, a new presentation, in the vernacular, so that it is appreciated and promoted among cultivated vernacular readers through a vernacular treatise: it is discussed by preceptors and in academies; this makes plausible the introduction of the new presentation at the level of university colleges and then the book is translated into Latin.

But what this new presentation consisted of? These vernacular publications appear as works written in a cultivated style, in humanistic French and in a compact and stylized form. They were usually short and light compared to earlier treatises.

---

[19]On this genre, see especially (Byrne 2010).

[20]For which scientific curriculum was expected at the time, which new books appeared in the sixteenth-century and for which public, see (Pantin 1995).

They strove for saying the minimum necessary to apprehend a topic, by providing the basic vocabulary and the crucial problems.[21]

However, Finé (1543) published also some books in French concerning the disciplines which did not have a classical tradition or a history of university teaching. In 1543, Finé published *Les Canons et documents très amples, touchants l'usage et practique des communs Almanachz que l'on nomme ephemerides. Briefve et isagogique introduction sur la judiciaire astrologie*, reprinted in 1551, in 1556 and after his death in 1557. Almanacs have their own special history. Present throughout the history of mathematics, they connected astronomers to people of basic level of literacy, university teaching to peasants. In fact, it was among the most well spread genres, stable in spite of its transformations: in the late Middle Ages, these started to include the position of celestial bodies, as opposed to the astronomical tables present in Ptolemy's almanac. So, while the genre of the almanacs belonged in fact to university tradition, its Renaissance version had an existence outside the universities. Notice that it was also one of the genres profiting from the printing press: it is a fact that is made clear by the fact that Gutenberg published the first printed almanac in Mainz in 1457, 8 years before the Bible.

What was the readership for these texts? Immediately we think of people who were not necessarily fluent in Latin, such as the members of the court in general, but also of the court academies (the ones studies by Yates)[22]; we think also of the aristocracy in general, including women and of cultivated merchants and their children, sent to school but not to university. However, there were also students who could appreciate mathematical books in the vernacular.[23] Astrology, for instance, had a vernacular audience and we know that readership of these works included medical students.[24] Furthermore, some of the sixteenth-century 'academies', I mean here the military colleges outside the university where young aristocrats were sent they did not intend to prepare for the university and they devised a curriculum entirely in the vernacular.[25] In this connection, it should be noted that part of the readership for cosmographical books was constituted by women[26]: while mathematics in our restricted sense does not seem to count female professionals at the time, cosmography included some women, at least at the level of calculators and almanacs writers. This aspect is particularly important here because it points to the open question of what kind of consistent education in mathematical sciences was offered to women at the time. Generally speaking, it probably consisted of sharing a preceptor's teaching with male relatives. Notice that preceptors or private teachers should in fact be seen as the most common opportunity for formal education in mathematics in sixteenth-century France, both for male and female pupils. In these

[21]On this point, see for instance (Dubourg-Glatigny and Vérin 2008).

[22]See the classical (Yates 1988).

[23]Eisenstein (1979).

[24]About the vernacular audience for astrological publications see (Schechner 1997).

[25]See the classical study (Chartier et al. 1976).

[26]Pritchard (1990).

teaching situations vernacular training in the "new" mathematical disciplines could be combined with Latin and Greek training for the Classics and classical disciplines.

In 1556 appeared posthumously another vernacular work by Oronce Finé: *La composition et usage du quarré géométrique, par lequel on peut mesurer fidèlement toutes les longueurs, hauteurs et profunditez.* This work belongs to the genre of practical geometry, of the type concerning instruments. While practical geometry in general had a university tradition, this aspect of it had been transmitted outside university teaching. Again, Finé produced a combination of the two traditions, and chose the vernacular to publish it.

In fact, bilingualism was most common status among sixteenth-century intellectuals. But bilingualism of authors took, especially in France, the form of systematic self-translation. Language can often indicate to us which kind of teaching was the target of the book. Latin was meant in general for traditional schools, such as colleges and universities, while French is for abacus schools – but they were rare in France-, and preceptors or private teaching, a very usual institution for mathematical sciences, if we remember that private teaching in mathematics could take the form of apprenticeship at the mathematician's household. Given that the change in public implies also a change in the reading process, what is good for classes (a written encyclopaedic reference book for oral Latin, like Finé's *Prothomathesis*) is not necessarily good for self teaching (mostly vernacular).

8.2.2 Peletier

After Finé's first steps in vernacular scientific publishing, we find again one of his disciples at the Collège Royal in Paris, Jacques Peletier du Mans, as one of the main promoters of the vernacular scientific book project. Peletier began when he was 27, in 1541, by encouraging the use of French in literature through his translation of Horace: the theory of imitation presented therein contributed to the *manifesto* of the *Pléiade.*[27] As he explains in *L'Art Poétique d'Horace, traduit en vers françois*, if the goal is being creative in science, writing in French is the precondition for being creative; the means by which the writing of science in French may be promoted are various, but the most basic and fundamental is the development of a new orthography. In the letter to the reader, he explains that, "just as the word signifies the thought similarly orthography signifies the word."[28]

This approach is clearly presented in the introduction to the *Dialogue de l'Ortografe et Prononciation Françoese* (1550).[29] The *Dialogue* itself represents a debate between Peletier's friends, Jean Martin, Théodore de Bèze, Denis Sauvage

---

[27]Peletier (1541).

[28]'Tout ainsi que la parole est significative de la pensée, semblablement l'orthographe de la parole'. (Peletier 1541, fol.2).

[29]Peletier (1550).

and Jean Paul Dauron, at the house of the publisher Vascosan.[30] Only two points about the *Dialogue* need to be mentioned here. First of all, Peletier's insistence on the notion that language comes from the people, and is transmitted by contact with the people. Across time and space, only writing can provide a substitute for the normal transmission of language, which is oral. However, writing is more than a mere substitute for speech. In the *Dialogue* he explains that

> This is why <writing> should not so much depend on speech but on understanding, given that the plus which we get from writing is the understanding of sense.[31]

In this way, Peletier states clearly that writing has autonomy with respect to speaking: writing has, in fact, an advantage on speaking, the reader will focus on the meaning. Furthermore, writing does not convey pronunciation, and this is why the study and improvement of orthography is so important as a tool for bridging the gap between written and spoken language.

The second point is at a different level, and concerns more directly the project which Peletier was to pursue throughout his life: the writing of scientific books in French. Peletier writes:

> Our mathematics has never been doing as well as now, nor in a better position to be understood in its perfection. And given that its true is manifest, infallible and constant, think of what immortality it could give to a language if it were written in it according to a good and true method. Let us look even to the Arabs, who in spite of being very far from us and almost in another world, yet some Europeans have wanted to learn their language, especially because of Astrology, and other secret things they have dealt with in their vernacular. This is indeed quite unfortunate for we know what sophistry they have mixed with Medicine and even with Mathematics. Still, they made their language necessary for the study of all this. We should think why we don't do not only as much but without comparison more of our language?

Peletier indicates quite clearly that his plan is to impose French as the language of science or, to paraphrase him, to make that language necessary for the study of science ('rendre notre langue requise en contemplation des sciences'). Peletier took this program seriously.

As we have seen, he had already produced, five years before the publication of the *Dialogue*, his edition (with commentary) of Gemma Frisius' *Arithmeticae practicae methodus facilis* (1545)[32]: from this prestigious beginning in Latin with a work that was meant to be used in teaching at the colleges in France, Peletier moved on to his own mathematical works, which are always marked by his concern for a perfection of the form of presentation and by an explicitly reasoned rhetorical, logical and orthographic strategy.

---

[30] The dialogue was actually written in Lyon, after the religious wars and the fragmentation of his circle made Peletier du Mans leave Paris. See the classical (Zemon Davis 1964).

[31] 'Voela commant elle ne doet point ètre tant sugette a la prolation qu'a l'antandemant, vu que le plus que nous retirons de l'Ecritture cét l'intelligance du sans.' (Peletier 1550).

[32] Frisius (1545).

Peletier's program in French became more precise with the four following works: first his own book on arithmetic, *L'Aritmetique departie en quatre liures, à Theodore de Beze* (1549),[33] then *Le dialogue de l'Ortografe* (1550),[34] third, his algebraic work *L'Algèbre* (1554)[35] and lastly *L'Art poétique* (1555).[36] Peletier's publications in French after the beginning in Latin with Frisius' *Arithmetica* underline his own personal transition from colleges where the *Arithmetica* could be used to an aristocratic cultural "*élite*" less familiar with Latin, that is, circles connected to the Court, as well as preceptor's teaching.

After this sketch of the larger programme of publishing vernacular scientific books according to Jacques Peletier du Mans, let us see how he put it to practice in his mathematical works and in what teaching situations they could be used.

8.3 Jacques Peletier's Works in Mathematics. Bilingualism and Self-Teaching

In 1549 Peletier published his own book on arithmetic, *L'Aritmetique.*[37] Here the new language means that after introducing a new topic at the French colleges by his edition of Frisius' book, Peletier introduced it to the court by means of a newly conceived book of arithmetic written in French. Peletier is responsible for many innovations in this and other mathematical texts, starting with devoting separate manuals to restructuring the domains of practical and commercial arithmetic on the one hand and algebra on the other: this was new in France. Through changes in the rhetoric of the manuals Peletier made them more acceptable to the wide audience of the court and to the *noblesse de robe* connected with it. In fact, Peletier's arithmetic contains many more commercial rules than contemporary books, including Frisius', and the four introductions to the four books orient the cultivated reader towards a new view of mathematics in which this newly conceived practical arithmetic (not only astronomical, but containing abacus arithmetic and algebra) is presented as a crucial part of the *quadrivium*.

For Peletier, abacus school's arithmetic and algebra were legitimate fields of speculative knowledge, in fact true sciences. In this sense, aristocrats should consider them positively and even practice them insofar as they represent the beneficial part of commerce. In this he was ahead of his time, since a positive view of commerce and interest was not common before the time of Richelieu. Here it is important to stress that he consciously theorized the features of a scientific book. This, I argue, is an indication of Peletier's increasing interest in the process of

[33]Peletier (1549).

[34]Peletier (1550).

[35]Peletier (1554).

[36]Peletier (1555).

[37]Peletier (1549).

printing and the possibilities of dissemination offered by it. This is confirmed by the fact that, once again, while his works were being printed, Peletier lived at the house of the publisher, in this case the Marnef family in Poitiers, even though the book had been written in Paris. But, more importantly, the Marnef were particularly aware of the ongoing orthographic reform, and ready to be leaders in the field, even competing with the main Parisian publishers of scientific books, such as Cavellat and Wechel. With the *Aritmetique* they became the disseminators of Peletier's orthographical reform. In fact, the *Aritmetique* is the first book that might have satisfied Peletier regarding its fidelity to his orthography.

The four books of *L'Aritmetique* contain a substantial introduction to what can be called Peletier's thought. In fact each of the *prooemia* of the four books justifies the novelty introduced in the book. The first prologue justifies the publication of a newly conceived book and comes before a book on the theory of numbers. Already from the beginning, Peletier detaches himself from the style and also the matter of practical arithmetic university texts, oriented to classification and exhaustiveness. Peletier privileges concision and chooses his priorities. He gives only the indispensable definitions for operations and introduces already at this first level the simplest form of the rule of three. In other words, Peletier simplifies the matter, pursuing in the process initiated by Gemma Frisius. In order to justify these choices his first *prooemium* concerns the principle of authority, the *auctoritates*. Peletier wonders why great authors of a discipline follow each others in a time and place as if they engendered one another, while other times and places, in spite of the efforts of some authors, remain without development and creativity. Peletier mentions the golden age of Greek philosophy and opposes it to the dark centuries where neither Porphyrius nor Boethius succeeded to improve the conditions of civilization. The only exceptions were Ptolemy and Galen, whereas the majority, despite their good will, managed only to produce "barbarousness in letters, sophistry in philosophy, darkness in mathematics, quackery in medicine, hypocrisy everywhere".[38] But the earth untilled for a long time becomes fruitful again:

> Which time has ever been more flourishing in Philosophy, Poetry, Painting, Architecture and new inventions in all things necessary to human life than ours?[39]

This is the moment, for mathematical sciences, to raise again, because everywhere mathematics is on the crest of the wave, it is the time of illustrious men in Germany, Italy, Spain and even France can make a name for itself. It is up to the sovereigns to encourage it, or else risk humankind falling back into hypocrisy. This *prooemium*, under an ordinary appearance, introduces the principle that some contemporary scientists are to be considered on the same footing as the major

[38]Peletier (1549, f.iii^r).

[39]"Quel temps s'est-il jamais trouvé plus fleurissant en Philosophie, Poésie, Peinture, Architecture et inventions nouvelles de toutes choses nécessaires à la vie des hommes que le notre?", *L'Aritmetique*, Ibidem.

authorities of the classical past. This shows Peletier's consciousness about the value of new discoveries and also of the "new disciplines", that is, of Arabic mathematics reformulated and acknowledged in vernacular languages of the Latin West.

This attitude seems to depend on Oronce Finé's teaching and practice, Peletier's mathematics teacher. Finé will be explicitly cited later, in the third book, in connection with the mathematically crucial theme of the extraction of roots. The section bears the meaningful title of *About the Use of Roots: where incidentally some excellent Mathematicians of our time are mentioned.*[40]

Finé is mentioned together with algebra authors, after Christoph Rudolff, Michael Stifel, Luca Pacioli "to whom we owe the fact of having enriched us of their labours".[41] Peletier goes on: "Comme en toutes parties des Mathématiques fait très bien son devoir notre renommé Oronce Finé lecteur du Roi en l'Université de Paris". Clearly Peletier is very aware of the mathematical renaissance: it is obvious to him that his commitment to *humanitates* does not imply to neglect present time concerns, discoveries and priorities, nor to subordinate it to a supposed golden age in science. The stake for Peletier is to understand the sequence of contributions by mathematicians in the course of time, contributions which are compatible and cumulative. This is explained in the fourth *prooemium.* While Oronce Finé, in his oration on the importance of mathematics had maintained that without mathematics nothing would exist, neither divine nor human, that mathematics is constituted once and for all in God or together with the Creation, Peletier chose not to found the justification for mathematics on it, and opted to enter immediately into the flesh of human creation of mathematical disciplines.

The constitution of sciences and of scientific traditions is the theme of the fourth *prooemium*, in which Peletier asks the question of why "the practice and use" of arithmetic have not been written by the ancients mathematicians, the Greeks.[42] The answer given by some authors, says Peletier, is that the ancients had avoided to assume what is certainly a burden: to collect so many rules, and rules which belong to the "mechanical people".[43] Some others, Peletier adds, maintain that the Greeks were interested in spiritual things and neglected or ignored practice. To this opinion Peletier opposes a thesis on the origin of demonstrative mathematics:

> I shall never be of these. For it seems to me to be a great mistake to believe that Theorems have been so well presented and in such a good order as we have them from the authors, without finding their certainty and experience by calculation.[44]

---

[40]"De l'Usage des Racines: là où incidemment est faite mention d'aucuns excellents Mathématiciens de notre temps".folio 003F.

[41]"Auxquels tous nous sommes grandement redevables pour nous avoir enrichis de leurs labeurs".

[42]This question will be taken up again by meaningful authors such as Descartes in connection with Greek geometrical analysis: how is it possible that the Greek did not transmit to posterity the method of analysis if they obtained their results by means of it?

[43]"gens mécaniques".

[44]"Je ne serai jamais d'avec ceux-ci. Car l'erreur me semble grand, de croire que les Théorèmes aient été si bien rédigés et en si bon ordre comme nous les avons, par les auteurs, que premièrerment ils n'en eussent trouvé en ratiocinant la certitude et l'expérience."

The second book introduces fractions and its *prooemium* elaborates on the literary themes of *brevitas* and its costs at the level of clarity. The question is whether one of the two should be preferred to the expenses of the other. This is of interest to us especially because of the reference to printing[45]:

> Among erudite people, my friend De Bèze, has been debated for a long time and is still to be decided which is more profitable to pursuing arts and disciplines: that the teachers, when they write them, deal with them clearly and extensively, or obscurely and shortly. Those who are of the latter party say that difficulty causes that one reflects for a longer time, or reads more attentively what one desires to know: that by reading it and rereading it the readers grasps it with greater certainty, and one formulates oneself the special doubts and lively reasons belonging to the matter, and by this means one comes to the knowledge of the points which are not explained, and acquires a habit and a firm resolution of what ones looks for. This does not happen when things are easy: for human mind being always curious to know, and to go on further, after having listened a passage, throws itself on the other, without taking the time to engrave <to print> nor to incorporate into memory what one needs to remember. And they say that since printing has been invented one has not seen so many people of learning or so solid as it was done in the past, because since people have a quantity of books at leisure, they want to embrace not only many authors of one profession, but also many different professions. By charging their mind of so many things, this is the cause of the fact that they are forced to let go a great part of each on the road, and in fact they are not satisfied of any.[46]

I have commented this passage elsewhere,[47] given its relevance for the definition of the new topic, algebra, by its two main features: brevity and clearness. Let us just notice here that Peletier does not say himself that printing should be considered a real risk for the avid reader, but rather that this is what some contemporaries maintain. They are those who think that easiness and availability of texts are a paradox: instead of providing "food for thought" they would produce passivity and superficiality in learning. On the opposite, difficult texts would be a good thing, because they force to make a voluntary effort. Peletier insists that clarity, together with brevity, are the tasks of the true author. But it is interesting for us to see that printing was not necessarily seen as a progress, and that spreading of information was seen also as an incentive to superficiality. Peletier show here to be particularly conscious of the challenges and the risks of the various forms of communication, as in the *L'Art Poetique* he shows to be conscious of the challenges and risks of the various styles of communication. Quantitative increase of the access to information does not by itself insure an increase of knowledge. This point is fundamental for us, insofar as a teacher plays the role of a guide in the universe of culture and a guide who makes sure that knowledge is actually acquired.[48] In fact, we see appearing here a reflection on the practice of self-teaching, as printing was connected to such a degree of availability of books that the chain of transmission through teachers as a

[45]This can be read as a dialectical question, as I have done in (Cifoletti 2001).

[46]*Ibid.,* folio 27v.

[47]Cifoletti (2004).

[48]On the problem of mastering information and having guides through the universe of culture, see (Blair 2010).

guarantee of a good interpretation could not be assured anymore. Obviously this has to do with the Reformation and its trust for the reader of Scriptures. Now also the reader of mathematical texts should be trusted to understand and master the matter.

Also about fractions Peletier applies his theories to his mathematical text. After having maintained that to combine clarity and brevity is possible, Peletier authorizes himself to reduce even more drastically the theory of decimal and astronomical fractions. This is all the more noticeable, given that astronomical fractions is precisely the field of interest for Peletier, in which he distinguished himself with his *Annotationes* to Gemma Frisius' *Arithmeticae practicae methodus facilis.* As we have seen above, his *Annotationes* on this topic were particularly appreciated in the sixteenth-century.

The third book is about roots extraction and proportions, both fundamental operations in astronomy and commercial arithmetic. It is certainly a great achievement among mathematical books of the time. Peletier introduces the book by the discussion of vocabulary and apologizes for his Hellenisms. He does not hesitate to simplify a great amount of matter: he selects the two themes and exemplifies them thanks to his experience with Gemma Frisius's *Arithmetica*, where both roots extraction and proportions are developed. Faced with a typically Euclidean topic, Peletier had an enormous tradition before him. He chose to rely on his teacher Oronce Finé, and precisely on his *Protomathesis.* The most evident change is obviously the brevity of Peletier's text with respect to his predecessor. Peletier also achieved the clarity his *prooemium* to the second book had promised.

Finally, the fourth book is the one which goes deeper into commercial arithmetic, again resting on some aspects of Gemma Frisius' results, but also independently. All commercial rules are explained, not only the various cases of the rule of three, but also the rule of false position and double false position. Its *prooemium* speculates on the missing written tradition of practical (in the sense of commercial) arithmetic. Peletier speculates on the connection between the rule of false and the rule of three:

> The Rule of False, which the Arabs call rule Catain is called in this way because from a false supposed case it teaches how to find the true. And among all the vulgar rules it is the one which is more beautifully and largely used. It has two parts, one of a single false Position, the other of two. The Rule of False of one position operates almost in the same way as the rule of Three, except that in the rule of Three we have three known terms: here we only have one (I mean, entering in the operation) in analogy of which we form the other two, one by multiplying, the other by dividing.[49]

After the treatment of the rule of double false position, Peletier stresses the role of Gemma Frisius's combination of the rule of false and the rule of three and in fact

---

[49] *Ibid.* folio 86: "La Regle de Faux, que les Arabes appellent la Regle Catain, est ainsi ditte, parce que d'un cas faux presupposé, elle enseigne à trouver le vrai. Et est celle de toutes les Regles vulgaires, de laquelle l'usage est plus beau et plus ample. Elle a deux parties, l'une d'une seule Position fausse: l'autre de deux. La Regle de Faux d'une position a presque pareille operation a celle de la Regle de Trois, excette qu'en la regle de Trois nous avons trois termes cognuz: ici nous n'en avons aucun (i'entens qui viene en operation) à la semblance duquel nous en formons deux autres, l'un multipliant et l'autre divisant."

the rule of algebra as will be presented by Stifel, explicitly mentioned by him: "On the amplification of the rule of false through the extraction of roots, very ingenious invention of Gemma Frisius".[50]

History of inventors, brevity and clarity, theory and practice, mathematics and language, writing of practical arts: the *prooemia* are very cultured presentations. The ideas are clearly based on ancient Greek philosophy and could introduce to a university text traditionally in Latin. Instead, they are the philosophical justification of a mathematical book of a new sort.

Peletier realized his programme by publishing several mathematical works. We shall look more closely at the ones he chose to translate. *L'Algebre,* printed in 1554,[51] is among the mathematical texts by Peletier the richest in philosophical remarks on the role of the mathematical author, the importance of language, and about algebra as the art of thinking in mathematics.[52]

What matters here is that Peletier chose to start with a French algebra book, importing in the French vernacular the mathematical contents already expressed in Latin by Stifel and Cardan. The book is written in the elegant style of the vernacular scientific books we described earlier: short, clear, with classical references and some philosophy. In 1560 Peletier published a Latin version of his algebra.[53] How should we interpret this fact? We can at first notice that this version of the book has been used at the level of university colleges, at least at the Collège Royal, as suggested by the annotated copy belonging to the royal professor Henri de Monantheuil. There are no similar indications concerning the French edition. Other hints connecting this work to its use in colleges are in the very content of this version of the book. For, there are important differences between the French and the Latin Algebra. First of all, all the passages concerning the linguistic reform and the use of algebra as an art of thinking are omitted. This is compensated by an important dedicatory letter, containing a critical analysis of the current mentality, with special reference to its lack of honesty and understanding of the past. A plausible conjecture is that in the colleges the mathematical content should be clearly distinguished from other, extra-curricular concerns. This would also avoid disagreements between the author and the teachers which did not bear on the actual mathematical statements. The two books differed also on other points: in the French version the structure and most of the content is borrowed from Stifel's *Arithmetica Integra,* whereas the Latin version includes a concern for demonstration of the rules of solution, and in this connection Peletier gives references to the relevant propositions in Euclid's *Elements.* The importance of proofs and the references to classical author are indeed typical features of college textbooks of the time, in all mathematical disciplines.

Jacques Peletier du Mans published another mathematical book in Latin and French: this is the *De Usu Geometriae* (1572) and its French translation *De l'Usage*

---

[50]"De l'amplification de la regle de Faux, qui est par extraction des racines, invention tres ingenieuse de Gemme Frisien".

[51]Peletier (1554).

[52]I have written on these topics elsewhere: (Cifoletti 2006a).

[53]Peletier (1560).

de la Géométrie (1573). This text is emblematic of Peletier's style in mathematical books: short, clear, in small format (in 8°), with few engravings, in his typical, ennobled French. Peletier published this book while he was employed by the Duke Emmanuel Philibert of Savoy as a preceptor of his son Charles Emmanuel. Peletier had served Emmanuel Philibert since 1570 and participated in his reform of the army and the public administration in Piedmont and Savoy. It is probable that French had been the language of Peletier's private lessons to Charles Emmanuel. In this case it is likely that a canvass of the French text was ready while the Latin text was printed. The dedicatory letters give a hint of Peletier's plans. The first, to Marguerite of Valois and her husband Emmanuel Philibert indicates that the purpose of the print edition in Latin was to give this text a widespread use in educational institutions, such as those founded by Emmanuel Philibert. The dedicatory letter of the French edition is to another eminent nobleman interested in military reform, that is Albert de Gondi, comte de Retz, *gouverneur et lieutenant general pour sa majesté*. The purpose of the French edition was to pursue the vernacular scientific book project of Peletier's youth and to open Peletier his return to Paris after the years in Lyon and Savoy.

Finally, we can recall another important publication by Peletier: his version of the first six books of Euclid's *Elements*. It appeared first in Latin, in 1557, and then translated by Jean II de Tournes at the beginning of the seventeenth-century.[54] This format, up to book VI, was typical of college teaching of Euclid. It is not a surprise, therefore, to find it in Latin at first, while by the beginning of the seventeenth-century it is possible to find other instances of teaching of Euclid in the vernacular, in the military "academies" for young aristocrats. In conclusion, we see Jacques Peletier as a crucial figure in the making of a mathematical literature both in Latin and in French. The use of language is consistent with the choice of public and, aside from the instance of practical geometry, so is the managing of the matter in the books.

8.4 Nunes' *Libro de Algebra*: In Connection with the French Algebraic Tradition[55]

Another sixteenth-century cosmographer was engaged in the vernacular scientific book project: the Portuguese Pedro Nunes. Here we shall deal only with The *Libro*

[54] Peletier (1557, 1611).

[55] My research on Pedro Nunes started with the study of Guillaume Gosselin in 1986. I have expressed various aspects of my reading of Nunes' algebra in three talks: "Gosselin's use and critique of Nunes' Libro de Algebra", conference *"Iberia in the Golden Age"*, 31/3-3/4 1993, Imperial College, London, in "*Subtilissima arte*. Pedro Nunes et l'algèbre de la Renaissance", conference *Pedro Nunes and the Science of His Time* 8/11-10/11 2002, U. of Coimbra, as well as in a talk with the same title in Paris, at the seminar of the Centre d'histoire des sciences et des philosophies arabes et médiévales (Paris-Villejuif), mars 2003. My review of Martyn's book on an algebraic ms attributed to Nunes appeared in (Cifoletti 2003b).

de Algebra en Arithmetica y Geometria (1567),[56] which has been an important book in its time. Again, we see a book in the vernacular devoted to a topic coming from the abacus tradition: against the wide Latin production by Nunes, the other two vernacular works concern navigation. The book has retained the attention of the historians of mathematics only up to a certain point before its recent revival connected with the edition of Nunes's Complete Works.[57] Nunes' book does not contain either absolute novelties, as H. Bosmans noticed in his two articles already at the beginning of the twentieth century.[58] He also suggested that the value of the text lies rather in the generality of his approach. According to Bosmans, the most important aspect of this generality, capable of transforming the discipline of algebra, consisted in translating a vast collection of merchant's problems -which constituted a traditional corpus at least since Fibonacci' *Liber Abaci*- in a set of problems on numbers. Aside from this achievement, Nunes is credited of some solutions adopted by later algebraists, such as the determination of the common divisor of polynomials. This justifies at least in part the enthusiasm expressed by mathematicians of the time. Even before publication of the actual book in Antwerp in 1567, Pedro Nunes's algebra circulated as a manuscript, and was appreciated. In France, Jacques Peletier du Mans mentions the book as an important algebra which he has not read yet, and this already in his *L'Algèbre* of 1554 and *In Euclidis Elementa* of 1557.[59] After its publication, the book was rapidly used as a reference. Certainly, part of the reason for this was Nunes's celebrity as royal cosmographer and as an author of important cosmographical works. However, it is also clear that this book responded to a need. The question is to know which need justifies the success of this book, as in the case of Frisius' book. For instance, we have also some contemporary explicit references and commentaries about the *Libro de Algebra.* On this basis, I claim that what the French algebraists were interested in, and in fact most contemporaries, was a text of algebra in which algebra could acquire or *recover,* as they put it, its role as an art of thinking, as the art of discovery that the ancients had hidden in their brilliant mathematical constructions.

As I have shown[60] elsewhere, mathematicians had a larger public across Europe, namely the readers of arithmetic and algebra in the sixteenth-century, who often had at least some level of a humanistic culture in the vernacular and at best a juridical, medical or theological training. For such a humanist public, mathematics was the form of thought closest to dialectic. Nunes wanted to recover algebra's original "ordre et méthode" which made it an art of thinking. The structure he tried to give to the art of algebra derived from his interpretation of Euclid, and in fact he ascribed algebra to Euclid. This attitude came from the humanist idea that considered every

[56]Nunez (1567).

[57]See (Leitao 2010), in particular Vol. VI. *Libro de algebra en arithmetica y Geometria*, which gives a view on the recent literature on the topic.

[58]Bosmans (1907–1908, 1908).

[59]Peletier (1554, 1557).

[60]Cifoletti (2003a).

discovery or invention, in mathematics as well as in poetry, as founded on the imitation of the great classics, on a translation and understanding of their works.[61] It is up to us, now, to reconstruct the reconstruction of the classics proposed by Nunes, because it is at this level that his work was considered profoundly innovative by sixteenth-century readers.[62] Nunes, more than any of his contemporaries, takes up the challenge to provide justifications for the solution formulas of first and second degree equations. Nunes calls this part *doctrina* and puts it at the foundation of the very subtle art of algebra. This is the way to see why this book has been considered pedagogically successful, and for this we need to enter a little into his mathematics.

Nunes follows Pacioli's for the names of the 'powers of the unknown': *Numero* is called a quantity meant as composed of units, either a integer, or a fraction, or a root, even when it is surd [in our terminology: irrational]. *Cosa* is called the root of any square. *Censo* is called the square which is born from this root.

Nunes reminds us that, according to Aristotle, we get the number by the division of continuum and that number is a collection of units. We must therefore imagine to obtain number and units in lines, units in surfaces, units in bodies, by dividing them into parts. We could divide in equal or unequal parts, Nunes says, but mathematicians divide in equal parts in order to establish relations between the parts and the whole, which are also represented by a number. Mathematicians have the habit to divide the plane surface in equal and square parts, any of which is called unit, in conformity to the linear unit it has as side. In order to calculate the area and the volume we establish that the value of the figure is represented by the product of linear units contained, for instance, by the two sides of the triangle. By root or cosa we should understand the rectangle which has the side of the square as a basis and the unit as height. Nunes claims that in any root of the square there are as many square units that there are roots constituting the square. (As in the figure: Here the strip of three vertical shadowed squares constitutes one root.)

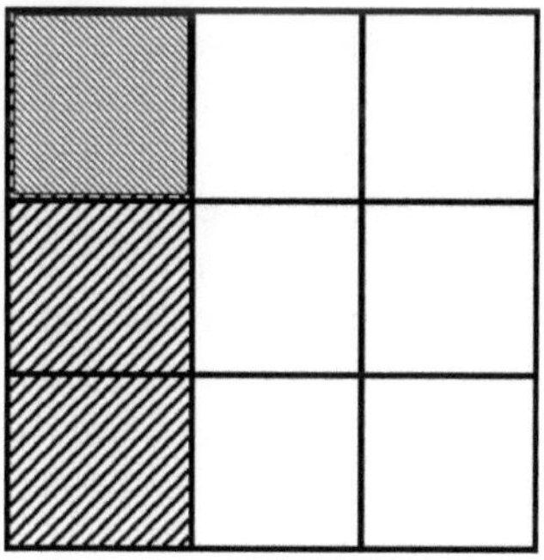

[61]Morse (1982).

[62]See (Cifoletti 2006a, b, c).

The number of square units contained in the square is therefore the value of the figure, whereas the value of the unknown is the number of linear units contained in the side of the square. As a consequence, we determine the solution of the equation, that is the value of the unknown, thanks to the equality posited, by counting the number of roots.

Nunes concludes by saying that many are mistaken by conceiving the root as a segment, and do not think that if the root was a straight line it could not be put as equal to a square.

Obviously Nunes is dealing here with the question of the homogeneity of the terms, but also of the correspondence, much more powerful than a simple illustration, that he wants to establish between geometrical quantities and monomials.

The reader is probably surprised by Nunes's statement, given our habit to see the root of a square precisely as a line, thanks to the introduction of the unit in the famous page of the *Géométrie* by Descartes. However, Nunes "intuitive" interpretation works.

Nunes states two rules:

1. There are as many units in any root, that there are roots in the square
2. The root of the square is a surface, but the side of the square is a line

Thus, the Root is different from the value of the unknown because Root is a geometrical entity, the stripe, while the value of the unknown is a number.

Note that if x^2 is censo, b the coefficient of the unknown is also number of units in the cosa, while x is also the number of the cosas (rule: number of units in cosa = number of cosa in censo).

In this framework, Nunes gives a proof a case of the solution formula for the second degree equation, based on the main references to Euclid.[63] What Nunes adds is the use of the *doctrina*, which establishes the correspondence between the two entities having a different geometrical meaning, but the same number: the number of square units in the root and the number of roots in the square. The value of the unknown is such number.

Nunes has attempted, with success, to advance in the geometrical interpretation of algebraic procedures. More than any author, Nunes succeeds to make explicit the sense of the algebraic operations. The geometrical sense of elementary operations has a heuristic role, but also what we would call a foundational role: at stake is the transition from the *ars rei et census* as it was presented in the abacus schools, to an art. Nunes stresses also that geometric progression, that is continuous proportion and the constant ratio between powers of the unknowns are the foundation of algebra. Nunes claims to have uncovered that *mathesis universalis* which is prior to arithmetic as well as to geometry, and that analytical doctrine which Euclid and Archimedes have used and hinted to without revealing it. Furthermore, that this art is in fact an art prior to arithmetic and to geometry guarantees that is has its place in the quadrivium but also that it can be legitimately used beyond practical calculation.

---

[63]I examine this point in detail, and compare it with Pacioli's treatment, in (Cifoletti forthcoming).

The reason why to start the book with the demonstrations of the solution formulas, then to structure it as an *algorismus,* and finally to get back to the theory of equations becomes clear: Nunes wanted to give algebra the foundations in the *doctrina.*

He actually gives a hint to this in the dedicatory letter to the King's son Henrique: the *doctrina de ciencia especulativa* is put at the very beginning because "a lot of time is usually spent in it, while in this way it can be learned in little time, easily and without the help of a teacher."[64]

In this sense, the theoretical summary at the beginning is especially important for some readers, that is for self-taught, and less for others, who could have a regular training in elementary arithmetic and geometry. Nunes should be acknowledged as the mathematician who actually connected directly the algebraic powers of the unknown to the notions given by Campanus and Zamberti in their interpretation of the arithmetical books of Euclid's *Elements* (founded on Theon of Alexandria) as well as to Jordanus' *Arithmetica* statements, in order to give foundation to algebra. Campanus and Zamberti had read the arithmetical books of the *Elements* (VII: VIII and IX) by associating them to the II and VI. This reading of Euclid became common, but few actually made this foundation explicit as Nunes did. This way is based on an interpretation of Euclid, each author proposing his own. Pacioli and Tartaglia had even provided a version of Euclid's *Elements*: Pacioli in Latin in 1505, based on Campanus's edition against the recent edition by Zamberti, whereas Tartaglia translated the *Elements* into Italian taking as basis both Campanus's and Zamberti's edition.

The dispute on Euclid was about main philological points aiming to establish the text of the *Elements*[65] but was also strongly philosophical insofar as it determined the interpretation of some fundamental aspects of the text: the connection between arithmetical and geometrical books, and most particularly the role of "geometric algebra", that is the theory presented in the books II and VI.[66] Pacioli's and Zamberti's dispute was overcome by the Lefèvre d'Etaples's edition (1516) and the Greek editio princeps together with Proclus's commentary (1533).

Nunes seems to echo what Tartaglia wrote in the frontispiece of his "Euclid"[67] promising:

> An extended presentation added again by the translator himself, so clear that any mediocre mind, without any notion oor help from any other science will be easily capable of understanding it.

---

[64]Nunez (1567, folio 3).

[65]The story of the contest between Pacioli and Zamberti has been described by Rose (1975) and more recently by Goulding (2010).

[66]I have sketched some of the philosophico-mathematical stakes of this debate in Cifoletti (2006c).

[67]"Con un'ampla espositione dello stesso tradottore di novo aggionta. Talmente chiara, che ogni mediocre ingegno, senza la notitia, over suffragio di alcun'altra scientia, con facilità sera capace a poterlo intendere." Tartaglia 1543.

8.4.1 The Dedicatory Letter

The dedicatory letter of the *Libro de Algebra* to his pupil the Infante Henrique, dated 1564, contains several indications as to Nunes' motivations and agenda and about the way he saw his work in the context of his time. It is the section in which Nunes gives more details about the social purpose of his work. Here I shall mention the themes we have already discussed for other authors, such as the relationship between theory and practice, the use of the vernacular in science (notice that the letter is in Portuguese, while the book is in Spanish), self-taught readership as well as Nunes' sources and the origins of algebra. Nunes writes:

> Algebra is an easy and short calculation to know the unknown quantity, in any problem of Arithmetic or Geometry, and in any other art making use of calculation and measure, such as Cosmography, Astrology, Architecture and Mercantile arts *(Mercantil)*[68]

Nunes states the possibility of using algebra in many mathematical sciences or rather in all those which are about calculation and measure: not only arithmetic and geometry, but also cosmography, astrology/astronomy,[69] architecture and the mercantile arts [Mercantil]. The potential universality of the use of algebra is a commonplace of all algebraic books of the time and Nunes includes, in fact, the mathematical sciences that he practices. Furthermore, Nunes adds here what can be considered an indication of the social role of algebraic books. He observes that in Italy there are many abacus schools, and in all towns there are teachers of counting in arithmetic and geometry who draw a salary form the town itself. But now also in the city of Lisbon mathematics is so wonderfully used in commerce and already financed by the King, given that 40 *contadores* are have a salary at the royal *fazenda* (*sua fazenda*). Therefore, the time has come, Nunes declares, that also in the city of Lisbon a book should appear without presupposing a doctrine of the speculative science, but so as to make it possible to learn it in little time, easily and without the need of a teacher.

Concerning the question of the use of the vernacular in science, he claims that he had written the majority of the text in Portuguese 30 years earlier. Jacques Peletier had already published this information in 1554 and in 1557.[70] However, this new version is, that the rest of the book is in Spanish in order to reach a wider public and because there are almost no books in Spanish devoted to Algebra.[71] He also mentions the conditions in which the book can be received: Nunes' remarks are very important as a witness of the situation of abacus schools discussed in the first chapter. It is evident from what Nunes says that in Portugal the abacus schools were not as present as they were in Italy. However there where 40 calculators paid to

[68]Nunez (1567, f.1).

[69]At the time astrology was computational astronomy and our astrology, whereas astronomy was natural philosophy, this is why I gave both terms in my paraphrasis: Nunes has *Astrologia*.

[70]In *L'Algèbre* and in *In Euclidis Elementa:* (Peletier 1554, 1557).

[71]Although, as to the presence of algebraic books in Spanish, it is surprising that Nunes does not take into account Marc Aurel *Libro primero de Arithmetica Algebratica.*

practise their art the court, which is a large number, and a new and growing public in the readers of vernacular books who had a sufficient general culture to learn a discipline without a master and wished to learn algebra outside the abacus schools. It was precisely this public who, according to Nunes, wanted the *doctrina speculativa* and convincing demonstrations in a compact form. Notice that Pacioli, one of the few authors explicitly mentioned and praised by Nunes, had already tried to found algebra on Euclid's *Elements*: in fact, he called *practica speculativa* that part of his treatise dealing with what we would call algebra and the demonstration of the solutions. Nunes sees this as a way to uncover the hidden practice of mathematics which Euclid and Archimedes presuppose.

Later we find a remarkable passage:

> How much better it would be if the Authors who wrote in the mathematical sciences had left their inventions written in the same way and with the same reasoning that they had used to get to them. And not, as Aristotle said in his *Mechanics* about artisans, who show only the external part of their machines and hide the *artificium* in order to be considered praiseworthy. In any art invention is very different from tradition and we should not think that all these propositions by Euclid and Archimedes were obtained in the same way that they used to transmit them.[72]

This explains the initial passage of the dedicatory letter, in which Nunes states that algebra is necessary for the practice of some books of Euclid and of Archimedes: for, here he says that algebra is the invention, whereas those books give us the tradition,[73] algebra gives the royal road or the *mathesis universalis* premises of arithmetic and geometry. Here Nunes deals with the *topos* of the hidden analysis of the Greeks, known from Peletier, Viète and Descartes among others: Nunes, not without irony, represents the great Greek mathematicians as highly specialised artisans, hiding their method of discovery in order to make their results look miraculous. This discussion on theory and practice could be incarnated in the contemporary time with the role of abacus schools versus humanistic college and university teaching of algebra.

8.4.2 *Motivation for the Foundations of Algebra: Nunes' Sources*

The background for Nunes' pedagogical foundations for algebra are his sources. As to contemporary authors, Nunes mentions from the beginning Luca Pacioli.[74]

[72]"O quan bueno fuera, si los Autores que escrivieron en las sciencias Mathematicas, nos dexaran escriptos los sus inventos por la misma via, y con los mismos discursos que hizieron hasta que pararon en ellos. Y no como Aristoteles dize en la Mechanica de los artifices, que nos muestran de la machina que tienen hecha lo de fuera, y esconden el artificio, por parescer admirabiles. Es la invencion muy differente de la tradicion en qualquier arte, ny penseis que aquellas tantas proposiciones de Euclides, y de Archimedes, furon todas halladas por la misma via que nos las han traido." (*Ibid*, f. 114).

[73]A different thesis is (Høyrup 1998).

[74]Henrique Leitao has developed the study of Nunes' sources, see (Leitao 2002, 31–58)

He says that Pacioli has published the first book on algebra. It is clearly the main source for Nunes in writing his own book. At the folio 323v we find the address of *The Author of this book to the readers*, which begins:

> I thought it a good thing to talk about the books of Algebra which arrived in Spain so far, so that you would read them with judgement and choose the most useful.[75]

In this connection Nunes develops also his criticism of Pacioli, explaining that Pacioli often uses the art without mentioning it, and when he does mention it he does so without getting to the point. This underlines once more that Nunes purpose in presenting the solutions of equations and their proofs at the very beginning of his treatise was to give a short and understandable foundation. A little later he repeats that Pacioli puts the conclusion at the beginning, which is not good for beginners. It is because of this great lack of pedagogy that Nunes has started to write the book and communicated it to some people.

Nunes also mentions a "summa" by Cardano, in emulation of Pacioli: here we should think of Cardano *Practica Arithmeticae* (1539). He comments that this book follows a certain order but it is not homogeneous. Furthermore, he writes, Cardano later edited a book on algebra – here we should think of the *Ars Magna* (1545) – which was completely without order (*que es un chaos*). Tartaglia's *Quesiti et Inventioni diverse* (1554) as well as his *General Trattato* (1557–60) are mentioned and praised for order, clarity and style. However, Nunes remarks:

> But it is not a selfcontained work [obra absoluta], because it refers the readers to to to others books of his, and presupposes some rules which he has not demonstrated and are not in the books of Euclid, from which all this doctrine comes.[76]

This is the clearest indication of Nunes purpose for his choice of starting his book with the doctrina.

In the course of the book Nunes mentions also Oronce Finé's *Arithmetica,* (by which he could have meant the 1532 *Protomathesis'* version or in the independent Parisian edition of 1542) and Jacques Peletier's *In Euclidis Elementa* (1557).[77] I have already recalled that Nunes does not mention Stifel's *Arithmetica Integra* 1544 and it is also noteworthy that Peletier *L'Algèbre* (1554) is not mentioned either.

He also mentions a few classical authors, and it is important to keep in mind that the editions of the classics occupied many mathematicians of the time and had a great influence on mathematical styles. First of all, of course, Euclid's *Elements*, for which Nunes most probably[78] had access to two editions: the first, based on Campanus' version, published by E. Ratdolt in Venice (1482), Zamberti's edition

---

75Nunez (1567, f. 323) I wish to thank Samuel Gessner for several discussions on Nunes' text and particularly on this section.

76Nunez (1567, f. 324).

77Peletier (1557).

78This is also the opinion of the editors of the volume (Leitao and Martins 2002).

of 1505 and also the edition including both Campanus and Zamberti's versions, published in Paris by Henri Etienne in 1516. Nunes explicitly mentions Archimedes, which he probably had access to in the Basel bilingual edition of 1544. The third mathematical authority mentioned is Jordanus Nemorarius, the thirteenth-century mathematician, for his *Arithmetica*. The edition Nunes could have access to is the 1514 Paris edition, edited by Jacques Lefèvre d'Etaples, the great humanist theologian who will edit two years later the edition of the *Elements*.

It goes without saying that there are many other contemporary sources for Nunes' work, even in the rather restricted circle of French humanistic mathematicians. Let us recall at least Elie Vinet, who was probably responsible for the contact between Peletier and Nunes.

Nunes also mentions some texts that are not exclusively mathematical, belonging to a larger humanist culture. It is important to keep them in mind not so much to show that he mastered such a culture, which is obvious, but to see how he considered them as pertinent sources in a book of algebra. The first is Marsilio Ficino's commentary on the Timaeus: we don't know whether Nunes had the original Florentine edition of the *Opera* (1491), the Venitian (1516) edition or the most recent and best selling edition in Basel (1561). The second is an encyclopaedia, Giorgio Valla's *De expetendis et fugiendis rebus*, edited in Venice in 1501. Ficino and Valla are the early masters of that neo-Platonic interpretation of Euclid which will take its final shape with Grynaeus' edition of Euclid's *Elements* and Proclus' Commentary.

In fact, the reading of Euclid's book II in terms of general algebraic quantity as proposed by Nunes seems be rooted in a neo-platonic and most particularly Proclus' reading. His *mathesis unversalis,* whose axioms are common to arithmetic and geometry, is, according to Nunes, nothing but algebra. In the use of Euclid's *Elements* in the book, Nunes' approach follows Campanus', but also Theon's reading proposed by Zamberti, with the goal of finding the correspondence between the theorems on segments of book II of the *Elements* and the theorems on numbers of books VII, VIII and IX. At the same time, Nunes could see, together with the French algebraists, the strong connection between this interpretation of Euclid and the work of Jacques Lefèvre d'Etaples and his programme on scientific manuscripts of late antiquity and with the related reform of logic.[79] Finally, Nunes could take advantage of Euclid's *Elements* in Grynaeus' editio princeps, which includes Proclus' commentary to the first book and the definition of *mathesis universalis*.

This had been already conveyed by Pacioli. But he lacked pedagogy. Therefore Nunes had been motivated to write. In fact, the reference to Pacioli is evident already in the title of Nunes' book. But also the title changes the balance of mathematical disciplines: Nunes wants to explain the meaning of algebra in arithmetic and in geometry, because he thought of algebra as a foundation to both, in the way of Proclus' *mathesis universalis*.

[79] See (Cifoletti 2006c).

8.5 Conclusion and Open Questions

A reference to teaching context is claimed by most sixteenth-century books of mathematical sciences. We have looked at some examples of texts, mostly belonging to the corpus of algebraic books of Renaissance Europe, and questioned the claim. For, it raises the question: which teaching and in what situation?

Teaching presupposes previous knowledge, wrote Aristotle and repeated Renaissance thinkers. But it presupposes also something new to be learned. The new could consist also, paradoxically, in the import of a classical text in the humanistic high culture: this could then imply a translation from Greek into Latin or a new better established Latin edition or even a translation from Latin into a vernacular language. An example of this is for instance the editions of Euclid's *Elements*, the various versions in Latin, culminating in Tartaglia's translation into Italian. A traditional and quite interesting way of answering the question would have been to study how sixteenth-century authors came to see ancient sources and interpreted them in or out a teaching context. The relation to ancient sources is the program underlying the Renaissance, there is a lot to be explored in this direction: but I hope the examples mentioned here can show to what extent sixteenth-century concerns and practices could influence the understanding of the past and of past sources. With this paper I insisted on another way to look at the connection between texts and teaching contexts in the Renaissance, that is looking at the vernacular scientific book project, bilingualism and self-translation as meaningful printing choices. For, another kind of "new knowledge" could consist of the import of a topic coming from abacus schools, when it reached the level of college or university teaching. This is the case of the various mathematical arts promoted by royal cosmographers such as Gemma Frisius and Oronce Fine. This is also the case of the algebras printed in the sixteenth-century, produced mostly by specialists trained at universities and working at universities. To what extent the content of these books was actually transmitted and learned by students is not easy to establish. However, the examples we have seen show an intentional effort to simplify the matter and to practice bilingualism. The authors we have seen are particularly careful about producing books which do not discourage the reader because of mathematical complexity: all the texts mentioned here were considered as self contained, not needing technical pre-requisites. The philosophical justification of the promotion of the vernacular was that creation could occur only in the mother tongue. While creativity could not be guaranteed, certainly the possibility of reading a book in both languages or even only in the vernacular led the reader to being autonomous, was the premise for successful self-teaching. Nunes follows this young printing tradition, but to the conscious pedagogical effort he adds his strong standpoint on cognition: mathematical knowledge was founded, and should be presented as founded, on algebra.

References

Blair, A. 2010. *Too much to know. Managing scholarly information before the modern age.* New Haven: Yale University Press.

Bosmans, H. 1907–1908. Sur le 'Libro de Algebra' de Pedro Nuñez. *Bibliotheca Mathematica* 8(1907–1908): 154–169.

Bosmans, H. 1908. L'Algèbre de Pedro Nuñez. *Annaes Scientificos da Academia Polytechnica do Porto* 3: 222–271.

Byrne, J.S. 2010. The mean distances of the Sun and commentaries on the Theorica planetarum. *Journal for the History of Astronomy* 42(2): 205–222.

Chartier, R., M.M. Compère, and D. Julia. 1976. *L'Éducation en France du XVIe au XVIIIe siècle.* Paris: SEDES-CDU.

Cifoletti, G.C. 2001. Matematica e fatti scientifici: polarità e simmetrie. Il ruolo degli algebristi francesi del Cinquecento nella costruzione dei fatti scientifici. *Quaderni Storici* 36(3): 771–798.

Cifoletti, G.C. 2003a. Mathematics and rhetoric. Princeton Univ.: Princeton dissertation, U.M.I.

Cifoletti, G.C. 2003b. Review of John R.C. Martyn (Editor and Translator). Pedro Nunes (1502–1578): His lost algebra and other discoveries. *Isis* 94(2): 369–371.

Cifoletti, G.C. 2004. The algebraic art of discourse. Algebraic Dispositio, invention and limitation in sixteenth-century France. In *History of science, history of text*, Boston studies in the philosophy of science, ed. K. Chemla, 123–135. Boston: Kluwer Academic Publishers/Springer/Collection.

Cifoletti, G.C. 2006a. *The art of thinking mathematically*, Special issue of early science and medicine, vol. 11, 4. Leiden: Brill.

Cifoletti, G.C. 2006b. Mathematics and rhetoric. Introduction. In (Cifoletti 2006, 369–389).

Cifoletti, G.C. 2006c. From Valla to Viète: The rhetorical reform of logic and its use in early modern algebra. In (Cifoletti 2006, 390–423).

Cifoletti, G.C. Forthcoming. *A mathematics of the human world.* To appear at Springer.

Dubourg-Glatigny, P., and H. Vérin (eds.). 2008. *Réduire en art, la technologie de la Renaissance aux Lumières.* Paris: Editions de la Maison des sciences de l'homme.

Echalier Serra, C. 2011. *L'Arithmétique au XVIe siècle: Enquête menée à partir de l'Arithmeticae practicae methodus facilis de Gemma Frisius (1540).* Memoir for "diplôme de l'EHESS", directed by G.C. Cifoletti.

Eisenstein, E. 1979. *The printing press as an agent of change: Communication and cultural transformation in early modern Europe.* Cambridge: Cambridge University Press.

Finé, Oronce. 1528. *La Théorique des Ciels, mouvements, et termes practiques des septs planètes.* Paris: S. du Bois.

Finé, Oronce. 1543. *Les Canons et documents très amples, touchants l'usage et practique des communs Almanachz que l'on nomme ephemerides. Briefve et isagogique introduction sur la judiciaire astrologie.* Paris: S. de Colines.

Finé, Oronce. 1556. *La composition et usage du quarré géométrique, par lequel on peut mesurer fidèlement toutes les longueurs, hauteurs et profunditez.* Paris: G. Gourbin.

Frisius, G. 1540. *Arithmeticae practicae methodus facilis.* Antwerp: G. Bonte.

Frisius, G. 1545. *Arithmeticae practicae methodus facilis. Cum Jacobi Peletarii Cenomani Annotationibus.* Paris: G. Richard.

Frisius, G. 1547. *Arithmeticae Practicae Methodus Facilis.* Antwerp: G. Bonte.

Frisius, G. 1567. *Aritmetica prattica facilissima ... Con l'aggiunta dell'abbreviamento de i rotti astronomici di G. Pelletario.* Trans. O. Toscanella. Venice: Giovanni Bariletto.

Goulding, R. 2010. *Defending Hypathia Ramus, Savile and the Renaissance discovery of the history of mathematical history.* Dordrecht: Springer.

Grafton, A., and L. Jardine. 1987. *From humanism to the humanities: The institutionalizing of the liberal arts in fifteenth- and sixteenth Europe*. Cambridge, MA: Harvard University Press.
Hallyn, F. 2008. *Gemma Frisius, Arpenteur de la terre et du ciel*. Paris: Champion.
Høyrup, J. 1998. A new art in ancient clothes. *Physis* 25: 11–50.
Leitão, H. 2002. Pedro Nunes, leitor de textos antigos e modernos. In *Pedro Nunes e Damião de Góis. Dois rostos do humanismo português,* Coord. Aires A. Nascimento, 31–58. Lisboa: Guimarães Editores.
Leitão, H. (ed.). 2010. *Obras de Pedro Nunes*. Lisboa: Academia das Ciências de Lisboa.
Leitão, H., and L. Azevedo Martins. 2002. *Pedro Nunes, 1502–1578: Novas terras, novos mares e o que mays he: novo ceo e novas estrellas.* Catálogo da Exposição Bibliográfica sobre Pedro Nunes. Comissário científico: Henrique de Sousa Leitão; coordenação técnica: Lígia de Azevedo Martins. Lisboa: Biblioteca Nacional.
Morse, J. 1982. The reception of Diophantus. Princeton Univ.: Princeton dissertation, U.M.I.
Nunez, P. 1567. *Libro de Algebra en Arithmetica y Geometria*. Antwerp: Heirs of A. Birkman.
Pantin, I. 1995. *La poésie du ciel en France*. Geneva: Droz.
Peletier du Mans, J. 1541. *L'Art Poetique d'Horace traduit en vers françois*. Paris: M. de Vascosan.
Peletier du Mans, J. 1549. *L'Aritmetique*. Poitiers: Marnef.
Peletier du Mans, J. 1550. *Dialogue de l'Ortografe et Prononciation Françoese*. Poitiers: Jean & Enguilbert de Marnef.
Peletier du Mans, J. 1554. *L'Algèbre*. Lyon: J. de Tournes.
Peletier du Mans, J. 1555. *L'Art Poetique*. Lyon: J. de Tournes.
Peletier du Mans, J. 1557. *In Euclidis elementa geometrica demonstrationum libri sex*. Lyon: J. De Tournes et G. Gazeau.
Peletier du Mans, J. 1560. *De Occulta parte numerorum, quae algebra et almucabala vulgo dicitur*. Paris: G. Cavellat.
Peletier du Mans, J. 1572. *De Usu geometriae*. Paris: G. Gourbin.
Peletier du Mans, J. 1573. *De l'usage de la Géométrie*. Paris: G. Gourbin.
Peletier du Mans, J. 1611. *Les six premiers livres des Éléments géométriques d'Euclide*, avec les démonstrations de Jaques Peletier, du Mans, traduicts en françois par Jean II de Tournes. Lyon: J. de Tournes.
Pritchard, R.E. 1990. The sixteenth-century: Excluding drama after 1550. *The Year's Work in English Studies* 70: 252–265.
Rose, P.L. 1975. *The Italian renaissance of mathematics. Studies on humanists and mathematicians from Petrarch to Galileo*. Geneva: Droz.
Schechner, S.J. 1997. *Comets, popular culture and the birth of modern cosmology*. Princeton: Princeton University Press.
Tartaglia, N. 1543. *Euclide Megarense philosopho Solo introduttore delle scientie mathematice*. Venise: V. Rofinelli.
Van Ortroy, F. (1920) 1960. *Bio-bibliographie de Gemma Frisius*. Amsterdam: Meridian.
Yates, F. 1988. *The French academies of the sixteenth century*, 1st publ. 1947. London: Routledge.
Zemon Davis, N. 1960. Sixteenth-century French arithmetics on the business life. *Journal of the History of Ideas* 21: 1.
Zemon Davis, N. 1964. Peletier and Beza part company. *Studies in the Renaissance* 11: 188–222.

Part IV
Zooming Approach

Chapter 9
Leonardo of Pisa and the *Liber Abaci*. Biographical Elements and the Project of the Work

Eva Caianiello

9.1 Introduction

Reconstructing the educational journey taken by Leonardo of Pisa, also known as Fibonacci, is no easy matter. Fibonacci received his scientific training in North Africa, as well as during his working visits to the eastern part of the Muslim world and Byzantium. He thus had the opportunity to draw directly on several Arab and Greek sources, which were not yet in the possession of his compatriots and over whom he therefore had an undoubted advantage.[1] The question of sources has, however, not yet been clarified and it would be very difficult to retrace his bibliographic steps with any degree of certainty: we do not have enough direct information on the works[2] (written in Latin) that Fibonacci could have studied and which subsequently inspired him in writing his own works. For the Arab sources, while the borrowings from al-Khwārizmī, Abū Kāmil and the tradition

The *Liber Abaci* (thereafter *L. A.*) is currently the subject of a digital critical edition, a translation into Italian and a historical commentary being undertaken by the "Project Fibonacci" team at the University of Naples Federico II, of which I am a member. The project and the critical edition of the text of the *Dedicatory Letter* to Michel Scott and of the *Prologue*, which make up the *Preface* to the *Liber Abaci* (ed. 1228), are presented in (Burattini et al. 2012). As a result, below, for that which relates to the *Dedicatory Letter* and the *Prologue*, I have used my French translation of the text established by the team (see annex), while, for that which relates to the body of the text of the *Liber Abaci*, I have used (Boncompagni 1857–1862, vol. I). There is an English translation of the *Liber Abaci* by L.E. Sigler in (Sigler 2002), but this has not been used here.

[1] Sesiano (1999, 99).

[2] For the available sources expressly cited by Leonardo cf. (Folkerts 2004).

E. Caianiello (✉)
Ecole des Hautes Etudes en Sciences Sociales (EHESS), Paris, France
e-mail: eva.caianiello@gmail.com

A. Bernard and C. Proust (eds.), *Scientific Sources and Teaching Contexts Throughout History: Problems and Perspectives*, Boston Studies in the Philosophy and History of Science 301, DOI 10.1007/978-94-007-5122-4_9,

of commercial arithmetic (*Mu'āmalāt)* of the Muslim East are incontestable, the influence of al-Karajī is, on the other hand, highly controversial.[3] Another difficulty in reconstructing his educational journey lies in the fact that the Pisan's life was split between two different worlds: for one part the commercial milieu he was born into and for the other, the most highly refined intellectual milieu of his time, that of the court of Frederick II, to which he was linked through close friendship with several prominent members of the court with whom he shared deep mutual respect.

Moreover, it would be impossible to understand the contribution the Pisan made to accounting and commercial arithmetic without invoking the influence that the socio-economic revolution, which took place at the turn of the thirteenth century, had on his education, with the passage from a feudal economy based on local retail trade to an economy based on credit, wholesale and overseas trade.

Similarly, if we try to reconstruct the teaching context at the time of the Pisan, in the sense of the transmission of knowledge from a scholar to a disciple in a school, we come up against the not insignificant obstacle of chronology. There is, in effect, a hiatus between the fairly broad knowledge we have of the Italian abacus schools from the second half of the thirteenth century, and the ignorance we have of Leonardo's "direct" educational context, that is, the type of school where he could have been taught and possibly later where he employed his own pedagogical skills. Indeed, he died post 1241, while the first known 'abacus master' is a certain Peter of Bologna who appeared as a witness in a deed of sale[4] in 1265 and the first abacus school to be established by the municipality of San Gemignano did not appear until 1276.[5]

Should we be put off by these difficulties? Obviously not. Indeed, Fibonacci seems to have been well recognized during his lifetime for his pedagogical abilities. The Pisan is referred to as *magister* in an edict[6] published by the municipality of Pisa in 1241 where the authorities recognized his doctrine and his skills, while congratulating him for his work as a consultant. We should remember that in the Middle Ages (twelfth to fourteenth centuries) the title *magister* was used for any person of noble disposition, who, after having studied, dedicated himself

---

[3]The borrowing from al-Karajī is inferred by (Woepcke 1853/1982) by the fact that in al-Karajī's *al-Fakhrī* there is a group of 2nd degree algebraic problems which are identical or quite similar to a certain number of problems in the 3rd section of chap. XV of the *L. A.* The same problems are found, however, in the Latin version of Abū Kāmil's Book of Algebra cf. (Sesiano 1993, 360–419). See also (Bortolotti 1929–1930; Levey 1966; Miura 1981; Rashed 2003, 55–57).

[4]Zaccagnini (1924, 283–284).

[5]Ulivi (2000, 87).

[6]Reproduced at the bottom of the page in the *Constitutum Pisanum legis et usus* and after the revision of 1233, among the additions made in 1241 under the podesta Ugolino (son) of Ugone Rossi, it was found by Bonaini (1857). See also (Maccagni 1988, 92).

generally to teaching; an intellectual committed to research and the transmission of knowledge, "...a master in the epoch of the time".[7] His having 'pedagogical experience', even if we do not know the details, is hardly in doubt. Other elements, taken from the *Preface* to the *L. A.* and even the structure of the work itself, can be compared to what we know later of the structure of teaching in the abacus schools. For example, if one studies the structure of the chapters making up the *L. A.* and then compare this with the structure and the teaching programs in some of the abacus schools,[8] one is struck by the resemblance of the structure of the content, which reveals indirectly the thread that unites Fibonacci to the teaching in the future abacus schools. The most widely aired opinion is that the *L. A.* was the precursor of most of the subsequent abacus textbooks. Jens Høyrup[9] offers a reverse perspective which is the subject of complex debate, which I will not discuss here.[10]

In the historical reconstruction we offer here, the presence of "voluntary testimony" (the autobiographical note in the *Prologue* to the *L. A.* and the structure of the *L. A.* itself) and "involuntary testimony",[11] such as the deliberation from the municipality of Pisa and other testimonies from his time, give us reason not to doubt the information we have invoking Leonardo's training in mathematics and his qualities as a teacher. Starting from this observation, our methodological approach will therefore attempt to overcome this lack of objective information by discovering all we can from a body of relevant clues, which are the milieu in which Fibonacci grew up and where he started out from. We shall draw here from a wide circle (and willingly widened) of *possible* elements for comparison, namely those which could better shed light on the very facture of Fibonacci's treatise, focusing particularly on its *Preface*. Any knowledge of past human acts is knowledge acquired from 'traces',[12] which become significant for those who know how to question them. For this reason that we will go on to draw from a later 'global context', even much later than the period under consideration, in our attempt to shed light on the object of our analysis. To do this, we will employ a method known as "comparison" which Henry Pirenne[13] summarizes in the pithy phrase: "If the paucity of our sources does not bring us conclusive proofs, probability will have to suffice for our edification". Historian Antonella Tarpino, in her *Postface* to the Italian edition of Pirenne's work,

[7]Fumagalli Beonio Brocchieri (1987, 204–205). However, in the vernacular (i.e. Italian), of the Italian municipalities, the terms "*maestro d'arte*" or "*maestro di bottega*" are used for a master craftsman and the place of learning is the craftsman's workshop.

[8]Arrighi (1968), Franci (2000), and Goldthwaite (1972, 420).

[9]Cf. Høyrup (2000, 2001, 2005, 2006, 2007, 30–44).

[10]See (Van Egmond 2009; Høyrup 2009a, b; Oaks 2009, 2010; Franci 2010; Folkerts 2011). Heeffer (2007, 6) appears to accept nearly all of Høyrup's theses.

[11]Following the terminology used by Marc Bloch (2009, 48–54).

[12]Bloch (2009, 44–45).

[13]Pirenne (1929, 14–15).

writes on this method the following: "Applied to the analysis of historical facts, 'comparison' can bridge space and time by highlighting the shifts in plans: the continuity and the breaks, the repetitions and the sequences ... ".[14]

This method should not be adopted as absolute nor theorized on as a mathematical postulate would be, but used with caution, assessing the conditions for its application case-by-case and as a support to research, where it is particularly useful when traces are inadequate, as in the case for certain periods of history.[15]

So, taking into account the areas of light and shade that have seeped into the Pisan mathematician's biography, and in order to progress in our investigation of the man in question and the objectives he pursues in the *L. A.*, I will divide this article into two parts. The first part offers some "biographical details" for Fibonacci which draws on the autobiographical passage in the *Prologue* – and at the same time sheds light on it – where Fibonacci describes his training in Béjaïa (lat. *Bugia*) as well as his travels within the Mediterranean basin. The content of this "extensive biographical investigation" is clearly an indispensable element in understanding the complexity of the "educational project" which structures Fibonacci's sophisticated *Preface*. In the second part, from some aspects of the *Dedicatory Letter* and the *Prologue*, we will try to explain the project and the work's meaning for Fibonacci, in the light of the biographical detail identified in the first part.

9.2 The Elements Which Allow the Reconstruction of Fibonacci's Life and Educational Journey

9.2.1 Documentary Sources that Directly or Indirectly Attest Fibonacci's Biography

The Rarity of Sources Directly Relative to Fibonacci's Life

With regard to biographical data, which are very scarce, we have the following documents:

1. the autobiographical note contained in the *Prologue* to the second edition of the *L. A.*
2. a deliberation by the municipality of Pisa dated 1241 which states: "In consideration of the honor and the advancement the city has received thanks to the teaching (*per doctrinam*)[16] and the unhesitating compliance of the perspicacious

[14] Pirenne (1985, 324). From the author's own translation.

[15] A personal communication (November, 04, 2011) from historian Marco Meriggi (University of Naples).

[16] The Latin word *doctrina* can also mean "knowledge, erudition, culture".

and learned master Leonardo Bigollo for his calculation of estimates and budgets (*abbacandis estimationibus et rationibus*)[17] we assign the annual sum of 20 lb in denari in recompense for the work he carries out as auditor in the process of consolidating the aforementioned estimates and budgets".[18] He was, therefore, undoubtedly an expert mathematician, a consultant to the officials of the Municipality – auditor of Pisa's "Libro di ragioni", according to Federigo Melis[19] and possibly also a teacher (magister). The epithet 'Bigollo', which appears not only in this document but also in the incipit of some of his other works means 'traveler' and not "fool" or "imbecile" as was believed for a long time.[20]

3. a notarial deed from 1226,[21] which gives information about his family, whose presence is attested in Pisa from the end of the eleventh century, as can be seen in a deed of sale dated January 11th, 1109.[22] Thanks to this notarial deed we know that his family name was Bonaccius, that his father was called Guglielmo and not Bonaccius, which was the name of his grandfather, and that his brother was called Bonaccingo.

Not much can be deduced from this thin documentary base, other than that Leonardo of Pisa, also known as Leonardo Fibonacci or simply Fibonacci, belonged to the merchant elite of his native city, Pisa, where he was born between 1170 and 1180. The exact dates of both his birth and his death are unknown and the above dates have been established on the basis of internal references in the texts of his works. Thus, after 1228, the year the revised edition of the *L. A.* was published, we know nothing more of the Pisan, except for the aforementioned deliberation by the municipality of Pisa in 1241, which shows that he was alive. It is not known whether he died in Pisa or elsewhere, nor exactly when. It is assumed that the date *terminus post quem* is the year 1241.

By contrast, the written work provides indirect testimony of Fibonacci's journey.

17 "*... qui eis tam per doctrinam quam per sedula obsequia discreti et sapientis viri magistri Leonardi Bigolli, in abbacandis estimationibus et rationibus [...] in recompensatione labori sui quem substinet in audiendis et consolidandis estimationibus et rationibus supradictis ....*". (Bonaini 1857, 241).

18 From the author's own translation.

19 Melis (1950, 591).

20 The term "*bigollo*" has two meanings: 1. "fool" or "good for nothing" (from "*bigolone*" and "*bighellone*"; 2. "spinning top" or "traveler" (from "*pigollo*"). Cf. *Tesoro della lingua italiana delle origini* http://tlio.ovi.cnr.it consulted 2012/02/08). See also (Boncompagni 1854, 16ff) and (Bonaini 1857, 242).

21 A deed of acquisition dated August 28 1226, (Milanesi 1867, 3–10).

22 Published by Mittarelli and Costadoni (1755–1760), according to the reconstruction by Mucillo (1997, 359–362). The founder of the family was probably Bonito, who died around 1100. Elisabetta Ulivi casts doubt on the line of descent from Bonito (Ulivi 2011, 253–254).

An Indirect Biographical Source: The Dating of Fibonacci's Written Work

Dating most of the works written by Fibonacci (except the two editions of the *L. A.* and the *Practica Geometriae*) is also inferred on the basis of internal references in the texts themselves. Thus the contested dates for the *Flos* and the *Liber Quadratorum* could go back to 1225, but there is a discrepancy between the date of Emperor Frederick II of Hohenstaufen's visit to Pisa, which could be dated in 1226[23] and the date (1225) of the *Liber Quadratorum* which was written after this visit. Because of this, Carlo Maccagni[24] argues that the date specified in the *Liber Quadratorum* is incorrect or that Fibonacci and the Emperor met before, perhaps in 1223. Between the two possibilities, he follows the first and suggests the following dating: *L. A.* (1202–1228); *Practica Geometriae* [1220/1221], *Flos* [1228/1234], *Epistola ad Magistrum Theodorum, phylosophum domini imperatoris* [post 1228], *Liber Quadratorum* [1226–1228]. Bear in mind that Leonardo dated his works according to the Pisan calendar[25] without indicating the month in which they were written, which could imply a margin of error of up to 1 year.

Leonardo also composed two works[26] which are now lost: *Libro di minor guisa*, a book of commercial arithmetic, quite likely in Italian vernacular and a commentary on book X of Euclid, where the author considers a numerical treatment of irrational numbers. The latter work, according to some authors, could have been inserted sequentially as "chapter XIV" in the second edition of the *L. A.*

This reconstruction of the dating and especially that of the dedicatees of these works is important for the reconstruction of Fibonacci's intellectual journey, as discussed below (Sect. 9.2.2).

---

[23]Stürner (2009, 787), based on Winkelmann (1880–1885, 259–260), appears to exclude the idea that Frederic visited Pisa before 1226. However, the Emperor's visit to Pisa can be assumed on the basis of analysis of the *Acta Imperii*. For local testimonies of the visit, cf. (Annales Placentini Gibellini 1863, XVIII: 469; Codagnello 1901, 83).

[24]Maccagni (1988).

[25]The calendar that was in effect in Pisa during the Middle Ages differed to those used in other Tuscan towns and the rest of Italy. Indeed, in Pisa, from the tenth century, it was decided that the start of the year would coincide with the Annunciation and therefore the Incarnation of Jesus Christ, 9 months before December 25 (Cappelli 1983, 8ff). See also (Ulivi 2011, 250–252). In this way, according to the Pisan calendar (*Ab Incarnatione Domini*), the first day of the year was March 25 and the year finished on the following March 24. Thus for the Pisans, any event taking place before March 25 belonged to the previous year and anything happening after March 25 was part of the following year. A non-Pisan dating must then be interpreted based on the time offset. Thus, if the Emperor was passed through Pisa on July 12, 1226, a Pisan would have noted the date as July 12, 1227.

[26]Franci (2002, 303) reports that in a fifteenth century manuscript in the National Library of Florence, they are mentioned as the *Libro di merchaanti detto di minor guisa et le Libro sopra il 10 di Euclide.* The first treatise (Lat. *Liber minoris guisae,* Eng. "Book in the smaller manner") is mentioned by Fibonacci in chapter XI (*L. A.*, 154): "*Est enim alius modo consolandi, quem in libro minoris guise docuimus*" (Indeed, there is another way to make the alloy, as we were taught in the 'book in the smaller manner'). As for the commentary on book X of Euclid, there is an allusion made by Fibonacci in his *Flos* (Boncompagni 1856, 3).

Fibonacci's Years of Travel Reflected in His Written Output

The *Prologue* [§.1] indicates that Fibonacci not only studied in Béjaïa, but that he also traveled throughout the Mediterranean region, "... *apud Egyptum, Syriam, Graecia, Siciliam et Provinciam* ... ". However, while Leonardo's stay in Béjaïa has been the subject of numerous investigations (see above), a systematic study of his educational travels and work in the countries mentioned above has yet to be carried out. However, his written work does provide some clues in this regard.

Thus, a trace of his sojourn in Byzantium could be deduced from some problems that the masters from Constantinople (with *magister* Muscus among them) submitted to him.[27] However, the fact that there are several references in the *L. A.* to Constantinople and the use of bezants could bear witness only to trade with the Byzantine world and not constitute proof of his presence in the city.

In Sicily, Fibonacci may have encountered a Greek to Latin translation of Euclid's *Elements*,[28] which was carried out by an unknown translator post 1160. Proof of this, or at least of his exposure to the Greek language lies in the use of several Greek terms in the *L. A.* which were only known from the direct translation of Euclid's *Elements*, such as: *epipedi* (*L. A.*, Chap. V, 30/Elements VII def.17); *riti* (*L. A.*, Chap. XIV, 356/Elements X def. 3); *riton et medium potens* (*L. A.* Chap. XIV, 357/Elements X.40); *media* (*L. A.* Chap. XIV, 356/Elements X. 21), *apothami* (*L. A.* Chap. XIV, 358/Elements X.92) etc.[29]

This very disparate documentary base may be supplemented by indirect information that we have for each of the contexts in which we know Fibonacci was active: The context of the Sicilian court (1.2), the Pisan context (1.3), the likely context of abacus schools in Italy at the time (1.4), and finally the Maghrebian context of the city of Béjaïa (1.5).

9.2.2 *The Context of Emperor Frederick II's Court*

Leonardo of Pisa's celebrity as a mathematician attracted the attention of the imperial court and of Frederick II himself. In his writings Fibonacci mentions numerous personages who made up the Emperor's entourage.

- Michael Scott,[30] addressee of the 1228 revised edition of the *L. A.* (*Dedicatory Letter* [§ 1]), astrologer, philosopher and scientific advisor to the Emperor, and Arabic and Hebrew translator.

---

[27]"*Questio nobis proposta a peritissimo magistro musco costantantinopolitano in Constantinopoli*" (*L. A.*, 249–250) (A problem proposed by a most learned master Muscus of Costantinople in Costantinople). See others problems in *L.A.* 188,190.

[28]Murdoch (1966) and Busard (1987).

[29]Busard ibid, (154, 211, 242, 228, 294). For an investigation of Greek testimonies in the complete works of Fibonacci see (Folkerts 2004).

[30]On this subject, cf. (Haskins 1924, 272–298; Thorndike 1965; Manselli 1979; Burnett 1994; Morpurgo 2005).

- Theodore of Antioch and John of Palermo,[31] were also very fine mathematicians: Theodore, a native speaker of Syriac, a philosopher, doctor and Imperial translator, had been a student of the great mathematician and astronomer Kamāl al-Dīn ben Yūnus (1156–1242). He appears in two works: as a formulator of questions in the *Liber Quadratorum* and as the addressee of a letter in the *Epistola ad Magistrum Theodorum, phylosophum domini imperatoris*. John of Palermo translated an Arabic treatise on conic sections into Latin.
- Master Dominic, possibly Dominic of Spain,[32] who was a scholar and translator at the School of Toledo before being moved to Frederick's court. Recipient of *Practica geometriae* written by Fibonacci in 1220 or 1221, it was he who presented Fibonacci to Frederick II in 1226 when the Emperor stopped in Pisa on his return to Sicily after the failure of the Diet of Cremona and received the mathematician in his palace.

During this audience, which is constantly invoked by the Pisan in his later writings, questions on number theory and indeterminate analysis – which Fibonacci solved brilliantly – were posed to him by John of Palermo and, possibly, later by Theodore of Antioch. This episode influenced Leonardo's later scientific work. Indeed, he went on to write the *Flos*[33] and the *Liber Quadratorum*,[34] where he developed the demonstrative methods that had brought him to the solutions to the aforementioned problems and the letter, addressed to master Theodore of Antioch, which can be found in the *Flos*.

Concerning the influence of the Swabian court on Fibonacci's journey, we know that scientific debate at the court of Frederick II invoked the great issues of the day thanks to the cultural exchanges the Emperor maintained with the Islamic world. This cultural milieu, where he encountered the masters of the most highly refined scientific traditions, very probably fueled Fibonacci's scientific journey.

However, if one examines the state of mathematical study at the Swabian court, as suggested by Carla Frova,[35] Fibonacci's works and particularly the books' dedicatory letters are practically the only source we have on the subject. Furthermore, while Frederick demonstrates a marked disposition for exact and applied sciences, his interest leant mainly towards natural science. The *L. A.* certainly aroused great interest, as the Pisan wrote in the incipit to *Liber Quadratorum*. The mathematician

[31]On Theodore: (Burnett 1995, 2000; Minervini 2005); on John: (Dall'Oco 2001).

[32]Very little is known about him, cf. (Arrighi 1970, 20–21; Stürner 2009, 825).

[33]The complete title of the work is: *Flos super solutionibus quarundam questionum ad numerum vel ad geometriam vel ad utrumque pertinentium*. Edited by (Boncompagni 1854, 1856, 1857–1862, 2: 227–252).

[34]For the edition of *Liber Quadratorum*, cf. (Boncompagni 1854, 1856, 1857–1862, 2: 253–83). For a further study of the work from a mathematical perspective cf. (Ver Eecke 1952; Picutti 1979; Sigler 1987).

[35]Frova (2005).

knew, from news emanating from the court, that the Emperor liked to revel in the subtleties of arithmetic and geometry from time to time.[36] But it seems that his interest did not go much further than this. Frova adds: "Despite the relationship with Michael Scott and the personal interest shown by Frederick II .... This is an experience that seems to be, in its entirety and complexity, rather alien to the intellectual tastes of that milieu."[37]

9.2.3 Pisa in the International and Political Context of the Day

International Trade

From the eleventh century, the fundamentals of social and economic development change: the rise of medieval Europe, the development of large towns, the growth in agricultural production accompanied by a multiplication and an extension of trade networks. Thus, trade with the Muslim East and Byzantium became more and more frequent: finely worked cloth, metals, gold and silver objects were exported; spices, silk, alum[38] – essential for linen dying – were imported. Italy found itself at the centre of this trade due to its geographical position and historical trading tradition: cities like Venice, Pisa, Florence, Genoa and Amalfi were key players in this phase of expansion.

Among the factors which determine this new context we must count the reconquest of the Mediterranean Sea. According to Melis, an economy of large trading blocs was established, that would expand in new directions, North–South and West–East, where "East" is the Near East and the Byzantine Empire.[39] The Mediterranean Sea, *mare nostrum,* again became the centre of the Christian world.

It is in this broader context that we must position Pisa both politically and economically, as we shall see.

The Context of Fibonacci's Birth: His Home Town

The city of Pisa was a maritime republic and was, in Leonardo's time, at the height of its trading powers: following a very aggressive war against Maghrebian pirates which started at the beginning of the eleventh century and against the Islamic world

[36]Boncompagni (1857–1862, 2: 54).

[37]From the author's own translation.

[38]This trade is well documented in chapters VIII and IX of the *L. A.* for that which concerns the links between Pisa and its trading partners.

[39]Melis (1989, 126).

in general, the city entered into a second phase, where a more peaceful commercial exchange took place between the former enemies, trade that was regulated by numerous treaties.[40]

Pisa had thus become the centre of a diverse, heterogeneous, open, society, frequented by Muslim traders and slaves, often of high rank, with whom friendly and cooperative relations were established. Quarters (called in the Muslim world funduqs[41]) set up in the very heart of the Mediterranean trading centers of Pisa, Venice and Genoa attracted many travelers, merchants and scholars in Maghreb, in Alexandria, Cairo, Constantinople, Acri and Antioch. Some of them, engaged in diplomatic missions, played a very important role of cultural mediation, as was the case of Burgondio, who learned Greek during his sojourn in Constantinople and "who may have discovered some manuscripts of the *Code of Justinian* in Constantinople during a long stay there (1137–1140), and took them to Pisa, studied and translated them".[42] And it is precisely to one of these funduqs, that of Béjaïa, that the Pisan scribe[43] Guglielmo (the title corresponding at the time to that of Consul) summoned his son Leonardo so that he could receive an appropriate training (possibly in 1180) that we read about in the *Prologue* §1 of the *L. A.*

The simple fact that Leonardo was born in Pisa is not negligible. It is very possible that the Pisan would have naturally adsorbed the mores, the orientation and the influence of Pisa's various legal and financial institutions. It is particularly necessary to mention that during the twelfth century, alongside the foundation of the municipality's political structure and its military and commercial expansion, Pisa adopted two legal codes[44] which are considered by historians to be among the most advanced, and anticipatory of Italian municipal law.[45] Such a set of legal rules, based on the discovery of Roman law and determined by the new socio-economic conditions could not but openly reject or ignore the hegemony claimed by the ecclesiastical power. This is especially true with regard to the highly controversial question of interest. Given that the practice of giving and receiving loans was widespread in most Italian municipalities despite prohibitory canonic laws, we can then explain, for example, how Fibonacci was able to speak openly about the charging of interest, which he called *usurae.*[46]

---

[40]Tangheroni (1994, 22–23).

[41]Valerian (2005).

[42]Volpe (1970, 154). According to Tangheroni (1994, 32), this sixth century code contained the fifty books of Justinian's *Digesta*, also known as the *Pandette*. The code is currently kept at the The Laurentian Library in Florence (*Biblioteca Medicea Laurenziana*), under the name *Lettera Pisana* or *Pisana* and it is proof, according to Tangheroni, that Pisan legal culture came directly from the Greek world, and is independent of the Bologna school of Law.

[43]The 'scribe' had a notarial and a diplomatic role. The title scribe preceded that of consul, which appeared later in Maghreb's funduqs. (Tangheroni 1994, 16; Aïssani and Valerian 2003, 9, note 2).

[44]The *Constitutum usus*, a code of common law concerning maritime legislation and the *Constitutum legis*, a code of Roman law.

[45]Salvatori (2002, 13–32).

[46]*L. A.*, 258–276. cf. (Caianiello 2004).

Fibonacci's Training in Mathematics

What training in mathematics could Leonardo have received in Pisa before he left for Béjaïa?

This question leads us to investigate the spread of the Hindu-Arabic numbering system in Italian cities during the second half of the twelfth century, the diffusion of the first Latin adaptations of *Indian arithmetic*[47] by al-Khwārizmī (also known as *Algorismi*),[48] and finally the organization of the educational system for future merchants. We ask ourselves if it was entrusted to the clergy or to laymen.

On the first two points, Leonardo went to Béjaïa when he was "*puer*" (see *Prologue* §1), therefore when he was between 7 and 14 years old.[49] It is thus possible that, like the other children of his age in Pisa, he learned to read, write[50] and count using an abacus[51] and the roman numbering system. Indeed, the diffusion of *algorisms* seems to have been limited to the circle of scholars and it is unlikely that it was part of the primary or secondary teaching programs. The Pisan also had to learn digital (finger) arithmetic and techniques for rapid memorization, which were part of the tradition of oral calculation.

Regarding the third point, the organization of the educational system, reconstructing the outline requires retrospective discussion that makes up the following part.

---

[47] Allard (1992).

[48] The oldest algorisms come from the twelfth century, the most important of them are *Dixit Algorizmi* (DA), *Liber Ysagogarum Alchoarismi* (LY), *Liber Alchorismi* (LA), *Liber Pulueris* (LP), which have been described by A. Allard (1992, 102ff), In the thirteenth century they were widespread, especially in universities, *Algorismus vulgaris* by Sacrobosco (John of Hollywood, also known as Johannes of Sacrobosco) and *Carmen de Algorismo* by Alexandre de Villedieu (Alexander de Villa Dei), described in (Karpinski and Waters 1928).

[49] *Pueritia* according to Isidor of Sevilla's classification in *Etymologiae Origines,* runs from the age of 7 until 14.

[50] In Italian cities set up as municipalities (twelfth to thirteenth centuries) written knowledge gradually acquired a hegemonic role in relation to other forms of knowledge. While in the feudal period preceding the establishment of the municipalities, not being able to write one's name was not seen as a sign of belonging to a lower class, it certainly was the case in the thirteenth century. (Cardini 1978, 152ff).

[51] The term *abacus*, which originally referred to the ancient calculating device, was so widespread throughout Europe in the Middle Ages – due to the fact that calculations were commonly carried out using an abacus – that the verbs "*abbacare, abbacizare*" ended up by meaning the "art of calculation". Likewise, starting from the eleventh century, arithmeticians were sometimes known as "*abacists*". Later, in Fibonacci's time, the term "*abacus*" was used for all types of arithmetic calculation (Smith 1925, 2: 177) while in subsequent abacus treatises – the *libri d'abaco* – the term indicated commercial calculations using Hindu-Arabic numerals. In the documents of the time – as is the case of any work written before the codification of modern spelling – the term "*abaco*" appears in a number of forms. The most common variation is the doubling of one or both of the consonants, as *abbaco* or *abbacco* and it is also common to find an h after the c, as in *abbacho* (Van Egmond 1988, 4–5). Although these different spellings are equivalents from a philological point of view, in order to avoid misunderstandings, Van Egmond (Van Egmond 1981, 5–6, 1999, 36), suggests using the word "abacus" when we refer to the calculating device and the word *abbacus* (with double b) when we refer to the art of calculation based on the Hindu-Arabic numerals.

9.2.4 *Some Elements for a Plausible Reconstruction of the Twelfth Century School Context in the City of Pisa*

All the information at our disposal on the organization of schools for merchants in medieval Italy relates to a period later then Leonardo's: it dates at best from the start of the fourteenth century. Supplemented by other commercial, political and institutional contextual elements, it allows us to reconstruct, to a certain degree, the school context in which Fibonacci developed, both as a pupil and later as a teacher. We will now review some of these elements and offer a plausible synthesis.

The Organization of Schools in Italy at the Start of the Fourteenth Century

We know that in Florence,[52] from the first half of the fourteenth century, around a half of all children learned to read and write and also learned the rudiments of Latin between the ages of 5 and 7.[53] It is also known that the teaching was carried out by primary school teachers called *doctores puerorum or doctrices puerorum*, if they were women. From 10 to 11 years of age, the sons of merchants (around 10 %), and often the sons of the nobility too, were orientated towards business, attending an abacus school, where they learned commercial arithmetic. At around 14 years of age they were ready for an apprenticeship in a trading house. A small number of them who had the possibility of going into international trade were sent abroad and traveled (as was the case for Leonardo).

There were, subsequently, schools called grammar schools where the Latin language, literature, rhetoric and logic were studied. This type of school was addressed to those (around 5 %) who were destined to civilian professions, but they were also attended by the pupils of the abacus schools who wanted to obtain a higher culture.[54]

An Indirect Testimony: The Circulation of Hindu-Arabic Numerals in the Twelfth Century

Hindu-Arabic numerals started to be used in place of Roman numerals between the end of the twelfth and the start of the thirteenth centuries: in the Laurentine Codex of Santa Croce,[55] the dating for the start, 1234, is in Arabic numerals; those at

---

[52]Thanks to a very well known passage by G. Villani in his *Cronaca fiorentina*.cf. (Davidsohn 1965, 4: 211).

[53]Davidsohn (1965, 4: 211–15), Grendler (1989, 77ff), and Høyrup (2007, 27–28). Firstly the children learned to read, then to write. Regarding this learning and the use of Latin, see (Black 2007, 43ff).

[54]Davidsohn, (ibid.); Goldthwaite (1972, 420), Ulivi (2000, 87), and Black (2007, 121–140, 226).

[55]Bartoli Angeli (2000, 230ff).

the end are in Roman characters. But already from the start of the twelfth century, there are documents which attest the use of Hindu-Arabic numerals for counting and not for arithmetic among some notaries in central and northern[56] Italy, among whom Raniero of Perugia appears.[57] It is probable that Hindu-Arabic and Roman numerals coexisted in teaching and in the practice of trade. From its introduction and for centuries thereafter (until the fifteenth–sixteenth centuries), there was much opposition, at different levels, to the use of the Hindu-Arabic numbering system.[58] Advocates of the Hindu-Arabic numerals were called *algorists*, while those who preferred the traditional abacus were the *abacists*.[59] It is known that in its 1299 statute, the *Arte del Cambio* forbade their use by its members and insisted on numbers written with letters.[60] There was also a ban in Padua some years later, when the University insisted that booksellers mark the prices of their books "*non per cifras, sed per literas claras*".[61] However, the new numbering system and the new arithmetic applied to trade answered the Italian merchants' urgent need for professional development.[62] Jacques Sesiano underlies that "The introduction of the new numerals and the new arithmetic was to be greatly facilitated by the introduction of account books, which favored the introduction of paper over costly parchment."[63] It is well-known by historians of accounting that Fibonacci gave the West its first (known) Latin accounting transcription.[64]

---

[56]Davidsohn (1965, 215).

[57]The case of Rainiero, active as a notary in Perugia from 1184 to 1211 is certainly the most significant: Rainiero numbers the pages of the document using an Arabic numeral followed by a period, which was written at the top right hand side of the page. The Roman numerals are marked with a period before and another after to indicate a change of register.

[58]Stone (1972) and Reynolds (1993).

[59]The possibility of easily changing the order of magnitude of a number, of adding or deleting a figure on the right as well as easier calculations, led to a fear of fraud and mistakes in calculations (Stone 1972, 348). It is for this reason that Fibonacci introduces the proof by nine, seven, eleven, and thirteen to check results in the section on arithmetic in the *L. A.* (Ambrosetti 2008, 233). Furthermore, the concept of zero, that is, a number whose status is "nothing", but can increase the value of a number tenfold, was hampered by conceptual difficulties on a religious and social level (Rotman 1987, 96–98). The church was against the democratization of arithmetic and against zero, the professional abacists feared the loss of their privileged positions (Ifrah 1994, 2: 367–68, 370).

[60]Camerani Marri (1955, 72–73).

[61]Cherubini (2006, 332).

[62]The mathematical tools available to Italian merchants at the start of the thirteenth century were largely inadequate: neither the very complicated Roman system nor arithmetic using the fingers or an abacus was sufficient when working on more complex problems. It is hard to imagine, but each city and each region had, more or less, different units of measurement for different goods and even different units of time. There were four different systems of currency, the most common being that established by Charlemagne, having livre, sou and denier (pound, shilling and penny) (Travaini 2007, 208–216). The amounts of money which were first recorded in units of account in contracts and other documents had to be converted into real money then converted into the different currencies, whose nominal values changed frequently. For trade between Pisa and Béjaïa, see (Aïssani and Valerian 2003).

[63]Sesiano (1999, 99).

[64]Melis (1950, 591). *L. A.*, 6.

Lay Schools and the Learning of Law

In regard to the question of laity in the schools intended for merchants, we know that in the twelfth to thirteenth centuries, episcopal and parish schools did not answer the needs of the nascent "bourgeoisie".[65] They were, over time, replaced by municipal schools, firstly private and later financed by local government, even if they were not yet "public" according to the current meaning of the word. They were run by the clergy, but often by laymen too, who worked principally as notaries. It seems that in most cases they acted as masters alongside the clergy in the municipalities.[66] This information could orientate research towards the study of the notarial archives and the municipalities' statutory documentation. In addition, as the schools were neither solely public nor private, but were also established by professional organizations and religious orders, it would be interesting to analyze the registers of these institutions.

However, although the documentation is fragmentary for instruction before the thirteenth century, it is unlikely that there were no lay schools in Pisa in the twelfth century, as there probably were in Pistoia[67] Genoa[68] and Florence. In Pisa, the only school attested is a probably lay law school, mentioned in the letter from a Victorine monk in 1124/1127.[69] This is a letter to his Abbot by a monk from St. Victor of Marseilles, traveling in Italy, where he asks permission to go to Pisa to devote himself to the study of law and where he remarks that many students from Provence were traveling to Italy to study law. It is therefore in notaries' offices and law schools that we should look for the beginning of lay teaching.

Conclusion: What Can Be Said About Leonardo's Education in Pisa

Armando Sapori[70] relates that Pirenne, starting from the conjecture that all trading, except for the most basic, presupposes a reasonably high level of education in those who are engaged in it,[71] wonders what the tools for training a merchant were. In the regions where Pirenne carried out his thorough research, Flanders, he went back to the mid thirteenth century and for Italy he advances the hypothesis that one can

[65]Pirenne (1929, 139ff) and Sapori (1937).

[66]Cardini (1978, 166ff).

[67]Chiappelli (1920).

[68]Schiaffini (1929).

[69]Dufour et al. (1979, 504).

[70]Sapori (1972).

[71]"How could it be accepted that, from this period Flemish merchants could both buy and sell large quantities of wool and linen in England and lend considerable amounts of money to all sorts of noble customers if they had to rely on their memory for the state of their debts and credit?" (Pirenne 1929, 19).

go back much further because in Italy, the level of instruction for thirteenth century merchants was far higher than that in more northerly regions. Sapori adds that in the absence of conclusive proof, what is known today allows this supposition to be seen as a "certainty".

According to Enrica Salvatori,[72] who takes up Pirenne's idea, the silence of the sources cannot be invoked as evidence of the non-existence of one or more schools. Indeed, the expansion of Pisa would be difficult to understand in the absence of two types of know how:

- Legal expertise, as borne out by the above letter from 1124 concerning a Victorine monk, as well as the enormous complexity of twelfth century regulation.
- A command of reading, writing and arithmetic.

While it is true that there is little direct evidence of the existence of schools for merchants before the thirteenth century, it is also true that "it is hard to imagine the hundreds of Pisan merchants scattered to the four corners of the world attending to their business without having mastered the notarial deed or the management of budgetary expenditure and income".[73]

In conclusion, we have underlined that throughout the twelfth century a strong legal tradition had taken root in Pisa. We have assumed that in medieval Italy and in Pisa there was a strong correlation between the study of law, notarial activity and the onset of lay teaching (see Sect. 9.2.4.3). One cannot doubt, therefore, the existence of one or more schools for merchants in Pisa. Leonardo may have studied the rudiments of commercial arithmetic and of Latin (a language he has shown to have mastered by the perfection of his writing) in such a school, or studied with a private tutor.

We can assume as well that, on his return to Pisa from Maghreb, Fibonacci, in his role as consultant (cf. Sect. 9.2.1.3), passed on to the officials of the Municipality his knowledge of Hindu-Arabic numerals that he demonstrates in the *L. A.*, in spite of the difficulty in the diffusion and acceptance of Hindu-Arabic numerals in institutional settings as we have seen (see Sect. 9.2.4.2). On the basis of these issues, we can reasonably believe that the *L. A.*'s real novelty was, above all, the diffusion of Hindu-Arabic numerals. Nevertheless, as Enrico Giusti stresses, "In the context of European mathematics, which has only recently become familiar with the great works of the past ... Fibonacci has given us a *summa* of the Arabic world's arithmetic and algebraic knowledge that is probably even broader than our current knowledge of the time allows us to understand; it may take at least a century before Fibonacci's works can be fully understood".[74]

---

[72]I would like to thank Enrica Salvatori (University of Pisa) who lead me to this passage through a personal communication (April, 04, 2009); from the author's own translation.

[73]Enrica Salvatori, personal communication (April, 04, 2009); from the author's own translation.

[74]Giusti (2002, 59).

9.2.5 The Maghrebian Context of Fibonacci's Stay in Béjaïa

The Political and Commercial Context

The city of Béjaïa, founded in 1067 and seat of an Almohad governor, was at the crossroads between the West, western Maghreb and the Sahara desert as well as a necessary route in the cultural exchanges between Spain and the Muslim East.[75] In the port district the city accommodated merchants, most of whom came from Italy: from Pisa, Venice, Genoa and probably from the south of Italy too. Scholars from all lands, but especially Muslim Arab scholars came to visit or settle in Béjaïa. A large Andalusian community also settled there.

Scientific and Mathematical Activity in Béjaïa

The sojourn at Béjaïa certainly had a profound influence on Fibonacci's mathematics. There, Leonardo learned Indian arithmetic, the fractional notation linked to the Maghrebian school, the fundamentals of algebra in the traditions of al-Khwārizmī and Abū Kāmil and thanks to his experience as a merchant's son, the fundamentals of the science of arithmetic as it applied to trade. The city, an intellectual centre for science of the highest level, was the seat of a school renown for the science of arithmetic and algebra.[76] We know, for example, that al-Qurashī (died 1184), a mathematician of great repute, worked there. His work has been lost, but, according to the testimony of Ibn Khaldūn,[77] he circulated a commentary on Abū Kāmil's algebraic treatise in northern Maghreb. There is also the testimony of the circulation in Béjaïa of works by Abū Bakr al-Hassār (twelfth century), Ibn al-Yāsamīn (died 1204) and Ibn Mun'im (died 1228). A work by al-Hassār, *The Book of Demonstration and Memorization,*[78] was used until the fourteenth century as a textbook in Maghreb. With regard to arithmetic,[79] Leonardo followed the fractional

[75] Aïssani and Valerian (2003).

[76] "The 'science of arithmetic' in the classical Arabic tradition encompassed not only Indian arithmetic, mental and digital (finger) arithmetic, but also procedures for resolving problems such as the four proportional number method, the single and double false position methods and algebraic algorithms associated with al-Khwārizmī's six canonic forms and a set of problems called "the arithmetic of transactions" (Djebbar 2005, 76 – author's translation). For texts on the science of arithmetic, see also (Djebbar 2001, 327–344).

[77] Ibn Khaldun (1958, 124).

[78] This work came to Europe, in a Hebrew translation by Moses Ibn Tibon in 1271 (Djebbar 2005, 109).

[79] Djebbar (1992, 223–245) and Aïssani (2004, 10).

notation of al-Hassār and al-Yāsamīn (in particular for continued fractions)[80] and he was the first Latin mathematician to introduce the '*virga*' (bar) for fractions.

Apart from the science of arithmetic, a tradition of pre-algebraic practical geometry is attested, which was popular with Latin translators.[81] The text by Abū Bakr, an almost unknown author of Andalusian origin who lived between the tenth and the twelfth centuries,[82] deals with surface areas using numerical algorithms and then proposes a solution in the same manner as al-Khwārizmī. Translated with the name of *Liber Mensurationum* by Gerard of Cremona, it may have been a source for Fibonacci's *Practica Geometriae.*[83] We do not find, by contrast, any references in the Muslim West to confirm the circulation of a subsequent work on algebra by Abū Kāmil, that refers to the tradition of Omar Khayyâm or al-Karajī. Sesiano affirms that

> While algebra in the Muslim East gradually freed itself from the tutelage of geometry and was to gain a degree of independence, algebraic reasoning being from then on regarded as valid in its own right, the Muslim West, which for geographical and also for political reasons had barely been aware of developments since the middle of the tenth century, remained outside this development. It remained faithful to the principles of Abū Kāmil's time.[84]

Given this cultural context, we can explain why Fibonacci does not appear within the current of research into algebra or number theory with his contemporaries in the Muslim East, but rather is part of the current of algebraic study in the ninth and first half of the tenth centuries.[85] Following the tradition of al-Khwārizmī and Abū Kāmil, the geometric validation of algebraic theorems was essential for the Pisan (see *Dedicatory letter* §3). However, it seems unlikely that al-Khwārizmī's Algebra or its Latin adaptations were direct sources for the *L. A.* There are, among other things, differences in the order of compound equations and in the choice of numeric examples.[86] This leaves the door open to other influences.

---

[80] In continued fractions the numerator and the denominator are made up of whole numbers (including zero for the numerator):

$$\frac{n_3 n_2 n_1}{m_3 m_2 m_1} = \frac{n_1}{m_1} + \frac{n_2}{m_2}\left(\frac{1}{m_1}\right) + \frac{n_3}{m_3}\left(\frac{1}{m_2 m_1}\right)$$

[81] A tenth century surveying book by al-Abdūn was found at Cordoba which describes plan and solid figures and solves second degree problems without reference to second degree equations or to the algebraic notation of al-Khwārizmī, using numeric algorithms which resemble the Babylonian type (Djebbar 2005, 81).

[82] According to Djebbar (2005, 123), he lived around the tenth century, according to Moyon (2008, 101) he lived sometime between the tenth and twelfth centuries.

[83] Busard (1968). See also (Høyrup 2007, 41, note 108).

[84] Sesiano (1999, 100–101).

[85] Rashed (1994).

[86] Leonardo, when using al-Khwārizmī's six canonic forms to solve a second degree equation, changes the order of the final three compound equations. The order used by al-Khwārizmī is (1) $ax^2 + bx = c$; (2) $ax^2 + c = bx$; (3) $bx + c = ax^2$: Fibonacci, however, presents and solves firstly

The School Context: Places and Teaching Activities

Did the young Fibonacci know Arabic? It is very likely he did. The archives in Pisa contain letters from the beginning of the thirteenth century written in Arabic by Maghrebian merchants to their Pisan partners.[87] The structure of the Muslim Arab scientific world in the Middle Ages was inseparable from the world of the ulemas,[88] the doctors of Islamic law, lawyers, and theologians. Al-Ġubrīnī relates 108 personages sojourning in Béjaïa, of whom 102 were sworn councilors and 5 expert mathematicians. It appears that the scholarly world of thirteenth century Béjaïa was composed of two groups: the independents and those connected to a personage.[89] We do not know to which group Fibonacci belonged.[90] The places of learning open to foreigners were the "House of wisdom" (*Beït al Hikma*) and the "City of science" (*Madinat al 'ilm*),[91] the method: "audiences by the princes of science." There may be an echo of these audiences when Fibonacci alludes to his oratorical jousts as being a fundamental stage in his education.[92]

In the first part we have, from the *Prologue* and from voluntary as well as involuntary testimonies (see the introduction) tried to throw some light on the milieu where Leonardo may have matured in Pisa, in Béjaïa and at the Swabian court. In the second part we will proceed with a selective commentary on the *Dedicatory Letter* and the *Prologue* in light of the elements identified in part 1. We will analyze the following aspects: the dedicatee, the pedagogical project, his readership, the relationship between theory and practice and in this way attempt to clarify the "facture" of the *L. A.* and the meaning of the book for Fibonacci in the context of his time.

---

type 1, then type 3, and then type 2. The coefficients Leonardo uses in the second and the third type are different from those that were chosen by al-Khwārizmī in his Algebra and in its Latin adaptations and those in Abū Kāmil's Algebra.

[87] Amari (1863, 48–62).

[88] Aïssani et al. (1994, 72).

[89] According to a statistical analysis by Urvoy (1976) based on a work by al-Ġubrīnī (died 1315) cf. (Aïssani 2004, 7).

[90] Aïssani claims that Fibonacci could have been part of mathematician al-Uṣūlī's group or the group which came from the town of Qalat Beni Hammad, the ancient Hammadid capital and an active cultural centre (Aïssani et al. 1994, 72; Aïssani 2004, 9).

[91] Aïssani (2004, 9).

[92] In the Muslim world, from the discipline of *kalām* – an Arabic term meaning, *strictu sensu*, "word" or "term" and, more broadly, a rational discourse on a field of investigation – develops a dialectic orientation, which leads to a kind of verbal sparring match. This is linked on one side to Arabic grammar and on the other to the process of *fiqh* (religious law). These are found in the presentation of problems in the form of questions and answers or objections and replies like at a trial (Urvoy 2006, 172–174). One can imagine the young Fibonacci applying himself to the oratory jousts provided by the educational methods of the time.

9.3 The Project of the Work: The Liber Abaci

9.3.1 The Dedicatory Letter of the Liber Abaci Addressed to Michael Scott: The Clue to a Special Architecture

Michael Scott, who was probably part of the Emperor's entourage from 1220, had a remarkable influence on the organization of scientific research at the imperial court. As Pietro Morpurgo points out, Scott urged Frederick II to ask questions to both Muslim and Christian scholars and masters on diverse philosophical and scientific problems in order to allow him to acquire knowledge that was a synthesis of different influences and a support for good government.[93] According to Frova, the *Dedicatory Letter* to Michael Scott is unconventional:

> it represents the presentation by Leonardo to Frederick's court of a book, a synthesis of different cultures, the author feels it to be a compendium of new mathematics, a tool that will allow the Latin peoples to put their knowledge to a level comparable with the most cultivated '... *apud Egyptum, Syriam, Graecia, Siciliam et Provinciam* ...' [*Prologue*, §2] and which Leonardo had become acquainted with during his travels.[94]

9.3.2 The Pedagogical Project

Fibonacci wants to transmit a complete doctrine of numbers (*Dedicatory Letter* §2) which, while taking account of the Euclidian legacy, was founded on Indian arithmetic and the method developed by the Arab tradition that Fibonacci had chosen as the best for this science and which he became acquainted with during his training in Maghreb and his travels in the Mediterranean basin. He certainly gives the content of his treatise a structure in 15 chapters that to a degree recalls the "science of arithmetic" textbooks of traditional Arabic mathematics (see Sect. 9.2.5.2). But, as Giusti points out, "Even in regards to his masters, Fibonacci wrote a unique work, if not in its originality, at least in its dimension ...".[95] In his text, a section on arithmetic, aimed at improving numeracy, including counting and operations and a large collection of problems for use by merchants, coexists with a section of practical geometry followed by a chapter on irrational numbers and finally a section of algebra, where problems are solved using equations and unknowns. In addition, a large number of problems of a practical nature are presented, relating to various aspects of everyday life and which are theoretical at the same time. There is also a section of fun and recreational problems.

---

[93] Morpurgo (1984, 287).

[94] Frova (2005).

[95] Giusti (2002, 60).

His Readership

In the *Prologue* (§ 4), Fibonacci addresses himself to the *gens latina,*[96] setting himself a very ambitious objective, namely to reach a broad audience. This is probably the reason for which he wrote the work in Latin.[97] In light of what was presented earlier in the first part (see Sect. 9.2.3.1), it appears that Fibonacci intended principally to teach mathematics skills to a high cultural milieu, if it is true that at the thirteenth century, most of the merchants in international "business" had a fairly high level of education and that Latin was the language of international trade. Belonging to this readership were the merchants,[98] the moneychangers, technicians, municipal officials, architects, everyone who needed mathematics in their work and not forgetting the professionals of the subject, such as the abacus masters and the lovers of mathematics. His audience was also made up of "intellectuals" belonging to highly educated circles, both ecclesiastical and lay, who knew him and regarded him with esteem. I am referring here to certain members of Frederick II of Hohenstaufen's court, with whom he maintained regular correspondence and to whom he dedicated most of his works (see Sect. 9.2.2).

Methodological Choices and Educational Aims

Fibonacci wanted to give his work a rigor based on convincing proofs and on the presentation of content which was accessible to the reader.

For the Pisan, following the tradition of Abū-Kamil's Algebra, the geometric justification of algebraic theorems was essential [*Dedicatory Letter* §3]. The geometric figures are used not only to represent the solutions to quadratic equations, but also to represent and solve the algebraic problems which follow them.[99] Even more than Abū Kāmil, Leonardo's processes are oriented towards a geometric proof.[100]

The treatise also provides innovations in arithmetic, which were not based on the erasure of numbers nor the use of the abacus, the ancient arithmetical tool, but on written calculations, where the remainders do not disappear but are kept in the memory or with the help of the fingers. We can thus explain the somewhat obscure passage, [*Prologue*, § 3] where he said he considered "the *Algorism* and the

[96]That is to say, to all peoples who share the Latin culture -which we might translate today as European peoples.

[97]Latin was the language of international trade, especially with the Germanic-speaking peoples (Pirenne 1985, 139 et al.).

[98]When we speak of merchants here, we do not mean the local storekeepers.

[99]See (*L. A.*, Chap. XV, section III, 410–459).

[100]Caianiello (2003, 87–98).

Pythagorean Arc ... practically as a mistake, with regard to the Indian method."[101] Fibonacci strongly emphasizes the quality of a learning method based on abstraction and supported through constant exercise and mnemonics. It is only through constant practice, he says, that the science becomes a frame of mind and that the memory and the intellect harmonize perfectly through the practice of arithmetic [*Dedicatory Letter* §5]. The above principles may be part of a vision of learning that has its roots in classical antiquity (e.g. Cicero and Quintilian among the Romans), where daily work and the continuous exercise of the memory and the intelligence were considered indispensable supports for student progress. Finally the desire to make the subject he presented available to his readers [*Dedicatory Letter*, §6], is present throughout the treatise. Thus Fibonacci presents things to his readers in a systematic and coherent manner. He arranges his examples from the simple to the more complicated. Throughout the book, charts and tables are often placed next to the text, to simplify or illustrate the explanation, or even make it easier to remember.

9.3.3 The Relationship Between Theory and Practice

"This book is certainly more oriented towards practice than theory" [*Dedicatory Letter*, §5]

If we ask ourselves whether the *L. A.* is more oriented towards theory rather than to practice, we encounter the first of a long series of difficulties of a textual nature. The manuscript tradition does not agree on this passage. Among the manuscripts we have examined thus far, F_1, R, N, S and V (see annex) bear witness, even if they use different lessons, to the *L. A.* "being more oriented to practice than to theory", while manuscript F (the Codex Magliabechiano) on which the Boncompagni edition is based, contends that the *L. A.* "looks more to theory than practice." So what was Fibonacci's true intention in the composition of the *L.A*? The question is controversial. On one hand, the theoretical aspect is always present: in the arithmetical part, Fibonacci uses Euclidean statements to verify the accuracy of the practical aspects;[102] he refers to Euclidean statements when he states that

[101]Fibonacci probably refers to the *Algorismus vulgaris* by Johannes de Sacrobosco and to *Carmen de Algorismo* by Alexander of Villedieu, which were widely circulated (especially at an academic level) in the twelfth and thirteenth centuries. The Pythagorean Arc is probably Gerbert of Aurillac's abacus (Burnett 2006, 19). Gerbert introduced a new type of abacus to the West where the beads were marked with numeric symbols (Indian figures), and were called *apices*. As G. Germano notes, see (Burattini et al. 2012, 79), R. E. Grimm's thesis (1973, 99–100), according to which it is necessary to correct *arcus in artem*, because *arcus* would be present clearly only in the Magliabechian Code (F) while the Riccardian Code (R) would be a questionable example, seems unconvincing. Indeed, after collating all the known witness manuscripts, it follows that R also clearly writes *arcus*. For a description of the manuscripts (see Annex).

[102]More precisely, he quotes Euclid 12 times (Folkerts 2004).

the propositions are taken from *The Elements,*[103] but also he himself shows many propositions in a Euclidean style, relating to the field of numbers. Not to mention the highly speculative content of chapters XIV and XV and in particular chapter XIV, which deals with quadratic and cubic radicals. On the other hand, it seems that the theoretical framework has a lesser weighting in the *L. A.* than in *Flos* and in the *Liber Quadratorum* (see Sect. 9.2.2). Moreover, as Concetta Carotenuto notes, Fibonacci does not only write that his book is practical, but that it "looks" and "relates to practice".[104] Indeed, he never loses sight of the practical application of the mathematics he deals with. While Marco Tangheroni stresses that "the *L. A.* cannot be considered to be a practical manual for use by merchants, because it offers no formulas for immediate application, but problems and their resolutions using mathematical procedures",[105] a large part is given over to arithmetical problems applied to trade. Tito Antoni[106] remarks that due to the volume of information on practices in the various markets, contained in chapters eight to twelve, we can compare the *L. A.* to an ante litteram "*Practica della Mercatura*".[107] Bearing these considerations in mind, the example of manuscripts F_1, R, N, S and V seems preferable.[108] The *L. A.* appears thus to be a hybrid. Leonardo was the first author in the Latin West to introduce large numbers of practical problems, but one can still find a theoretical analysis in his treatise, which was neglected in most later abacus textbooks. Thus in referring to Giovanni Sfortunati's treatise,[109] Ugo Tucci[110] observes that Sfortunati fails to explain the definition of the number in the manner of Euclid, Boethius and Sacrobosco, even though he was well acquainted with the definition, because he considered it too laborious and difficult for the merchants (*per essere cosa al mercante laboriosa e difficile*).

9.4 Conclusion

Firstly, we studied the conditions which led Fibonacci to write his work, by describing the world in which he matured, the milieus he frequented, the contexts of his training and his scientific production. Thanks to this brief historical excursion, we have caught a glimpse of Pisa's expansion in the context of the Mediterranean

---

[103](*L. A.*, 356).

[104]See (Burattini et al. 2012, 131). From the author's own translation.

[105]Tangheroni (1994, 25).

[106]Antoni (1994, 46).

[107]The earliest "*Practica di Mercatura*" to have reached us was written by Francesco di Balduccio Pegolotti, factor to the Bardi family in 1340.

[108]This interpretation is shared by mathematician E. Giusti (University of Florence) (personal communication, October, 10, 2010).

[109]Sfortunati (1545).

[110]Tucci (1994, 55).

world[111] in the twelfth to thirteenth centuries as well as its relations with the Islamic world. We have also discussed of the role played by the city of Béjaïa in mathematics and the processes by which knowledge spread throughout the Mediterranean basin, and the context of the scholars at the Swabian court. Finally a clear picture of Fibonacci and his complex character appears, as well as his role as mediator between Arabic science and Latin science. This "extended biographical investigation" led us to understand and comment on some aspects of the project of the *L. A.*, those that are articulated in its *Preface*. The pedagogical issues in the treatise are, in our opinion, clearly explained. Similarly, it is easy to see that *learning* and *teaching* are present in equal measure in Fibonacci's pedagogical goals (see Sect. 9.3.2.2). However, if we want to answer the questions of whether, in his lifetime, Fibonacci practiced as a master or "taught the *L. A.*", we must remain cautious. On one hand, the deliberation dated 1241 (see Sect. 9.2.1) does not give any positive information in this regard, while on the other, the *L. A.* was probably too difficult for his contemporaries and "went too far beyond the interests and requirements of the merchants of his time[112]" as well as the Swabian court (see Sect. 9.2.2). Aware, maybe, of the difficulty of his work, Fibonacci wrote another book with simpler arithmetic for merchants, the *Libro di minore guisa* (see Sect. 9.2.1.2) which is lost. In conclusion, as Menso Folkerts highlights[113] the *L. A.* was rather "the ring of conjunction between the theoretical mathematics taught successively at the University and the practical mathematics for use by merchants."

Appendix

The text was collated and translated into Italian by Concetta Carotenuto, it was reviewed and critically established by Giuseppe Germano, translated here into French by myself. For the critical analysis prepared by G. Germano and the commentary, refer to (Burattini et al. 2012).

Collated and translated manuscripts that contain the entire or part of the *Preface* of the *Liber Abaci*:

F FLORENCE, National Central Library (Biblioteca Nazionale Centrale di Firenze), ms. Conv. Soppr. C. 1. 2616, parchment, XIV century, f. 1r

F_2 FLORENCE, National Central Library (Biblioteca Nazionale Centrale di Firenze), ms. Fond. Prin. II. III. 25 [Magl. XI. 22], paper, XV or XVI century, f. 1r

F_1 FLORENCE, National Central Library (Biblioteca Nazionale Centrale di Firenze), ms. Magl. XI. 21, parchment, XIV century, f. 1r

---

[111] Aïssani and Valerian (2003, 9–31).

[112] Frova (2005).

[113] Folkerts (2001, 320).

R FLORENCE, Riccardian Library (Biblioteca Riccardiana), ms. 783, paper, XV century, ff. 1r/v
A MILAN, Ambrosian Library (Biblioteca Ambrosiana), ms. I. 72 Sup., parchment, XIII century, f. 1r
N NAPLES, National Library (Biblioteca Nazionale di Napoli), ms. VIII. C. 18, paper, XVII century, ff. 3r–4r
S SIENA, Civic Library of the Intronati (Biblioteca Comunale degli Intronati), ms. L. IV. 20, parchment, XIII century ex.-XIV century in., f. 1r
V VATICAN CITY, Vatican Apostolic Library (Biblioteca Apostolica Vaticana), ms. Palat. Lat. 1343, parchment, XIII century ex.-XIV century in., f. 8r

Leonardo Fibonacci of Pisa, The Book of the Abacus

Leonardo of Pisa to the Great Philosopher Michael Scott

(1) You wrote to me, my Lord and Master Michael Scott, so that I might transcribe for you my book on numbers which I composed long ago, it is for this reason that, in complying with your request and by the means of more subtle research on the book, I have corrected it in your honor and for the utility of many others. In this correction, I have added certain necessary points, and cut out other superfluous ones. (2) I present here the entire doctrine of numbers, following the Indian manner, which I have chosen as the most valid in this same science. (3) And given that the science of arithmetic and the science of geometry are related and mutually subsidiary, it is not possible to pass on an entire doctrine on numbers if one does not encounter notions of geometry or deal with geometry, which is practiced here solely according to numerical methods; such methods derived from many proofs and many demonstrations made with geometric figures.

(4) It is thus in another book, consecrated to applied geometry, that I described a great many problems related to geometry and to several other issues, by developing them one by one by the presentation of geometric proofs. (5) This present book is certainly oriented more towards practice than theory. It follows that those who would like, through it, to acquire a good knowledge of the practice of this science, should do so through the diligent application of continuous use and sustained exercise. Because once the science has become second nature through practice, the memory and the intelligence attune to such a point with the hands and the numbers, that they harmonize together spontaneously towards the same goal, in every way, like a single breath or a single impulse, at the same moment. Thus when the student has trained until it becomes natural, he will, little by little, be able to attain perfection in such practice.

(6) For the doctrine to be easier to acquire, I have divided this book into 15 chapters, in the aim of making it easier for the reader who is looking for something in particular.

Prologue

1. In my childhood, when my father was promoted by the state to be the scribe responsible for Customs in Béjaïa to deal with the Pisan merchants who met there, he sent for me and suggested keeping me there for some time in order that I learn the abacus, recognizing the use and benefit I could draw from this in the future.
2. Here, initiated into this art by exceptional teaching by means of the nine Indian numerals, I was enthralled by this above all things and I excelled to such a point that eventually, through diligent study and oratorical jousts, that I learned everything that was studied on the subject in Italy, Syria, Greece, Sicily and Provence, countries that thereafter I visited for my work.
3. But in the end, I considered all this, as well as the algorism and the Pythagorean Arc almost for a mistake, with regard to the Indian method.
4. That is why, embracing not only this Indian method, but also by studying it thoroughly, and adding certain developments of my own, derived from the subtleties of the art of Euclidian geometry, I tried to make the whole subject of this book as accessible as I could by differentiating 15 chapters. I have basically done all this, supported by conclusive proofs, so that, in this perfect way among all the others, those who are interested have access to this science, and those peoples using the Latin language cease to neglect it, as has been the case until now.
5. If, by chance, I have overreached myself in parts in an unjustified way or without reason, I call on your indulgence, indeed, as no-one is immune from imperfection, we must take certain precautions everywhere.

References

Aïssani, D. 2004. Béjaia à l'époque médiévale: Centre de Savoir et de Transmission Méditerranéen (IX–XV) siècles. *Manifestation Béjaia, Ville Des Sciences,* 1–13. Ethnomathematics Digital Library (EDL). www.ethnomath.org/resources/aissani2002.pdf. Consulted 8 Feb 2012.

Aïssani, D., D. Valerian. 2003. Mathématiques, commerce et société à Béjaia (Bugia) au moment du séjour de Leonardo Fibonacci (XII–XIII siècle). *Bollettino di storia delle scienze matematiche* 23: 9–31.

Aïssani, D., et al. 1994. Les mathématiques à Bougie médiévale et Fibonacci. In *Leonardo Fibonacci: Il tempo, le opere, l'eredità scientifica*, dir. M. Morelli and M. Tangheroni, 67–82. Pisa: Pacini.

Allard, A. 1992. *Muhammad Ibn Musa Al-Khwarizmi. Le Calcul Indien (Algorismus).* Paris: Blanchard.

Amari, M. 1863. *Diplomi arabi del R. Archivio fiorentino*. Florence: Successori Le Monnier.

Ambrosetti, N. 2008. *L'eredità arabo-islamica nelle scienze e nelle arti del calcolo dell'Europa medievale*. Milan: LED.

Angeli, Bartoli. 2000. I notai e i numeri. In *Scienze matematiche e insegnamento in epoca medievale, Atti del convegno internazionale di studi.* Chieti, 2–4 May 1996, dir. P. Freguglia, L. Pellegrini, and R. Paciocco, 225–254. Naples/Rome: Edizioni Scientifiche Italiane.

Annales Placentini Gibellini. 1863. In *Scriptores rerum germanica rum in usum scholarum ex Monumentis Germaniae historicis separatim editi*, ed. G.H. Pertz. XVIII, 457–581. Hannover 1863.

Antoni, T. 1994. Leonardo Pisano detto il Fibonacci e lo sviluppo della contabilità mercantile del '200. In *Leonardo Fibonacci: Il tempo, le opere, l'eredità scientifica,* dir. M. Morelli and M. Tangheroni, 45–50. Pisa: Pacini.

Arrighi, G. 1968. Un 'programma' di didattica di matematica nella prima metà del quattrocento (Dal Codice 2186 della Biblioteca Riccardiana di Firenze). In *Atti e Memorie della Accademia Petrarca di lettere, arti e scienze di Arezzo*, 117–128.

Arrighi, G. 1970. La fortuna di Leonardo Pisano alla corte di Federico II. In *Dante e la cultura sveva*(Melfi 1969), 17–31. Florence: Olschki.

Black, R.D. 2007. *Education and Society in Florentine Tuscany: Teachers, Pupils and Scholars c. 1250–1500*, Education and Society in the Middle Ages and Renaissance, vol. 29. Leiden/Boston: Brill. books.google.it/books?isbn=9004158537. Consulted 8 Feb 2012.

Bloch, M. 2009. *Apologia della storia o mestiere dello storico*, 2nd Italian ed. (translated from the 1998 French original edition by Giuseppe Gouthier, contributions of Étienne Bloch, foreword by Jacques Le Goff). Turin: Einaudi. 739.

Bonaini, F. 1857. Memoria unica sincrona di Leonardo Fibonacci nuovamente trovata dal prof. F. B. *Giornale Storico degli Archivi Toscani* 1(4): 239–246.

Boncompagni, B. 1854. *Intorno a tre scritti inediti di Leonardo Pisano, pubblicati da Baldassarre Boncompagni secondo la lezione di un codice della Biblioteca Ambrosiana di Milano*. Florence: Tipografia Galileiana.

Boncompagni, B. 1856. *Opuscoli di Leonardo Pisano secondo la lezione di un codice della Biblioteca Ambrosiana di Milano*, 2nd ed. Florence: Tipografia galileiana.

Boncompagni, B. 1857–1862. *Scritti di Leonardo Pisano matematico del secolo decimo terzo. 2 vols.: I. Il Liber abaci pubblicato secondo la lezione del Codice Magliabechiano C. 1., 2616, Badia Fiorentina, n. 73 da B. Boncompagni; II. La Practica geometriae. Opuscoli: Flos, le Questiones avium e il Liber quadratorum, pubblicati da B. Boncompagni*. Rome: Boncompagni Tipografia delle Scienze matematiche.

Bortolotti, E. 1929–1930. Le fonti arabe di Leonardo Pisano. *Memorie dell'Accademia delle scienze dell'Istituto di Bologna* 8: 1–30.

Burattini, E., E. Caianiello, C. Carotenuto, G. Germano, and L. Sauro. 2012. Per un'edizione critica del Liber Abaci di Leonardo Pisano, detto il Fibonacci. In *Forme e modi delle lingue dei testi tecnici antichi*, ed. G. Matino, 51–138. Naples: D'Auria.

Burnett, C. 1994. Michael Scot and the Transmission of Scientific Culture from Toledo to Bologna via the Court of Frederick II Hohenstaufen. *Micrologus* 2: 101–126.

Burnett, C. 1995. Master Theodore, Frederick II's philosopher. In *Federico II e le nuove culture. Atti del XXXI Convegno storico internazionale (Todi, 9–12 ottobre 1994)*, 225–285. Spoleto: Centro Italiano di Studi sull'Alto Medioevo.

Burnett, C. 2000. Antioch as a link between Arabic and Latin culture in the twelfth and thirteenth centuries. In *Occident et Proche-Orient: Contacts scientifiques au temps des Croisades*, ed. I. Draelants, A. Tihon, and B. van den Abeele, 1–78. Turnhout/Leuven: Brepols.

Burnett, C. 2006. The semantics of Indian numerals in Arabic, Greek and Latin. *Journal of Indian Philosophy* 34(1–2): 15–30.

Busard, H. 1968. L'algèbre au moyen age: Le 'Liber mensurationum' d'Abū Bakr. *Journal des Savants* 2: 65–125.

Busard, H. 1987. *The mediaeval Latin translation of Euclid's "Elements": Made directly from the Greek.* Wiesbaden: F. Steiner Verlag.

Caianiello, E. 2003. *Le Liber Abbaci de Leonard de Pise. Préliminaire à une étude historique.* D.E.A. memoir, EHESS.

Caianiello, E. 2004. Interest and Usury in Fibonacci's Liber Abaci. A first comparison of the mathematical tradition of the western Muslim world. In *Actes de l'8ème Colloque Maghrébin sur l'Histoire des Mathématiques Arabes*, 93–122. Tunis: Association Tunisienne des Sciences Mathématiques.

Camerani Marri,G. 1955. *Statuti dell'arte del cambio di Firenze (1299–1316)*, ed. G. Camerani Marri. Florence: Olschki.

Cappelli, A. 1983. *Cronologia, cronografia e calendario perpetuo*. Milan: Hoepli.

Cardini, F. 1978. Alfabetismo e cultura scritta nell'età comunale: alcuni problemi. In *Alfabetismo e cultura scritta nella storia della società italiana*, dir. A. Bartoli Angeli, 147–193. Perugia: Università degli Studi.

Cherubini, P. 2006. Il numero come elemento di disturbo. In *Lo scaffale della biblioteca scientifica in volgare (secoli XIII–XVI) – Atti del Convegno*, Matera, 14–15 October 2004, dir. R. Librandi and R. Piro, 313–339. Florence: Sismel – edizioni del Galluzzo.

Chiappelli, L. 1920. Maestri e scuole in Pistoia fino al sec. XIV. *Archivio Storico Italiano* 78: 161–214.

Codagnello, J. 1901. Annales Placentini. In *Scriptores rerum germanica rum in usum scholarum ex Monumentis Germaniae historicis separatim editi*, XXIII. Hannover/Leipzig: O. Holder-Eggar.

Dall'Oco, S. 2001. Giovanni da Palermo. In *Dizionario Biografico degli Italiani*, 56. Rome: Istituto della Enciclopedia Italiana Treccani. www.treccani.it/enciclopedia/giovanni-da-palermo_(Dizionario_Biografico)/. Consulted 8 Feb 2012.

Davidsohn, R. 1956–1968. *Storia di Firenze.* ed.it, 8 vols. Firenze: Sansoni.

Djebbar, A. 1992. Le traitement des fractions dans la tradition mathématique arabe du Maghreb. In *Histoire de fractions, fractions d'histoire*, ed. P. Benoit, K. Chemla, and J. Ritter, 223–245. Basel: Birkhäuser.

Djebbar, A. 2001. Les Transactions dans les mathématiques arabes: Classification, résolution et circulation. In *Commerce et mathématiques du Moyen Age à la Renaissance, autour de la Méditerranée, Actes du Colloque International*, 327–344. Toulouse: Editions du C.I.H.S.O.

Djebbar, A. 2005. *L'algèbre arabe. Genèse d'un art*. Paris: Vuibert.

Dufour, J., et al. 1979. L'attrait des 'Leges'. Note sur la lettre d'un moine victorin (vers 1124/27). *Studia et Documenta Historiae Iuris* 45: 504–529.

Folkerts, M. 2001. Le discipline matematiche. In *Storia della scienza*, vol. IV, 313–323. Rome: Istituto dell'Enciclopedia Italiana Treccani.

Folkerts, M. 2004. Leonardo Fibonacci's knowledge of Euclid's elements and of other mathematical texts. *Bollettino di storia delle scienze matematiche* 24: 93–113.

Folkerts, M. 2011. Jacopo da Florence's Tractatus Algorismi and early Italian abacus culture. *Annals of Science* 68(2): 282–284.

Franci, R. 2000. L'insegnamento dell'Aritmetica nel Medioevo. In *Scienze matematiche e insegnamento in epoca medievale, Atti del convegno internazionale di studi.* Chieti, 2–4 May 1996, dir. P. Freguglia, L. Pellegrini, and R. Paciocco, 111–132. Naples/Rome: Edizioni Scientifiche Italiane.

Franci, R. 2002. Il Liber Abaci di Leonardo Fibonacci: 1202–2002. *Bollettino della Unione Matematica Italiana* 5-A: 293–328.

Franci, R. 2010. The history of algebra in Italy in the 14th and in 15th centuries. Some remarks on recent historiography. *Actes d'Història de la Ciència i de la Tècnica NOVA ÈPOCA* 3(2): 175–194.

Frova, C. 2005. Matematica. In *Enciclopedia Federiciana*. Rome: Istituto dell'Enciclopedia Italiana Treccani. www.treccani.it/enciclopedia/matematica_(Federiciana)/. Consulted 8 Feb 2012.

Fumagalli Beonio Brocchieri, M. 1987. L'intellettuale. In *L'uomo medievale*, 1st edit. it, dir. J.L. Goff, 201–233. Bari: Laterza.

Giusti, E. 2002. Matematica e commercio nel *Liber Abaci.* In *Un ponte sul Mediterraneo: Leonardo Pisano, la scienza araba e la rinascita della matematica in Occidente,* dir. E. Giusti, with the collaboration of R. Petti, 59–120. Firenze: Il Giardino di Archimede. http://php.math.unifi.it/archimede/archimede/fibonacci/catalogo/giusti.php. Consulted 8 Feb 2012.

Goldthwaite, R. 1972. Schools and teachers of commercial arithmetic in renaissance Florence. *The Journal of European Economic History* 1(2): 418–433.

Grendler, P. 1989. *Schooling in renaissance Italy: Literacy and learning, 1300–1600.* Baltimore: Johns Hopkins University Press.

Grimm, R.E. 1973. The autobiography of Leonardo Pisano. *The Fibonacci Quaterley* 11(1): 99–104.
Haskins, C.H. 1924. *Studies in the history of medieval sciences*. Cambridge, MA: Harvard University Press.
Heeffer, A. 2007. *Humanist repudiation of eastern influences in early modern mathematics*. pdf 1–60.
Høyrup, H. 2000. Jacobus de Florentia, Tractatus algorismi (1307). *Centaurus* 42: 21–69.
Høyrup, H. 2001. The founding of Italian vernacular algebra. In *Commerce et mathematiques du moyen âge à la reinassance*, 129–156. Toulouse: Editions du C.I.S.H.O.
Høyrup, J. 2005. Leonardo Fibonacci and Abbaco culture: A proposal to invert the roles. *Revue d'Histoire des Mathématiques* 11: 23–56.
Høyrup, J. 2006. Jacopo da Florence and the beginning of Italian vernacular algebra. *Historia Mathematica* 33: 4–42.
Høyrup, J. 2007. *Jacopo da Florence's Tractatus Algorismi and early Italian abacus culture*. Basel/Boston/Berlin: Birkhäuser.
Høyrup, J. 2009a. *A response to Van Egmond on Høyrup, Jacopo da Florence's Tractatus Algorismi and Early Italian abacus Culture*. www.ircps.org/sites/ircps.org/files/aestimatio/6/2009-14_Hoyrup.pdf. Consulted 8 Feb 2012.
Høyrup, J. 2009b. Answer to Jeffrey Oaks' review of Jens Høyrup, Jacopo da Florence's Tractatus Algorismi and Early Italian abacus culture. In *CSHPM/SCHPM bulletin*, 45. Basel: Birkhäuser. faculty.umf.maine.edu/michael.molinsky/public.www/cshpm/archives/bulletins/AnswerToOaks.pdf. Consulted 8 Feb 2012.
Ibn Khaldun, A.-R. 1958. *The Muqaddimah: An introduction to history*, dir. F. Rosenthal. New York: Pantheon Books.
Ifrah, G. 1994. *Histoire universelle des chiffres*. 2nd ed. 2 vols. Paris: R. Laffont.
Karpinski, L.C., and E.G.R. Waters. 1928. A thirteenth century algorism in French verse. *Isis* 11: 45–84.
Levey, M. 1966. *The algebra of Abū Kāmil*. Madison: University of Wisconsin Press.
Maccagni, C. 1988. Leonardo Fibonacci e il rinnovamento delle matematiche. In *L'Italia ed i paesi mediterranei*: *Atti del Convegno internazionale di studi* (Pisa, 6–7 June 1987), 91–115. Pisa: Nistri-Lischi e Pacini.
Manselli, R. 1979. La corte di Federico II e Michele Scoto. In *L'Averroismo in Italia. Atti del Convegno dell'Accademia Nazionale dei Lincei* (Rome, 18–20 aprile 1977), 63–80. Rome: Atti dei Convegni Lincei.
Melis, F. 1950. *Storia della ragioneria. Contributo alla conoscenza e interpretazione delle fonti più significative della storia economica*. Bologna: C. Zuffi.
Melis, F. 1989. *Industria e Commercio nella Toscana Medievale*, ed. B. Dini. Firenze: Le Monnier.
Milanesi, G. 1867. Documento inedito sconosciuto intorno a Lionardo F. *Giornale arcadico, n.s.* LII: 3–10.
Minervini, L. 2005. Teodoro di Antiochia. In *Enciclopedia Federiciana*. Rome: Istituto della Enciclopedia Italiana Treccani. www.treccani.it/enciclopedia/teodoro-di-antiochia_(Federiciana)/. Consulted 8 Feb 2012.
Mittarelli G.B., and A. Costadoni. 1755–1760 (s.d.). *Annales camaldulenses ordini Sancti Benedicti*, III Venetiis 1755–1760, col. 212, App., n. CXLVI.
Miura, N. 1981. The algebra in the Liber Abaci of Leonardo Pisano. *Historia Scientiarum* 21: 57–65.
Morpurgo, P. 1984. Il 'Sermo suasionis in bono' di Michele Scoto a Federico II. *Rendiconti dell'Accademia dei Lincei* 38: 287–300.
Morpurgo, P. 2005. Michele Scoto. In *Enciclopedia Federiciana*. Rome: Istituto della Enciclopedia Italiana Treccani. http://www.treccani.it/enciclopedia/michele-scoto_(Federiciana)/. Consulted 8 Feb 2012.
Moyon, M. 2008. La géometrie pratique en Europe en relation avec la tradition arabe, l'exemple du mesurage et du découpage: Contribution à l'étude des mathématiques médievales. Ph.D. thesis, Lille: Université des Sciences et technologies de Lille. http://ori.univ-lille1.fr/notice/view/univ-lille1-ori-12110. Consulted 8 Feb 2012.

Mucillo, M. 1997. Fibonacci, Leonardo. In *Dizionario Biografico degli Italiani*, 47. Rome: Istituto dell'Enciclopedia Italiana Treccani. www.treccani.it/enciclopedia/leonardo-fibonacci_(Dizionario-Biografico)/. Consulted 8 Feb 2012.

Murdoch, J. 1966. Euclides greco-latinus, a hitherto unknown medieval translation of the *elements* made directly from the Greek. *Harvard Studies in Classical Philology* 71: 249–302.

Oaks, J. 2009. Essay review: Medieval Italian practical mathematics. In *Jacopo da Florence's Tractatus Algorismi and early Italian abacus culture*, CSHPM/SCHPM bulletin, vol. 45, ed. Jens Høyrup. Basel etc.: Birkhauser.

Oaks, J. 2010. Response to Jens Høyrup's. In *Answer to Jeff Oaks' review of Jens Høyrup Jacopo da Florence's Tractatus Algorismi and early Italian abacus culture*. Basel etc.: Birkhauser, 2007 Available on J. Oak's website. http://pages.uindy.edu/~oaks/Articles/ResponseHoyrup.pdf. Consulted 8 Feb 2012.

Picutti, E. 1979. *Il Libro dei quadrati di Leonardo Pisano e i problemi di analisi indeterminata nel Codice Palatino 557 della Biblioteca Nazionale di Firenze*. Florence: Olschki.

Pirenne, H. 1929. L'instruction des marchands au moyen age. *Annales d'histoire economique et sociale,* tome 1(1): 13–28.

Pirenne, H. 1985. *Storia economica e sociale del Medioevo.* Postface by Antonella Tarpino, with a bibliographical and critical appendix by Hans van Werveke. 2nd Italian ed. translated from the 1963 French edition by Leonardo Cammerano. Milan: Garzanti.

Rashed, R. 1994. Fibonacci et les mathématiques arabes. *Micrologus* 2: 145–160.

Rashed, R. 2003. Fibonacci et le prolongement latin des mathématiques arabes. *Bollettino di Storia delle Scienze Matematiche* 23(2): 55–73.

Reynolds, B.E. 1993. The Algorists vs. The Abacists: An ancient controversy on the use of calculators. *The College Mathematics Journal* 24: 218–223.

Rotman, B. 1987. *Signifying nothing. The semeiotics of zero*. London: Macmillan Press Ltd.

Salvatori, E. 2002. Pisa in the Middle Ages: The dream and the reality of an empire. In *European Perspective*, dir. E. Steven, 13–32. Pisa: Plus University of Pisa. www.stm.unipi.it/Clioh/tabs/libri/6/02-Salvatori_13-32.pdf. Consulted 8 Feb 2010.

Sapori, A. 1937. La cultura del mercante italiano. *Rivista di storia economica* 2: 89–125.

Sapori, A. 1972. *La mercatura medievale>il mercante italiano dell'età eroica> l'istruzione.* Florence: Sansoni (Scuola aperta). http://www.retimedievali.it. Consulted 8 Feb 2010.

Schiaffini, A. 1929. Il mercante genovese nel Dugento. Extracted from revue "A Compagna", October 1929, 10:1–20. www.librianticchicavallero.com. Consulted 8 Feb 2012.

Sesiano, J. 1993. La version latine médiévale de l'Algèbre d'Abū Kāmil. In *Vestigia mathematica. Studies in medieval and early modern mathematics in Honour of H.L.L. Busard*, ed. Menso Folkerts and Jan P. Hogendijk, 315–452. Amsterdam/Atlanta: Rodopi.

Sesiano, J. 1999. *Une introduction à l'histoire de l'algèbre: Résolution des équations des Mésopotamiens à la Renaissance*. Lausanne: Presses polytechniques et universitaires romendes.

Sfortunati, G. 1545. *Nuovo Lume, libro de arithmetica*. Venice: Bernardino Bindoni.

Sigler, L.E. 1987. *The book of squares of Leonardo Pisano Fibonacci. An annotated translation into modern English by L.E. Sigler*. Boston: Academic.

Sigler, L.E. 2002. *Fibonacci's Liber Abaci. A translation into modern English of Leonardo Pisano's book of Calculation*. New York: Springer.

Smith, D. 1923/1925. *History of mathematics*. 2 vols. Boston: Ginn and Company.

Stone, W.E. 1972. Abacists versus Algorists. *Journal of Accounting Research* 10: 345–350.

Stürner, W. 2009. *Federico II e l'apogeo del suo impero*. Rome: Salerno Editrice.

Tangheroni, M. 1994. Fibonacci, Pisa e il Mediterraneo. In *Leonardo Fibonacci, il tempo, le opere, l'eredità scientifica*, dir. M. Tangheroni and M. Morelli, 15–34. Pisa: Pacini.

Thorndike, L. 1965. *Michael Scott*. London/Edinburgh: Nelson Ltd.

Travaini, L. 2007. *Monete e storia nell'Italia medievale*. Rome: Istituto Poligrafico e Zecca dello Stato.

Tucci, U. 1994. Manuali d'aritmetica e mentalità mercantile tra Medioevo e Rinascimento. In *Leonardo Fibonacci, il tempo, le opere, l'eredità scientifica,* dir. M. Tangheroni and M. Morelli, 51–65. Pisa: Pacini.

Ulivi, E. 2000. Le scuole d'abaco e l'insegnamento della matematica a Firenze nei secoli XIII–XVI. In *Scienze matematiche e insegnamento in epoca medievale, Atti del Convegno internazionale di studio, Chieti, 2–4 May 1996*, dir. P. Freguglia, L. Pellegrini, and R.Paciocco, 85–110, Naples/Rome: Edizioni Scientifiche Italiane.

Ulivi, E. 2011. Su Leonardo Fibonacci e sui maestri d'abaco pisani dei secoli XIII–XV. *Bollettino di storia delle scienze matematiche* XXXI: 247–286.

Urvoy, D. 1976. La structuration du monde des ulemas à Bougie au VII/XII siècles. *Studia Islamica* 43: 87–107.

Urvoy, D. 2006. *Histoire de la pensée arabe et islamique*. Paris: Du Seuil.

Valerian, D. 2005. Les marchands latins dans les ports musulmans méditerranéens: une minorité confinée dans des espaces communautaires? In *Revue des mondes musulmans et de la Méditerranée* [On line], 107–110. September 2005, put on line 11 January 2012. http://remmm.revues.org/2828. Consulted 8 Feb 2012.

Van Egmond, W. 1981. Practical mathematics in the Italian renaissance: A catalog of Italian abbacus manuscripts and printed books to 1600. In *Annali dell'Istituto e Museo di Storia della Scienza*, xliv-442 (Supplement 1980). Florence: Giunti Barbera.

Van Egmond, W. 1988. *The commercial revolution and the beginnings of western mathematics in renaissance Florence, 1300–1500.* Ann Arbor: UMI Dissertation Information Service.

Van Egmond, W. 1999. Abacus, algorisme, abacus: Methods of reckoning in the merchant culture of Mediterranean. In *Actes du colloque international du Centre International d'Histoire des Sciences Occitanes, Beaumont de Lomagne*, 13–16 May 1999, 21–53. Toulouse: Editions du C.I.H.S.O.

Van Egmond, W. 2009. Iacopo da Florence's Tractatus Algorismi and early Italian abacus. In *Science networks-historical studies,* ed. Jens Høyrup, 34. *Aestimatio* 6: 37–47.

Ver Eecke, P. 1952. *Léonard de Pise. Le livre des nombres carrés.* Bruges: Desclée de Brouwer & Ge.

Volpe, G. 1970. *Studi sulle istituzioni comunali a Pisa.* Florence: G.C. Sansoni.

Winkelmann, E.A. 1880–1885. *Acta Imperii Inedita Saeculi XIII [et XIV]. Urkunden und Briefe zur Geschichte des Kaiserreichs und des Königreichs Sicilien*, I, re-edition Aalen 1964. 2 vols. Innsbruck: ed. E.A. Winkelmann.

Woepcke, F. 1853/1982. *Extrait du Fakhrī, traité d'algèbre par Abou Bekr Mohammed Ben Alhacan Alkarkhi.* Reprint of the 1853 original ed. Hildesheim/Zurich/New York: Georg Olms Verlag.

Zaccagnini, G. 1924. L'insegnamento privato a Bologna e altrove nei secc. XIII e XIV. In *Atti e Memorie della R. Deputazione di Storia Patria per le Provincie di Romagna, 4th ser*, vol. XIV, 254–301.

Chapter 10
Didactical Dimensions of Mathematical Problems: Weighted Distribution in a Vietnamese Mathematical Treatise

Alexei Volkov

10.1 Introduction

Interpretations of pre-modern East and Southeast Asian mathematical texts offered by modern historians of science are often based on implicit assumptions concerning the circumstances under which these texts were compiled and the ways in which they were used.[1] Interpretations of a single text may vary considerably depending on whether it is considered as a "mathematical treatise" addressed primarily to the members of a community of contemporaneous mathematicians, a "mathematical classic", that is, the starting point of a commentarial tradition, or a "mathematical textbook" used for educational purposes. Since the extant texts often lack explicit mention of the intended goals and targeted audiences, and since these goals and audiences may have changed considerably throughout the history of the texts, an analysis of the circumstances under which they were compiled and used would be a highly relevant part of their study. However, historians often do not pay enough attention to this part, probably tacitly assuming (or at least making the reader assume so) that the content of a given text was dealt with in one and the same way throughout its entire history; equally tacitly, they often assume that this very way is correctly guessed and adequately treated in their own reconstructions. Changes

[1] The "pre-modern" Asian mathematical traditions mentioned here include the mathematical practices and related texts that either antedated the encounter with Western mathematics that happened in the seventeenth century in China or remained uninfluenced by it, as in the case of traditional Japanese mathematics *wasan* 和算 of the seventeenth to nineteenth centuries.

A. Volkov (✉)
Center for General Education and Institute of History, National Tsing Hua University, Hsinchu, Taiwan
e-mail: alexei.volkov@gmail.com

A. Bernard and C. Proust (eds.), *Scientific Sources and Teaching Contexts Throughout History: Problems and Perspectives*, Boston Studies in the Philosophy and History of Science 301, DOI 10.1007/978-94-007-5122-4_10,

to these assumptions may alter considerably the interpretations of even those mathematical texts that previously were believed to have been thoroughly studied.

The most fundamental questions a historian of science has to answer when working with a given pre-modern mathematical text can thus be posed as follows: Was the text originally compiled and used exclusively for educational purposes or was it a mathematical treatise written for a relatively small circle of peers? Were there commentarial traditions that considered the text to be a "classic"? Certainly, these questions are not expected to be (and in many cases cannot be) answered with simple "yes" or "no", since one and the same text may have been used for *different* purposes in different social and historical settings, notwithstanding the original intention(s) of its compiler(s). For instance, the Euclidean *Elements*, whatever this treatise was meant to be when being compiled, it eventually became a "classic" as well as *the* textbook of geometry for generations of later readers, in its original form or in its numerous adaptations. Another example: K. Chemla recently argued that the Chinese mathematical treatise *Jiu zhang suan shu* 九章算術 (Computational Procedures of Nine Categories) was a "classic" (Chinese *jing* 經) that drew special attention from later commentators, in particular, Liu Hui 劉徽 (fl. AD 263)[2]; she also warned against "des modes de lecture hâtive ou trop sommaires, qui verraient en cet ouvrage un avatar d'un quelconque manuel d'école", that is, against simplifying interpretations portraying the treatise as a school textbook.[3] However, the treatise was in fact one of the keynote textbooks in the curricula of the state-run educational institutions in China (see below) for many centuries, and its interpretation as a textbook thus would be *also* fully legitimate.

An ideal methodology for a study of a given pre-modern mathematical text therefore should include a thorough investigation of the social context in which it was used; in the cases where this information is insufficient or unavailable, the very content and the structure of the text may, to a certain extent, help offer informed conjecture concerning the modalities of its functioning.

As far as extant Chinese mathematical texts are concerned, it can be shown that a considerable number of them were used in an educational context, that is, were originally designed, or at least adapted later, specially for mathematical instruction. The mathematical tradition that existed in pre-modern Vietnam historically originated from the Chinese tradition, and therefore there are strong reasons to believe that the Vietnamese mathematical treatises were also compiled and used primarily for educational purposes. Certainly, one cannot entirely deny the "scholarly" function of some pre-modern mathematical texts, which, although styled as collections of problems, did contain new methods devised by their authors and therefore were

[2]In this paper, the *pinyin* transliteration system is used for Chinese terms (including titles of treatises and personal names), the Quốc Ngữ transliteration is used for Vietnamese terms, and the Hepburn romanization system is used for Japanese terms. The so-called "traditional Chinese characters" (and not their simplified versions used in the People's Republic of China and, in certain cases, in Japan) are used everywhere in this paper except the Bibliography.

[3]Chemla and Guo 2004: 4.

addressed, at least to a certain extent, to peers and not exclusively to the learners.[4] However, the present paper takes as its point of departure the assumption that a considerable number of pre-modern mathematical texts, especially those styled as collections of arithmetical problems, even though produced in a variety of historical and cultural settings in the Chinese world and beyond, were originally compiled or adapted at subsequent stages to be used in an educational context and thus are much closer, as far as their content and application are concerned, to modern mathematical textbooks rather than scholarly monographs.[5]

Even when the information about the use of a given text for educational purposes is insufficient or unavailable, its very structure, and, in some cases, the numerical parameters of the problems (as in the case of the present study), may provide sufficient material for a sensible reconstruction of at least a hypothetical use of the text in an educational context, while the same elements would appear superfluous or irrelevant if the text was designed and used exclusively for communication between professional mathematicians.[6]

Below I will discuss how this approach can be applied to a study of a series of arithmetical problems on weighted distribution from the Vietnamese mathematical text *Ý Trai toán pháp nhất đắc lục* 意齋算法一得錄 (A Record of What Ý Trai Got Right in Computational Methods) compiled in 1829 by Nguyễn Hữu Thận 阮有慎; I will conjecture that the numerical values in these problems may have been chosen to be used as a didactical tool to introduce new theoretical concepts and reasoning strategies to the learners.

10.2 Numerical Parameters and the Order of Problems: Preliminary Remarks

A considerable number of pre-modern mathematical texts belonging to various traditions contained groups of problems supposed to be solved with one and the same algorithm (or with generic algorithms apparently perceived by the compilers as variants of the same algorithm). Even though the numerical parameters of the problems often may have been chosen by the compilers randomly, there are reasons to believe that in some cases the parameters could hardly be random. The choice

---

[4]In traditional Japan, professional mathematicians published collections of problems *idai* 遺題 (lit.: "problems left [for solution of the reader]") without providing their solutions to challenge fellow mathematicians, see Horiuchi 1994: 40, 65, 114–116, 158–167.

[5]The present collection of papers contains contributions providing argument for, as well as against, this assumption in various historical contexts; see, for example, the papers of Christine Proust and Anne-Marie Chartier.

[6]What is known about the social positions of those who studied and taught mathematics, for instance, in traditional China from the second half of the first millennium to the early second millennium AD amounts to a fragmented picture of a rather complex network of individuals and institutions, including the state-run School of Computations (see below).

of the parameters may have pursued various goals. For instance, the compilers may have arranged the numerical parameters in a series of problems in such a way that all the problems would have had the same answer;[7] theoretically, this may have been done in order to minimize the time necessary for instructors to evaluate the solutions submitted to them by a large number of students, that is, the choice of the parameters did not have any particular mathematical or educational meaning, even though the choice of parameters was the result of the use of the text in an educational context. The choice of the parameters could have also been related to the intention of the compilers to assign relatively small integer values (or common fractions with relatively small denominators) to the parameters the learners were supposed to operate with at all steps when solving the problems, as well as to their answers; when solving such problems, the learners thus would not have been distracted by complex computations and would focus on the algorithms instead. In some cases there are strong reasons to suggest that the parameters were meticulously selected to highlight some particular elements of the methods introduced; the sequence of problems discussed in this paper belongs to this category.

As I will try to make clear in this paper, the order in which generic problems were listed in a given mathematical treatise also may have played an important role, and therefore when interpreting such a series one needs to make two assumptions:

1. the instructors administered the problems to the learners in the order in which they were listed in the treatise, and
2. the order of problems in the received editions of the treatise is the same as in its original edition.

As far as assumption (1) is concerned, my observations of present-day educational activities suggest that the actual order in which the problems are administered to the learners by some teachers may differ considerably from that found in the textbooks. Unfortunately, the available descriptions of the instruction procedures in pre-modern mathematical traditions do not specify how closely the order of the problems in the textbooks was followed by the instructors. As for assumption (2), it is not always certain whether the order of problems in the original edition(s) of a given textbook was identical to that found in its received editions; in some cases there are reasons to believe that the original order of problems may have been altered considerably.[8] Moreover, it remains unclear to what extent one can apply the very

[7]Series of Babylonian algebraic problems all having the same answer were discussed in Vaiman 1961: 174–175; Høyrup 1990: 330 also mentions such problems and claims, without providing references though, that "it was recognized already in the early 1930s that Babylonian 'algebra' problems were constructed from known solutions." I am thankful to Dr. Evgenii I. Slavutin (Moscow) who kindly drew my attention to this fact in a series of inspiring discussions in the 1980s.

[8]This is the case of the series of problems on weighted distribution from Chapter 3 of the *Jiu zhang suan shu*: an analysis of a source dated of the late first millennium AD shows that the order of the problems in all the extant editions differs from that found in an earlier (and now lost) edition of the text (Volkov 2011).

concept of the "established order of problems" in the case of those mathematical treatises that existed exclusively in manuscript form since the order of problems may have varied across the copies of one and the same treatise.[9]

10.3 State Mathematics Education in China: A Historical Overview

In China, the state-run School of Computations (*Suan xue* 筭學) was established no later than the Northern Zhou 北周 dynasty (557–581)[10]; the School was reopened during the Sui 隋 (581–618) and Tang 唐 (618–907) dynasties. Information about the School during the Northern Zhou and Sui periods is scarce. The extant descriptions of the School during the Tang dynasty provide descriptions of the textbooks used and of the instruction procedures.[11]

Twelve mathematical treatises used as textbooks in the School of Computations were edited by a team of high-ranking scholars and officially approved in 656.[12] Virtually all the extant mathematical treatises from the late first millennium BC – first millennium AD such as, for instance, the aforementioned *Jiu zhang suan shu* 九章筭術 (Computational Procedures of Nine Categories) were among the textbooks used during the Tang dynasty.[13]

The instruction included oral explanations provided by instructors, quizzes conducted every 10 days, and yearly examinations. A quiz included two questions on memorization and one question on the "general meaning" of an excerpt from a mathematical textbook, while a yearly examination included ten questions on "general meaning" answered orally.[14] Those students who successfully graduated from the School were allowed to take the examination for the doctoral degree

[9]The earliest printed editions of the Chinese mathematical treatises used for educational purposes are dated of the late eleventh century; one can therefore only speak about their established versions starting from that time. As far as the extant Vietnamese mathematical treatises are concerned, the majority of them are handwritten, and it is unknown whether their printed versions ever existed.

[10]Sun 2000: 138; Lee 2000: 515, n. 230.

[11]See des Rotours 1932: 139–142, 154–155; Siu 1995: 226; Siu and Volkov 1999; Volkov 2013.

[12]It is possible that earlier editions of at least nine of these textbooks had been used for instruction in the School of the Northern Zhou dynasty (Volkov 2014). Descriptions of the mathematics curriculum of the Tang State University and the identification of the textbooks with the extant mathematical treatises as well as information about their authors can be found in Siu and Volkov 1999; Volkov 2012, 2014.

[13]The extant treatises of this period not used in the School are the recently unearthed mathematical texts of the late first millennium BC as well as the mathematical manuscripts found in the Dunhuang caves (Gansu Province, China). However, it is quite possible that these texts were also used for educational purposes.

[14]Des Rotours 1932: 141–142, for a discussion of the procedure see Siu and Volkov 1999; Volkov 2012.

in mathematics (*ming suan* 明筭). The latter examination was made up of two parts, the first part consisted of writing essays in answering ten questions, while the second part was a test of memorization.[15] It has been argued that in the first part the candidates were asked to solve mathematical problems similar to those solved in the treatises studied yet with modified numerical parameters.[16] The change of parameters in a given problem may have required a modification of the algorithm already known to the candidates.[17] To solve the problems, the candidates thus had to use algorithms which were not mere replicas of those found in the textbooks but design generic versions suited to the modified parameters of the problems.

The School was reopened for short periods of time in the late eleventh and early twelfth centuries. After the transfer of the capital in 1127 the School ceased to exist as an independent unit, yet mathematics instruction continued in the State Astronomical Bureau; it was combined with mathematical astronomy and calendrical computations.

As far as the teaching materials are concerned, a set of textbooks based on the treatises used for instruction during the Tang dynasty was edited and block-printed for the first time in 1084. Later, the textbooks were reprinted in 1200–1213 on the basis of the still extant 1084 prints and other sources. The exact number of textbooks printed in 1084 and reprinted in the early thirteenth century is not known; Cheng Dawei 程大位 (1533–1606) suggested that it was ten.[18] The available descriptions suggest that, in the early second millennium AD, only some of these texts (including the *Jiu zhang suan shu*) were used for admission and internal examinations.

The only source of information concerning the mathematics instruction at that time is the chapter entitled "Xi suan gang mu" 習算綱目 (Master list [of topics] for exercising computations) of the mathematical treatise *Suan fa tong bian ben mo* 算法通變本末 (Alpha and Omega of the Continuity and Transformation of the Counting Methods, 1274) by Yang Hui 楊輝 (active in 1261–1275). The "Master list" contains a detailed description of the topics supposed to be known by those who studied mathematics and the number of days necessary to study each topic.[19] However, it is not certain whether Yang Hui was describing the instruction in the governmental School of Computations in the early thirteenth century, since he mentions two mathematical treatises not listed in the received descriptions of the School.

[15]On the memorization test see des Rotours 1932: 30–31, 141, n. 2; Siu and Volkov 1999: 91, n. 41; Lee 2000: 142.

[16]Siu and Volkov 1999; a piece of evidence supporting this hypothesis is found in a Vietnamese mathematical treatise; see Volkov 2012.

[17]This conjecture was made in Siu and Volkov 1999 and discussed in great detail in Siu 1999, 2004.

[18]*CDW* 1990: 991.

[19]*YH* 1993: 1048–1050; for English and French translations see Lam 1977: 11–14 and Friedsam 2003: 58–63, respectively; see also Zhou 1990: 397.

During the (Mongolian) Yuan 元 dynasty (1279–1368), certain mathematical subjects were taught at the Astronomical Bureau (*Si tian jian* 司天監 or *Si tian tai* 司天臺) and at the Academy of the Great Astrologer (*Tai shi yuan* 太史院),[20] yet nothing is known about the *syllabi*.[21] The rulers of the Ming 明 dynasty (1368–1644) made attempts to restore state mathematics education;[22] however, no information exists about any systematic attempts to rebuild curricula, reprint mathematical textbooks, and re-establish the examinations. Traditional Chinese state-run mathematics education thus ceased to exist no later than the late fourteenth century.

10.4 Vietnamese Mathematics: Education and Treatises

Vietnam was a province of the Chinese Empire from the second century BC until the tenth century AD. During this time, young Vietnamese males could participate in the provincial and national degree examinations and were eligible, theoretically at least, to study at the (Chinese) metropolitan School of Computations; the mathematical texts used for preparation for the exams that circulated in the province therefore would have been compatible with those used for instruction in the capital. After Vietnam separated from China in 939, the Vietnamese rulers established a system of national education emulating the contemporaneous Chinese model. The state-run mathematics education in Vietnam outlived its Chinese counterpart: there are historical records mentioning the examinations in "counting" (Viet. *toán* 算) that took place in Vietnam in 1077, 1179, 1261, 1363, 1404, 1437, 1477, 1507, 1711, and 1762.[23] It appears reasonable to conjecture that the Vietnamese teaching methodology and examination procedures imitated the Chinese ones, at least at the initial stages after the split in 939. No actual essays produced during mathematical examinations have been found so far, yet a "model" examination paper does exist, suggesting to the learners how an ideal examination answer was supposed to look. This "model" examination essay contains a detailed solution to a problem (interestingly enough, a problem on weighted distribution) found in the Vietnamese treatise *Chỉ minh lập thành toán pháp* 指明立成算法 (Guidance for Understanding of the *Ready-made Computational Methods*) compiled by Phan Huy Khuông 潘輝框 in 1820.[24] This piece of evidence demonstrates rather unambiguously that by the early nineteenth century at least *some* Vietnamese mathematical treatises were designed as textbooks to be used by candidates preparing for the state mathematics examinations.

[20] Sun 1996: 517; Lee 2000: 521; Sivin 2009: 147.

[21] Sun 1996: 517; Yang 2003, vol. 2: 819–820.

[22] Li 1955: 279–280.

[23] Volkov 2002: 373.

[24] A translation and an analysis of this model examination paper can be found in Volkov 2012.

The number of extant Vietnamese mathematical treatises amounts to 22. One of them, the *Toán pháp đại thành* 算法大成 (Great Compendium of Mathematical Methods) is conventionally dated as coming from the fifteenth century, yet it is more likely that it was compiled as late as the eighteenth or nineteenth century.[25] Another book, the *Cửu chương lập thành tính pháp* 九章立成併法 (Ready-made Methods of Addition [sic] of Nine Categories) was block-printed in the early eighteenth century. Three more books were published as late as 1909; the dates of publication for the rest of the mathematical treatises are uncertain, yet a preliminary analysis suggests that most likely they were compiled between the late eighteenth century and 1900.[26] All these books are written in the style of traditional Chinese mathematical treatises, i.e. having the format "problem – answer – solution," and nine of them contain voluminous quotations from the Chinese mathematical treatise *Suan fa tong zong* 算法統宗 (Systematic Treatise on Methods of Computation, 1592) by Cheng Dawei 程大位 (1533–1606).[27] The remaining 13 books, including the treatise discussed in the present paper do not, to the best of my knowledge, contain such large *verbatim* quotations from the extant Chinese mathematical treatises.

10.5 Nguyễn Hữu Thận and His Treatise

Below I will argue that in pre-modern mathematical treatises certain methods may have been introduced, and their rationale, explained, with sequences of problems with especially selected values of the involved parameters. To provide an example supporting this hypothesis, I will discuss the Vietnamese mathematical treatise *Ý Trai toán pháp nhất đắc lục* 意齋算法一得錄 (A Record of What Ý Trai [=Nguyễn Hữu Thận] Got Right in Computational Methods) by Nguyễn Hữu Thận 阮有慎 (to be referred to as *A Record* hereafter).[28] In this section I will briefly introduce the treatise and its author.

The reference manual by Trinh Khac Mạnh and the *Catalogue of the Books in Han-Nom* by Trần Nghĩa and François Gros both suggest that Nguyễn Hữu Thận lived from 1736 to 1813;[29] however, the preface of *A Record* is signed by Nguyễn Hữu Thận in the 10th year of the Minh Mệnh 明命 era (1829),[30] and it is therefore

[25]Volkov 2002, 2005, 2006.

[26]Volkov 2008, 2009.

[27]*CDW* 1990.

[28]*YT* 1829. The title of the treatise contains an allusion to the Chinese proverb (*chengyu*) "*qian lü yi de*" 千慮一得 ("Out of one thousand ideas [of a silly man] one is right").

[29]Trần and Gros 1993, 2: 51; Trinh 2007: 509. I am grateful to Nguyễn Thị Dương who kindly sent me a translation of the biographical notice on Nguyễn Hữu Thận from Trinh 2007.

[30]See also the notice no. 4505 in Trần and Gros 1993, 3: 708–709.

plausible to suggest that Nguyễn passed away some time after 1829.[31] Nguyễn Hữu Thận had the style name (*tự* 字) Chơn Chi 真之 and pseudonym (*hiệu* 號) Ý Trai 意齋,[32] occupied the position of Vice-Minister *Thị lang* 侍郎[33] of Ministry of Finance *Bộ Hộ* 戶部 during the Tây Sơn 西山 period (1778–1802) and the position of Minister *Thượng thư* 尚書 of Ministry of Civil Service Affairs *Bộ Lại* 吏部 during the Nguyễn Dynasty (1802–1945).[34] Besides the mathematical treatise, Nguyễn Hữu Thận authored the *Tam Thiên Tự Lịch Đại Văn Chú* 三千字歷代文註 (Commentaries on the [*Canon of*] *Three Thousand Characters* through generations)[35] and *Minh Mệnh Công Văn* 明命公文 (Official documents of the Minh Mang [era])[36] as well as a memorial appended to the *Hiếu Sử Lược Thuyên* 孝史略詮 (Commentaries and Historical Examples of Filial Piety).[37] The preface of *A Record* suggests that Nguyễn visited China some time between 1809 and 1813 where he obtained the Chinese astronomical treatise *Li xiang kao cheng* 曆象考成 (Completion of the Research on Calendar and [Celestial] Bodies, 1713) and discussed questions related to calendrical computations with local experts.[38]

A Record contains eight chapters. The first chapter features the multiplication table $9 \cdot 9$ as well as other tables, including magic squares; chapter 2 deals with the calculation of the areas of flat figures; chapter 3 is devoted to weighted distribution; chapter 4 focuses on the "extraction of square roots" (that is, in modern terms, quadratic equations); chapter 5 deals with right-angled triangles; chapter 6 is devoted to calculation of volumes; chapter 7 discusses so-called "difficult problems"; and chapter 8 is related to the "extraction of cube roots" (that is, the numerical solution of cubic equations).

10.6 Didactical Variables and Representations Consistent with Parameters

The experimentally observed phenomenon of the modification of mental representations and reasoning strategies by the learners in response to modifications of the settings of mathematical problems has been discussed by modern educators in terms of "didactical variables", that is, those numerical and textual elements of the

---

[31] Some Vietnamese authors recently suggested 1757–1831 for his lifetime.

[32] Trinh 2007: 509–510; see also Trần and Gros 1993, notices 1707, 3091, 4505 (2: 50–51; 3: 22, 708–709); Liu et al. 2002, notices 0123 (1: 22), 1954 (1: 384), and 2772 (1: 526).

[33] This interpretation of the term 侍郎 used in the biography is based on its Chinese equivalent; see Hucker 1985: 426–427, notice 5278.

[34] See Hucker 1985: 306–307, notice 3630; 410–411, notice 5042.

[35] Trần and Gros 1993, 3: 22, notice 3091.

[36] Trần and Gros 1993, 2: 286, notice 2165.

[37] Trần and Gros 1993, 1: 776–777; Liu et al. 2002: 21.

[38] *YT* 1829, preface: 1b.

Fig. 10.1 Sharing the amount S = 12 among n = 3 sharers

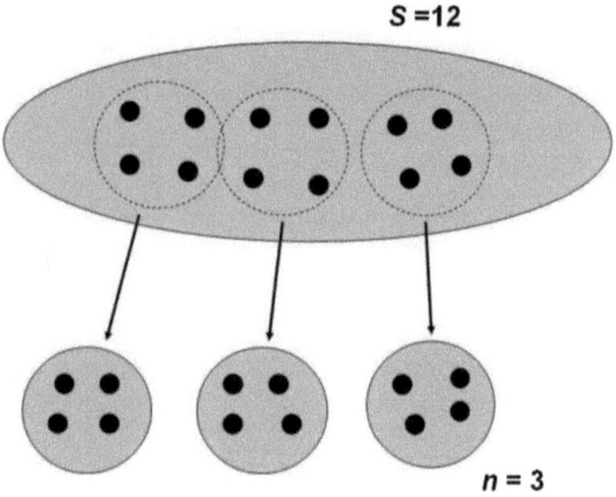

problems that may shift the learners' attention in various directions and thus result in different approaches to "mathematically identical" problems.[39]

The values of the parameters involved, the choice of the units of measurement, and even of the wording of a given mathematical problem may impose particular restrictions on the values of the parameters involved and suggest certain reasoning strategies (not necessarily relevant or efficient) to the learners. For example, if we deal with a problem of the flat-rate distribution of *S* objects among *n* sharers, the choice of units for the value of *n* is restricted to positive integers, and the choice of units for the value of S determines whether *S* can be an integer or a fraction, a positive or a negative number. However, the solution of the problem in all cases can be expressed by the formula $S \div n$ which does not impose any restriction on the values of *S* and *n* (except that *n* cannot be equal to 0).

If a problem is worded as dealing with the sharing of a given number of (raw) eggs *S* among a given number, *n*, of peasants one can expect that the original number of eggs and the number of peasants to be positive and integers, and, moreover, each peasant will obtain an integer number of eggs. These conditions thus set specific restrictions on the involved parameters *S* and *n*, as well as on the answer $S \div n$, and at the same time specify a model which can be visualized, for example, as shown in Fig. 10.1.[40]

If, however, the answer obtained is in conflict with the type of the numerical parameters involved (e.g., the result is fractional while the unit of measurement

[39]See, for example, Brousseau 1997: 176. By "mathematically identical problems" I mean the problems whose generic solutions can be written with the same mathematical formula.

[40]This interpretation of the division $S \div n$ is quite naturally dubbed "distribution division" in the literature on didactics (Gravemejier 1997: 335), since several objects are (equally) distributed among several sharers.

is such that the result can only be an integer, as in the case of raw eggs),[41] the imposed representation cannot be used any longer and should be replaced by a modified one, or some *ad hoc* additional strategy has to be applied to safeguard the original representation. For example, when solving the problem of sharing eggs, one can apply division with a remainder, thus leaving the original representation unchanged, or, if the division with a remainder does not fit into the context, switch to representations featuring, for example, beaten eggs or fried eggs. The change in representation is thus related to the change in the numerical values of the parameters set by the author(s), and can lead to a change in the degree of complexity of the problem and, in some cases, can help the learners establish links with other problems and other methods.

10.7 Weighted Distribution in *A Record*

Chapter 3 of *A Record* is devoted to weighted distribution and contains 27 problems. In this section I will show how the author used these problems to introduce the method of weighted distribution. In my analysis I will focus on the first eight problems in the chapter; it will be argued that this series of problems was designed specially to introduce the concept of weighted distribution.

This discussion below requires a short introduction of the necessary mathematical terminology. "Weighted distribution" (or "proportional distribution/sharing") is the distribution of a given amount S among a number of "sharers", n, according to given "weights" assigned to the "sharers". Symbolically, one has to find the values $x_1, x_2, \ldots, x_n$ given that $x_1 + x_2 + \cdots + x_n = S$ and $x_1 : x_2 \ \ldots \ x_n :: k_1 : k_2 : \ldots : k_n$ for some weighting coefficients $k_1, k_2, \ldots, k_n$. In the case when $k_1 : k_2 : \ldots : k_n ::$ $1 : 1 : \ldots : 1$, that is, when all the weights are equal, I will use the term "flat-rate distribution". A problem on flat-rate distribution corresponds to the search for the arithmetical mean $M = \frac{S}{n}$, since in this case the solution to the distribution problem is provided by $x_i = M$ for $i = 1, \ldots, n$. The solution of a weighted distribution problem in general case is given by the formula $x_j = \frac{Sk_j}{\sum_{i=1}^{n} k_i}$.

Problem 1 of Chapter 3 reads as follows:

平分。

說。有錢十七貫五陌。分三十六人。

問。每人[得]幾何。

法。以人數三十六人。為法。錢數十七貫五陌。通成錢積。一萬零五百文。為寔。法除寔。得二百九十一文。為每人所分錢數。外餘二十四文。記之。[42]

[41]Problems with such conditions are found in various traditions; Chinese mathematical treatises, for example, contain problems in which fractional numbers of individuals are supposed to be obtained, see Volkov 2011.

[42]*YT* 1829, 3: 2a. The original manuscript *YT* 1829 contains commentaries written in smaller characters, and since the first page does not contain any mention of commentator(s), I assume that

Level sharing.

It is said: there is money, 17 *quan* 5 *bách*. Share [it among] 36 people.

Question: how much will each person get?

The method is: take the number of people, <36 people,> this will be the divisor. [Take] the amount of money, <17 *quan* 5 *bách*,> convert it [into the smallest monetary units], establish the integral amount of money, <10,500 *văn*>. [This] is the dividend. Divide the dividend by the divisor. <Obtain 291 *văn*.> This is the amount of money each person will obtain as a share. Besides, there is a remainder, 24 *văn*. Write it down.

Here one has to share 17 *quan* 5 *bách* [= 10,500 *văn*] among 36 people.[43] The answer is given as an integer amount of money (291 *văn*) obtained by each person, and a remainder equal to 24 *văn*.[44] This classical flat-rate distribution problem explicitly suggests the model of "sharing a given integer among a given number of sharers" similar to that shown in Fig. 10.1. In this model the divisor n is supposed to be an integer; it is equal to the number of groups into which the tokens representing the dividend are to be organized. Since *văn* is the smallest monetary unit, the answer is also supposed to be an integer. Thus the setting of the problem assumed that the divisor, the dividend, and the quotient are integer numbers, and in the case of the dividend not being a multiple of the divisor, the "sharing" is represented with the operation of division with a remainder. The underlying representation thus suggests that the number of sharers, n, has to be an integer; if it is not, the interpretation of the operation of division has to be radically modified.

Problem 2 reads as follows:

平分有半人。
說有錢十八貫一陌三十二文。分三十六人半。
問每得幾[何]。
法。以人數三十六人半。半即五分。為法。錢數十八貫一陌三十二文。通成錢積。一萬零八百九十二文。為寔。定位在百位。半人不可計為寔多法一位之數。法除寔。得二百九十八文。為全人。三十六人。所受錢數。折半。得一百四十九文。為半人所受錢數。外餘錢十五文。記之。要還原依法。[45]

Level sharing with a half-person.

It is said: there is money, 18 *quan* 1 *bách* 32 *văn*. Share [it among] 36 and a half people.

Question: how much each person will get?

The method is: take the number of people, <36 and a half people, a half is 5 parts>, this will be the divisor. [Take] the amount of money, <18 *quan* 1 *bách* 32 *văn*,> convert it

the commentaries were added by the author himself; some of the commentaries quoted below are consistent with this hypothesis. When quoting the original treatise I reproduce the commentaries in a smaller font size, in translation they are placed in angular brackets <like this>. The text quoted below contains several corrections made on the basis of comparison of *YT* 1829 with another manuscript copy of the treatise preserved in the library of Han-Nom Institute, Hanoi (call number VHv.1184). Added characters in the text and added words in the translation are placed in square brackets. The translation is mine.

[43]The Vietnamese monetary units used here are *quan* 貫, *bách* 陌, and *văn* 文; they had the following relative values: 1 *quan* = 10 *bách* = 600 *văn*.

[44]The mathematical identity justifying this answer is $(10{,}500–24) \div 36 = 291$.

[45]*YT* 1829, 3: 2a–b.

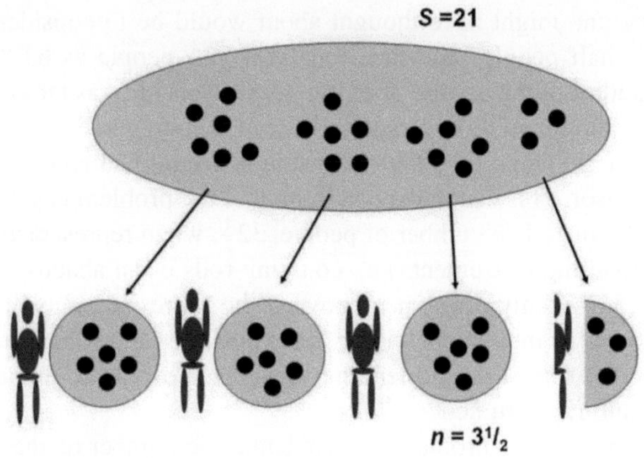

Fig. 10.2 Sharing 21 objects among $3\frac{1}{2}$ people

> [into the smallest monetary units], establish the integral amount of money, 10,892 *văn*. [This] is the dividend. Fix the [highest] position [of the divisor] in the position of hundreds. <Half a person cannot be counted, to [modify] the dividend increase [it and] the divisor; the [increment] value is of 1 position.> Divide the [modified] dividend by the [modified] divisor, to obtain 298 *văn*. For the "entire" people, <36 people,> [this is] the amount of money obtained [by each person]. Cut [this amount] in half, to obtain 149 *văn*. [This] is the amount of money obtained by the half-person. Besides, there is a remainder, 15 *văn*. Write it down. [If you] want to "return to the source",[46] rely upon the method.[47]

Textually, the problem is almost identical to Problem 1, yet it contains an important modification. According to the conditions, one should share a given amount of money (18 *quan* 1 *bách* 32 *văn* = 10,892 *văn*) among a *fractional* number of sharers ($36\frac{1}{2}$ people). The connection between the two problems is made quite clear: one should pass from 32 people to $32\frac{1}{2}$ people, whatever that could mean.[48] This small modification of one parameter, mathematically (but not didactically!) insignificant, radically changes the associated representation. It is difficult to imagine how the diagram shown in Fig. 10.1 would have been changed in order to account for this modification; one of the possibilities is shown in Fig. 10.2.[49]

[46]"Return to the source" is a standard term meaning a verification of the numerical solution.

[47]This instruction may have prescribed performing the operations in reverse order. See a verification of this type provided in Problem 3 below.

[48]The Vietnamese author is not the first to suggest sharing an amount among a fractional number of sharers, it had been done many times well before him. One of the earliest examples can be found in Problem 18 of Chapter 1 of the *Jiu zhang suan shu*: $6^1/_3$ coins are to be shared among $3^1/_3$ people (Chemla and Guo 2004: 167).

[49]The visual representations shown in Fig. 10.2 as well as in the other figures below are *not* reconstructions of the representations of the Vietnamese learners (for which no evidence exists); they are provided for the modern reader to depict the underlying models of distribution imposed by the choice of parameters. The actual representations employed by learners in the traditions discussed, theoretically, may have differed considerably from the suggested here.

The strategy one might have thought about would be to consider each person as two virtual "half people" and thus represent $32\frac{1}{2}$ people as 63 "half people". However, the author of the treatise does not discuss this idea, as far as the (probably corrupted) text allows us to understand; instead, he suggests interpreting $32\frac{1}{2}$ as 325, multiply the dividend by 10, and dividing the modified amount of money by the modified divisor. This can make one think that the problem is solved by formal analogy to Problem 1. The number of people, $32\frac{1}{2}$, when represented as a decimal fraction on a counting instrument (i.e., counting rods or an abacus),[50] would look exactly like 325: the only difference between the representations of the numbers 32.5 and 325 on either instrument will be in the position of the units. In other words, the method used to solve this problem may have been explained to the learner with the use of a counting instrument.

Conversely to the latter problem, in Problem 3 the number of sharers cannot be represented as a (mixed) decimal fraction. The problem reads as follows:

平分有一人。受幾分之幾。

說有錢三貫一陌十五文。分三十人。內有一人受七分之四。

問。各得幾何。

法。於全人三十人內。減出一人。餘二十九人。以母分七乘之。得二百零三分。又以分子四加入。共二百零七分。為法。錢數三貫一陌十五文。通成錢積。一千八百七十五文。又以分母七乘錢積。得一萬三千一百二十五文。為寔。法除寔。得六十三文。餘八十四文。謂之假餘。以分母分七歸之。為外餘錢十二文。其除得六十三文。為全人。二十九人。每人所得錢數。又以六十三文分子四乘之。分母七除之。得三十六文。為內有一人受四分所得錢數。又以全人二十九人。與每人受錢六十三文。相乘。得一千八百二十七文。為全人所受總錢。又加內有一人受錢三十六文。并外餘。十二文。得原錢積。一千八百七十五文。[51]

Level sharing with one person obtaining a fractional [share].[52]

It is said: there is money, 3 *quan* 1 *bách* 15 *văn*. Share [it among] 30 people. Among [them] there is one person obtaining $^4/_7$.

Question: how much will each person get?

Method: subtract one person from the "entire" people <30 people>,[53] the remainder is 29 people. Multiply this [number, i.e. 29] by the denominator, <seven>, to obtain 203 parts. Also add the numerator, <four>. Together this will be 207 parts. This is the divisor. [Take] the amount of money, 3 *quan* 1 *bách* 15 *văn*, convert it [into the smallest monetary units], establish the integral amount of money, 1,875 *văn*. Also, multiply [this] integral amount of money by the denominator, seven, to obtain 13,125 *văn*. [This] is the dividend. Divide the dividend by the divisor, <obtain 63 *văn*>. [Take] the remainder, <84 *văn*, [we] call it fake/borrowed remainder>, divide it by the denominator, <seven>, [this] will be the remaining money left aside, 12 *văn*. This division yields 63 *văn* for the "entire" people, <i.e., for 29 people >, [this] is the amount of money each [of those] people [will] get.

[50]On the use of the counting rods in Vietnam see Mikami 1934; Volkov 2009, 2013.

[51]*YT* 1829, 3: 2b–3b.

[52]The author is using the expression 幾分之幾, literally meaning "a number [m] of parts obtained by division of a unit by an integer number [n]", an equivalent of the modern expression "m/n".

[53]The author uses here the term 全人 he probably coined himself; it can be understood as "entire/complete person" as well as "the person [having a share expressed with] integer [number]". The ambiguity of the term may have been intentional and was meant to reflect the possibility of the two interpretations discussed below.

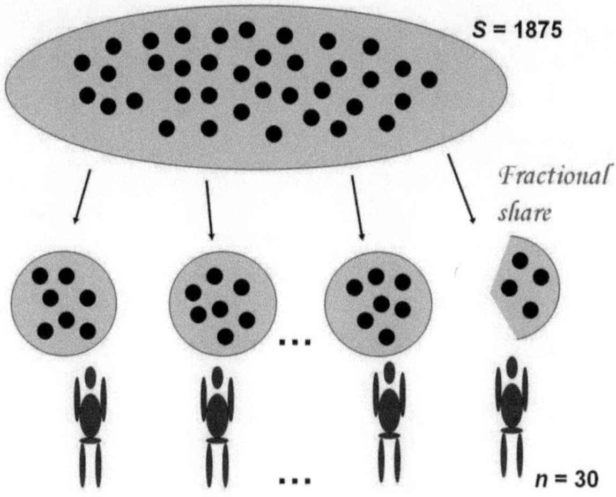

Fig. 10.3 Problem 3 and the re-interpretation of the concept of share

> Also, take 63 *văn* and multiply it by the numerator, <four>, divide it by the denominator, <seven>. Obtain 36 *văn*, this is the amount of money obtained for four parts received by [this] one man in [the group who obtained $^4/_7$ of a share]. Also, take the number of the "entire" people <i.e., 29 people>, multiply by the amount of money obtained by each person <63 *văn*>, to obtain 1,827 *văn*, this will be the total amount of money obtained by [all] the "entire" people. Also, add [to this amount] the money obtained <by the one person> in [the group who obtained $^4/_7$ of a share] <36 *văn*>, add the remainder left aside <12 *văn*>, to obtain the original integrated amount of money, <1,875 *văn*>.

The amount of 3 *quan* 1 *bách* 15 *văn* (that is, 1,875 *văn*) is to be shared among an integer number of people, 30 people, yet one of them obtains $^4/_7$ of a share while 29 people obtain 1 share. Here the representation is changed again: the number of people sharing money is an integer, as in Problem 1, yet the shares are not equal to each other any longer. Problem 2 thus plays the role of a "go-between": on the one hand, it shows that a fractional number of sharers can be interpreted in the context of the flat-rate distribution of Problem 1; on the other hand, the reader can see that the "fractional amount of sharers" ($\frac{1}{2} = {}^4/_8$) is a fraction quite close to the fraction used in Problem 3 ($^4/_7$), yet the latter fraction cannot be represented on the counting device as a decimal fraction. The fractions used suggest a transition from the concept of "one person – one share" to the concept of "a person obtaining a fractional share"; see Fig. 10.3.

The solution to Problem 3 provides a method of dealing with the case when one share has the form $^p/_q$, $p < q$ (a common regular fraction). The solution runs as follows:

To obtain the "divisor", the number of people having one share (29) is to be multiplied by the denominator of the fractional share (seven) thus yielding 203; and the numerator (four) is to be added to this product yielding 207. This number is referred to as the "divisor". The amount of money to be shared, 1,875 *văn*, is to be multiplied by the denominator to obtain a "dividend". The "dividend" is to be divided by the "divisor" yielding the quotient 63 *văn* and the remainder 84 *văn*.

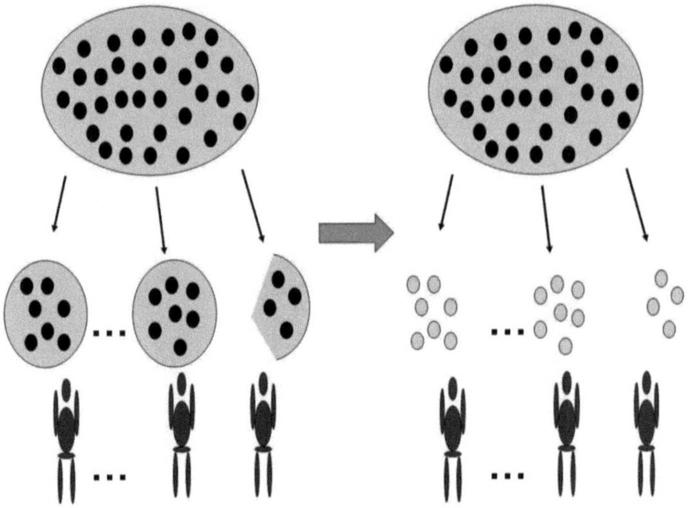

Fig. 10.4 The transition to "new shares"

However, the Vietnamese author stresses that the remainder is not a "true" remainder (he calls it a "fake" or "borrowed" remainder): since the total amount of money was multiplied by seven, to obtain the "true" remainder one has to divide the "fake" one by seven. The 63 *văn* obtained is the amount to be given to each person having one share; as for the person with $^4/_7$ of a share, he will obtain $(^4/_7) \cdot 63 = 36$ *văn*. The last part of the algorithm is devoted to checking the result obtained.

The rationale of the method again is not stated explicitly, but the very sequence of operations may suggest that the original shares were replaced by *smaller* shares, such that each "regular" person had seven of them, and the only person with a fractional share, three of them. This replacement corresponds to the transition from 1 to 7, and from $^4/_7$ to 4, as shown in Fig. 10.4, and the amount to be shared, 1,875 *văn*, therefore must be increased proportionally, that is, multiplied by seven, too.[54] The original problem of distributing 1,875 *văn* among 29 $^4/_7$ sharers is thus transformed into a new problem of distributing $7 \cdot 1{,}875$ *văn* = 13,125 *văn* among $7 \cdot 29\,^4/_7 = 207$ sharers.

The problem is followed by a commentary which reads as follows:

此俗所謂通分。納子法也。内有一人受七分之四。則七分為母。四分為子。三十人之中。二十九人。各受七分。故為母。内有一人受四分。故為子。以七乘人、乘錢。故曰通分。以子加入末位。故曰納子。又以分母七乘錢者。每人既通為七分。則錢每文亦通為七文。乃可相當以除之也。外餘以七歸除。成寔分者。原一文。而通為七文。以當七分。今分之不盡。故以七除。然後復為一文也。此法俗算書所載算術。多有謬妄。今正之。[55]

[54]This multiplication by seven may have been perceived by the reader of the treatise as analogous to the simultaneous multiplication of the "dividend" and "divisor" by ten in the previous problem.

[55]*YT* 1829, 3: 3b–4a.

> This is what is called in a vulgar way "the method of connecting fractions and inserting numerators". The reason is: among [the sharers] there is one person obtaining $^4/_7$ [of a share], thus 7 parts is a denominator, 4 parts is a numerator. Among the 30 people, 29 obtain seven parts each, this is why [seven] is the denominator. One person among [them] obtains four parts, this is why [four] is the numerator. [The procedure prescribes] multiply the people and money by seven, this is why it is called "connecting fractions". [The procedure prescribes] take the numerator [= four] and enter it in the last position [= add in the position of units], this is why it is called "inserting numerators". Also, [it is prescribed] multiply the [shared amount of] money by the denominator, seven; [the meaning of it is that] since each person was "connected" [= proportionally increased] with [the coefficient of] seven parts, therefore, in the [amount of] money each *văn* also [should be] "connected" [= proportionally increased] to become seven *văn*, and only after that [their amounts would] correspond to each other [and one could] be divided [by the other]. [Take] the remainder left aside, divide it by seven, establish the "actual parts"; [the meaning of it is that each] one original *văn* [through the operation of] "connecting" became seven *văn*, in order to correspond to seven parts. Now, the distribution of it [= the amount of money] is not completed [since there is a remainder]. This is why after [this] division by seven [we] return to "one *văn*". This method is written in vulgar mathematical books as "computational procedure" [= an algorithm lacking explanations], and many of them contain mistakes and absurdities. Now [I] corrected it.

This commentary provides a justification for the procedure using the standard Chinese terms of operations with fractions, 通分 (Chinese *tong fen*) and 納子 (Chinese *na zi*). The procedure mentioned here is identical to the one described in the treatise, yet the terms differ. The "vulgar books" the author refers to are, apparently, mathematical texts that circulated prior to the compilation of his treatise; their origin and content are impossible to identify. The author does not dismiss the explanation he quoted, he only remarks that these "vulgar" books contained mistakes that he corrected; one can thus understand that this commentary contains a corrected version of the justification of an algorithm found in mathematical books compiled prior to the 1820s.

In Problem 4, the situation is made even more complex: a given amount of money, 75 *quan* (= 45,000 *văn*), is to be shared among 239 people, 234 of whom obtain one share each and five people obtain $^7/_9$ of a share. The original text reads as follows:

> 平分内有多人。受幾分之幾。
>
> 説有錢七十五貫。分二百三十九人。内有五人受九分之七。
>
> 問各得幾何。
>
> 法於人數。二百三十九人。減内有[人]五人。餘為全人。二百三十四人。以母分九乘之。得二千一百零六分數。又以子分七。乘内有人五人。得數三十五分。兩乘積相併。得二千一百四十一分。為法。錢數七十五貫。通成錢積四萬五千文。又以母分九乘之。得四十萬五千文。為寔。法除寔。得一百八十九文。為全人。二百三十四人。後者凡稱全人者。倣此推之。每人所得錢數。所餘三百五十一文。以母分九除。得寔餘錢。三十九文。爰以全人所得錢一百八十九文。以子分七乘。以母分九除。得一百四十七文。為内有人五人每人所得錢。又以五人乘之。得七百三十五文。内有五人總錢。再以全人二百三十四人。與每人錢一百八十九文相乘。得四萬四千二百二十六文。為全人總錢。併兩總錢。并外餘錢。即原問錢積錢數。[56]

[56] *YT* 1829, 3: 4a–5a.

Level sharing with numerous people obtaining fractional [shares].

It is said: there is money, 75 *quan*. Share [it among] 239 people. Among [them] there are five people who obtain $^7/_9$ [each].

Question: how much will each [person] get?

Method: from the [total] number of people <239 people> subtract the "people inside [the group obtaining fractional shares]" <i.e., five people>, the remainder will be the "entire people" <234 people>. Multiply this [amount] by the denominator <nine>, to obtain 2,106, the number of shares. Also, multiply the [number] of 'people inside [the group obtaining fractional shares]' <five people> by the numerator <seven>, to obtain the number of 35 shares. Add both products, to obtain 2,141. This will be the divisor. Take the amount of money, 75 *quan*, convert [it into the smallest monetary units] and establish the aggregated amount, 45,000 *văn*. Again, multiply this by the denominator, nine, to obtain 405,000 *văn*. This will be the dividend. Divide the dividend by the divisor, to obtain 189 *văn*, this is the amount of money obtained by each person among the "entire people", <234 people; hereafter, in all cases for the so-called "entire people" proceed similarly.> The remaining 351 *văn*, divide it by the denominator <nine>, to obtain the remaining money of the dividend, 39 *văn*. Then multiply the [amount of] money obtained by [each] "entire person", 189 *văn*, by the numerator, <seven>, and divide by the denominator, <nine>, to obtain 147 *văn*. This will be the [amount] of money obtained by each person among the "people inside [the group obtaining fractional shares]," <five people>. Also, multiply it by <five people>, to obtain 735 *văn*. It will be the total [amount] of money [obtained] by the "people inside [the group obtaining fractional shares]," five people. Once again, multiply the "entire people," <234 people> and the [amount] of money [obtained] by each person <189 *văn*>, to obtain 44,226 *văn*, this is the total [amount of] money obtained by the "entire people." Add the two total [amounts of] money, add the [amount of] money remaining outside, then this will be the aggregated amount of money in the original money-[sharing] problem.

According to the algorithm, the first operation to be performed is to multiply the number of people with one share (234) by nine to obtain 2,106. At the next step, the number of people obtaining the fractional share, five, is to be multiplied by the numerator of the fractional share, seven, to obtain 35. These two products are to be added to yield 2,141; this number is to be kept as the "divisor". The "dividend" is obtained *via* two successive multiplications: the amount to share (75 *quan*) is multiplied by 600 to obtain 45,000 *văn*, and then by the denominator, nine, to obtain 405,000. The result of the division $405{,}000 \div 2{,}141$ is not an integer number; the integer quotient is 189, and the remainder, 351 *văn*. According to the author, this remainder, however, is not the "true remainder", as in the previous problem; to produce the "true" one, 39, it has to be divided by nine (the denominator of the fractional share). The answer, 189 *văn*, is clearly stated only for the aforementioned "entire people" (that is, for the 234 people having 1 share). To calculate the amount of money to be obtained by each of the five people getting $^7/_9$ of a share (147 *văn*), one should multiply 189 by $^7/_9$. The last part of the procedure is the verification of the result obtained; it consists only of the summation of all the calculated amounts yielding the original total, 75 *quan*.

Problem 5 belongs to the same category as Problem 4 and is designed, according to the author's commentary, in order to offer another way of solving this type of problem. It reads as follows:

平分内有多人。受幾分之幾。另有算法。不用通分納子。

說有錢二十五貫一陌三十文。分二百四十九人。

内有十七人受七分之三。問各得幾何。

法。以人數二百四十九人。減内有十七人。餘二百三十二人。為全人。以母分七乘之。得一千六百二十四分。為全人分。又以内有人十七人。與子分三相乘。得五十一分。為内有人分兩分積。相併。得一千六百七十五分。為總分積。為法。錢數通成錢積。一萬五千零九十文。為寔。法除寔。得九文。為每分所得錢。外餘十五文。記之。爰以每分所得九文。與母分七乘之。得六十三文。為全人二百三十二人。每人所得錢。又以全人二百三十二人乘之。得全人總錢。一萬四千六百十六文。再以每分所得錢九文。與子分三相乘。得二十七文。為内有十七人。每人所得錢。又以内有十七人乘之。得内有十七人總錢。四百五十九文。要還原依法。此另有一法。或用前條通分納子法算之。其所分錢數。亦相符合。[57]

Level sharing with numerous people obtaining fractional [shares] [solved] with another computational method not using the [procedure of] “reducing fractions [to a common denominator] and inserting [the denominator] into the numerator.”

It is said: there is money, 25 *quan*, 1 *bách* 35 *văn*. Share [it among] 249 people. Among [them] there are 17 people obtaining $^{3}/_{7}$ [each].

Question: how much will each [person] get?

Method: from the [total] number of people <249 people> subtract the “people inside [the group obtaining fractional shares],” 17 people, the remainder is 232 people, those will be the “entire people”. Multiply this [amount] by the denominator, seven, to obtain 1,624, the [number of] shares. [Those] are the shares of the “entire people.” Also, multiply the [number] of the “people inside [the group obtaining fractional shares]” <i.e., 17 people> by the numerator, three, to obtain 51 shares. These are two integrated amounts of shares [of the “entire people”] and the “people inside [the group obtaining] fractional [shares].” Add [them] to each other, to obtain 1,675 shares. This will be the total integrated amount of shares. It will be the divisor. Take the amount of money, convert [it into the smallest monetary units] and establish the aggregated amount, 15,090 *văn*. This will be the dividend. [Divide the dividend by the divisor], to obtain 9 *văn*, this is the [amount of] money obtained by each [person]. The remaining 15 *văn*, write [this amount] down. Then multiply nine *văn* obtained for each share by the numerator, <seven>, to obtain 63 *văn*. This will be the [amount] of money obtained by each person among the “entire people,” <i.e., 232 people>. Also, multiply it by [the number of] the “entire people” <232 people>, to obtain the total [amount] of money [obtained] by the “entire people” <232 people>, 14,616 *văn*. Again, multiply the [amount of] money obtained for each share, nine *văn*, and the numerator, three, to obtain 27 *văn*, [this] will be the [amount of] money obtained by each person not in the “people inside [the group obtaining fractional shares],” 17 people. Also, multiply [this amount of money] by [the number of] the “[people] inside [the group obtaining fractional shares],” obtain the total [amount of] money of the “[people] inside [the group obtaining fractional shares],” 17 people, 459 *văn*. [If you] want to “return to the beginning” [i.e., to verify the answer], rely upon the method. <Here is one more method. Alternatively, [to solve this problem one can use] the method of “reduction of fractions [to a common denominator] and the introduction [of denominators] in numerators” [described] in the previous problem. The values of distributed money [obtained] will also coincide [with those obtained here].>

In this problem the given amount of money, 25 *quan* 1 *bách* 30 *văn* (that is, 15,090 *văn*), is to be shared among 249 people, of which 17 people should obtain $^{3}/_{7}$ of a share, while the others receive one share each. According to the algorithm

---

[57] *YT* 1829, 3: 5b–6a.

provided, the first operation is to multiply 232 (the number of people with one share) by seven (the denominator of the shares of each of the remaining 17 people) to obtain 1,624 ("new"?) shares or "parts" (here the Chinese character *fen* 分 can be understood in both senses). Next operation is to multiply 17 by three to obtain 51 ("new") shares/parts. Adding these two products one obtains 1,675 shares which will play the role of "divisor". In the next step, the amount of money to share, 15,090 *văn*, is divided by the divisor to obtain nine *văn* and a remainder, 15 *văn*. Each "share" thus is equal to nine *văn*. Now, to calculate the amount of money obtained by each "entire person" the value of one share, nine *văn*, has to be multiplied by seven since each "entire person" has seven ("new") shares, $1 = 7 \cdot (^1/_7)$, and the amount of money obtained by each of the 17 people with "fractional shares" is equal to $3 \cdot 9 = 27$ *văn*. At the end the author makes a remark stating that in order to complete the solution one has to carry out a verification (he himself does not perform it). A commentary states that this new method differs from the old one used in the previous problem, and that the old method also can be used to solve the problem and to obtain the same result.

The difference between the two methods is as follows: if S is the amount of money to be shared and the group of sharers consists of N people having one share and M people having $^p/_q$ of a share ($p < q$), then the former procedure (Problem 4) prescribes calculating the divisor $\mathrm{D} = \mathrm{Nq} + \mathrm{Mp}$ and dividing the product Sq by D thus yielding the amount of money to be obtained by each person with one share. In the latter procedure (Problem 5), the same divisor $\mathrm{D} = \mathrm{Nq} + \mathrm{Mp}$ is used to divide the amount of money S, and the amount of money obtained by each person with one share is calculated as the result of the multiplication of the quotient by q (apparently, on the basis of the fact that one "old" share of each person is interpreted as q "new" shares of $^1/_q$ each). Both approaches are therefore based on the same idea of choosing the $^1/_q$ as a newly established share, yet the implementation differs.

In Problem 6 the amount 204 *quan* 1 *bách* 30 *văn* (that is, 122,490 *văn*) is to be shared among 345 people; 289 of them obtain one share, 12 obtain five shares, 19 obtain seven shares, 16 obtain $^6/_8$ of a share, and nine obtain $\frac{1}{2}$ of a share. The parameters thus include a combination of integer weighting coefficients (equal to one and greater than one), and two different fractional coefficients represented by proper common fractions. The only type of coefficient not presented in the problem is a mixed fraction. However, mixed fractions as coefficients are found in Problem 8: here the amount of 94 *quan* 4 *bách* (that is, 56,640 *văn*) is to be shared among 54 people; 12 people obtain $8\,\frac{1}{2}$ shares, 18 obtain $7\,\frac{1}{2}$ shares, and 24 obtain $6\,\frac{1}{2}$ shares.

10.8 Discussion

This presentation of the sequence of problems found in the Vietnamese mathematical treatise shows how a choice of parameters may have been related to the reasoning strategies suggested. The simplest "sharing" representation of the division suggested in Problem 1 by the very nature of the data (integer amounts of the monetary units and people) is radically modified when the number n of the sharers becomes fractional. This happens in Problem 2 where one has to share an integer amount of monetary units among $36\frac{1}{2}$ people. Since a "person" cannot be cut in half, it is reasonable to suggest that the choice of "people" as the units in this position was made in order to draw the attention of the reader to a different interpretation. Even though the problem may have been solved either by formal analogy with Problem 1 or by applying a simultaneous multiplication of the divisor and dividend by 10 suggested by the very structure of the counting device (an abacus or counting rods) used, the "half-person" may have also been perceived, in this particular context, as a person having only a half of a share of a regular person. That is, the choice of "people" as the unit for the number of sharers and the choice of the fractional number for the number of people may have suggested to the learner a new interpretation: seeing "people" *not* as actual individuals, but as "packages of shares". In this case, the number of actual people in this problem may still have been perceived as 37, except that one of them obtained only one half of the amount obtained by the others. This re-interpretation is further developed in the next problem: the number of sharers is equal to 30, yet one of them obtains $^4/_7$ of a share, which could have been understood as a placeholder for an arbitrary common fraction (and not only a decimal fraction, as in the previous problem); as we have noticed, at the same time the fraction used, $^4/_7$, remained close to the fraction $\frac{1}{2}$ ($= {}^4/_8$) used in the previous problem. Once this interpretation was established, not only one, but several people may have had "shares" expressed with common fractions. This representation contains, in a nutshell, the general concept of weighted distribution: each sharer is given a weight, a number of shares, which specifies the amount she or he is supposed to obtain. Problems 6 and 8 conclude the group; they depict the cases in which the sharers have "shares" (weights) expressed with combinations of integers and (proper) fractions as well as with mixed fractions. The problem which would logically conclude this sequence would have contained weights expressed with integer numbers, proper common fractions, and mixed fractions; the reason why such a problem is not found in the treatise cannot be answered here.

10.9 Conclusions

In the problems discussed above the numerical parameters, the choice of units (monetary units and "people"), and the order of the problems itself were all apparently related to the elucidation and justification of the method of solution of the problems on weighted distribution taking flat-rate distribution as a starting point. It remains unknown though whether in doing so the author of the treatise followed earlier Vietnamese mathematical textbooks that were presumably available to him or if he designed this sequence of problems by himself.[58]

---

[58] The Vietnamese treatises available to me do not contain sequences of problems similar to those found in *A Record*.

One more question is related to the aforementioned concept of "didactical variable". Its definition is not limited solely to numerical values or the wording of mathematical problems but also may be understood as including certain elements of the context in which the problems were posed, for example, whether the problems were posed orally or in written form, what was the actual order in which the problems were administered to the learners, and so on. In this paper I do not discuss these "extra-textual" elements, mainly for the reason that virtually nothing is known about the process of mathematics instruction in the Vietnamese state-run educational institutions, and even less is known about the interaction between instructors and students in the context of private small-scale schools. The extant Vietnamese mathematical treatises, especially those arguably used for educational purposes, are more reliable sources of information, and in this paper, when discussing "didactical variables," I referred exclusively to identifiable textual elements, such as the numerical parameters and the wording of problems.

The principal difficulty in interpreting ancient texts with the help of the concept of "didactical variable" is related first of all to the very nature of the phenomenon this concept is supposed to discern: the "didactical variable" is not, strictly speaking, a purely textual element; since it is the element of the text of a problem which, when modified, may result in a modification of the strategies of the learners (or at least *some* of the learners), its identification and study would necessarily involve experiments with the learners. In recent studies certain elements in mathematical problems were identified by the researchers as potential "variables" prior to the experiments, and series of problems with various "values" of the "variables" were administered to groups of learners; the outcomes of the experiment were analyzed, and interviews with the learners were conducted in some cases to identify the impact of the changed parameters on their reasoning strategies. No part of this experimental methodology, for obvious reasons, can be applied to the study of the ancient mathematical textbooks. In particular, it is impossible to know how certain mathematical concepts were, in fact, perceived at the time and what related difficulties the learners experienced in the context of the contemporaneous mathematical culture they all shared: it is quite possible that the elements perceived as "difficult" by modern learners (as well as by the modern historian mathematics analyzing the text) may have been identified as "easy" by the ancient learners, and vice versa. However, it cannot be doubted that the variation of certain parameters in mathematical problems resulted in the changes of the level of difficulty of the problems and of the reasoning strategies employed by learners, and these changes certainly did not remain unnoticed by the instructors in traditional China and Vietnam.

Acknowledgment I would like to express my gratitude to the Institute for Advanced Study (Princeton, NJ, USA) where my research on Vietnamese mathematical treatises was conducted in 2007; I also gratefully acknowledge the support obtained in the framework of the project "Multiculturalism in Monsoon Asia" (季風亞洲與多元文化) of the National Tsing-Hua University (Hsinchu, Taiwan) throughout the work on the paper. I would like to thank Christine Proust and Alain Bernard for their encouragement and numerous suggestions, as well as the participants of the Workshop "Textes et instruments scientifiques anciens élaborés dans un contexte d'enseignement: situations, usages, fonctions" (Paris, France, December 15–16, 2008) for their comments on the first draft of the paper; my thanks also go to Richard Kennedy who patiently polished the English of the final version.

Reference

Primary Sources

CDW. 1990. Cheng Dawei 程大位. *Suan fa tong zong jiaoshi* 算法統宗校釋 [Unified origin of counting methods, with emendations and explanations], ed. Mei Rongzhao 梅荣照 and Li Zhaohua 李兆华. Hefei: Anhui jiaoyu.

YH. 1993. Yang Hui 楊輝, *Cheng chu tong bian ben mo* 乘除通變本末 [Alpha and omega of the continuity and transformation of the counting methods]. In Guo Shuchun 郭書春 ed. *Shu xue juan* 數學卷 [Mathematical section] of Ren Jiyu 任繼愈 ed. *Zhongguo kexue jishu dianji tonghui* 中國科學技術典籍通彙 [Comprehensive collection of primary materials on Science and Technology in China]. Henan Education Publishers 1: 1047–1072.

YT. 1829. Nguyễn Hữu Thận 阮有慎 *Ý Trai toán pháp nhất đắc lục* 意齋算法一得錄 [A eecord of what Ý Trai [=Nguyễn Hữu Thận] got right in computational methods]. Manuscript A.1336 preserved in Han Nom Institute, Hanoi, Vietnam.

Secondary Sources

Brousseau, Guy. 1997. *Theory of didactical situations in mathematics*. Dordrecht/Boston: Kluwer Academic Publishers.

Chemla, Karine and Guo Shuchun. 2004. 九章算術. *Les Neuf chapitres. Le classique mathématique de la Chine ancienne et ses commentaires.* Paris: Dunod.

Des Rotours, Robert. 1932. *Le traité des examens, traduit de la Nouvelle histoire des T'ang (chap. 44–45).* Paris: Librairie Ernest Leroux.

Friedsam, Manfred. 2003. L'enseignement des mathématiques sous les Song et Yuan. In *Éducation et instruction en Chine*, Les formations spécialisées, vol. 2, ed. C. Despeux and C. Nguyen Tri, 49–68. Paris/Louvain: Editions Peeters.

Gravemeijer, Koeno. 1997. Mediating between concrete and abstract. In *Learning and teaching mathematics: An international perspective*, ed. T. Nunes and P. Bryant, 315–345. Hove: Psychology Press.

Horiuchi, Annick. 1994. *Les mathématiques japonaises à l'époque d'Edo*. Paris: Vrin.

Høyrup, Jens. 1990. Algebra and naive geometry. An investigation of some basic aspects of Old Babylonian mathematical thought. *Altorientalische Forschungen* 17(2): 27–69, 17(2): 262–354.

Hucker, Charles O. 1985. *A dictionary of official titles in Imperial China*. Stanford: Stanford University Press.

Lam Lay Yong. 1977. *A critical study of the Yang Hui suan fa*. Singapore: NUS Press.

Lee, Thomas H.C. 2000. *Education in traditional China: A history*. Leiden/Boston: Brill.

Li Yan 李儼. 1955. Tang Song Yuan Ming shuxue jiaoyu zhidu 唐宋元明數學教育制度 [The system of mathematics education [in China] of the Tang, Song, Yuan, and Ming [dynasties]]. *Zhong suan shi luncong* 中算史論叢 [Collected works on the history of Chinese mathematics]. Beijing: Keji Publishers. 4: 238–280.

Liu Chun-Yin 劉春銀, Wang Xiaodun 王小盾, Trần Nghĩa 陳義 (eds.). 2002. *Yuenan Han Nan wenxian mulu tiyao* 越南漢喃文獻目錄提要 [Annotated catalogue of the Vietnamese materials in Han and Nom]. Taibei: Institute of Chinese Literature and Philosophy, Academia Sinica.
Mikami Yoshio 三上義夫. 1934. Annan-no ichi sansho-ni tsuite 安南の一算書に就て [On one mathematical book from Annam [= Vietnam]]. *Gakko sugaku* 學校數學 14: 3–11.
Siu Man-Keung. 1995. Mathematics education in ancient China: What lesson do we learn form it? *Historia Scientiarum* 4(3): 223–232.
Siu Man-Keung. 1999. How did candidates pass the state examinations in mathematics in the Tang dynasty (618–917)? – Myth of the 'Confucian-Heritage-Culture' classroom. In *Actes du Colloque "Histoire et épistémologie dans l'éducation mathématique"*, 321–334. Louvain: Université catholique de Louvain.
Siu Man-Keung. 2004. Official curriculum in mathematics in ancient China: How did candidates study for the examination? In *How Chinese learn mathematics: Perspective from insiders*, ed. Fan Lianghuo, Wong Ngai-Ying, Cai Jinfa, and Li Shiqi, 157–185. Singapore/River Edge: World Scientific.
Siu Man-Keung, and Alexei Volkov. 1999. Official curriculum in traditional Chinese mathematics: How did candidates pass the examinations? *Historia Scientiarum* 9(1): 85–99.
Sivin, Nathan. 2009. *Granting the seasons*. New York: Springer Science + Business Media LLC.
Sun Hongan. 1996. *Zhongguo gudai kexue jiaoyu shilüe* [Brief history of science education in ancient China]. Shenyang: Liaoning jiaoyu Publishers.
Sun Peiqing 孫培青 (ed.). 2000. *Zhongguo jiaoyu shi* 中國教育史 [History of education in China]. Shanghai: Huadong Shifan Daxue Publishers.
Trần Nghĩa, and François Gros. 1993. *Catalogue des Livres en Han-Nôm*. Hanoi: Nhà xuất ban khoa học xã hội.
Trinh Khac Mạnh. 2007. *Nhà xuất ban Văn hóa thông tin (Tái ban có chinh lý và bổ sung)*. Hanoi: Viện khoa học xã hội Việt Nam, Viện Nghiên cứu Hán Nôm.
Vaiman, Aizik Abramovich (Вайман, Айзик Абрамович). 1961. *Шумеро-вавилонская математика III—I тысячелетия до н. э.* [Sumerian and Babylonian mathematics from the third to the first millennium BC, in Russian]. Moscow: Oriental Literature Publishers.
Volkov, Alexei. 2002. On the origins of the *Toan phap dai thanh* [Great compendium of mathematical methods]. In *From China to Paris: 2000 years transmission of mathematical ideas*, ed. Y. Dold-Samplonius, J.W. Dauben, M. Folkerts, and B. van Dallen, 369–410. Stuttgart: Franz Steiner Verlag.
Volkov, Alexei. 2005. Traditional Vietnamese mathematics: The case of Lương Thế Vinh (1441–1496?) and his treatise *Toan phap dai thanh* (Great compendium of mathematical methods). *Traditions of knowledge in Southeast Asia,* vol. 3, ed. U Kyi Win, 156–177. Yangon: Myanmar Historical Commission.
Volkov, Alexei. 2006. State mathematics education in traditional China and Vietnam: Formation of the "mathematical hagiography" of Luong The Vinh (1441–1496?). In *Nho Giao o Viet Nam/Confucianism in Vietnam*, ed. Trinh Khac Manh and Phan Van Cac, 272–309. Hanoi: Social Sciences Publishing House.
Volkov, Alexei. 2008. Mathematics in Vietnam. In *Encyclopedia of the history of science, technology, and medicine in non-western cultures*, 2nd ed, ed. H. Selin, 1425–1432. Dordrecht/Berlin: Springer.
Volkov, Alexei. 2009. Mathematics and mathematics education in traditional Vietnam. In *The Oxford handbook of the history of mathematics*, ed. E. Robson and J. Stedall, 153–176. Oxford/New York: Oxford University Press.
Volkov, Alexei. 2011. Weighted distribution in Vietnamese mathematics. Paper delivered at the 13th international conference on the history of science in East Asia, Hefei, P.R. of China, 25–29 July 2011.
Volkov, Alexei. 2012. Argumentation for state examinations: demonstration in traditional Chinese and Vietnamese mathematics. In *History and historiography of mathematical proof in ancient traditions*, ed. K. Chemla, 509–551. Cambridge: Cambridge University Press.

Volkov, Alexei. 2013. An early Japanese work on Chinese mathematics in Vietnam: Mikami Yoshio's study of the Vietnamese mathematical treatise *Chi minh toan phap* 指明算法. In *Seki, Founder of modern mathematics in Japan: A commemoration on his tercentenary*, eds. E. Knobloch, H. Komatsu, and D. Liu, 149–172. Tokyo etc.: Springer.

Volkov, Alexei. 2014. History of mathematics education in Oriental Antiquity and Middle Ages. In *Handbook on the history of mathematics education,* ed. A. Karp and G. Schubring. New York etc.: Springer (forthcoming).

Yang Xuewei 楊學為 (ed.). 2003. *Zhongguo kaoshi shi wenxian jicheng* 中國考試史文獻集成 [Collected materials on the history of examinations in China]. Beijing: Gaodeng jiaoyu Publishers.

Zhou Dongming 周東明. 1990. 'Xi suan gang mu' yu Yang Hui de shuxue jiaoyu sixiang 《习算纲目》与杨辉的数学教育思想 [The 'syllabus for studying mathematics' and Yang Hui's philosophy of mathematics education]. 华中师范大学学报(自然科学版) [*Journal of Central Normal University, Natural Sciences Series*] 24(3): 396–399.

Chapter 11
Learning and Teaching Medicine in Late Imperial China

Florence Bretelle-Establet

Describing the processes and pedagogical activities that took place within the framework of medical apprenticeship in late imperial China raises a number of issues that have already been addressed in this volume. Firstly, we know very little on the practical processes of teaching in pre-modern traditions and even less on how pupils or beginners actually learned. While teaching schedules were at times defined by masters, we have very little evidence of how teachers and beginners put them into practice. Secondly, when sources related to a teaching/learning situation do exist, a second problem is one that derives from attempting to understand these sources and the activities they support from the point of view of our modern experience of learning and teaching. Against this latter problem, the posture Pierre Hadot adopted in his analysis of the philosophical texts of Antiquity, illustrated by his well-known injunction "toute oeuvre doit être placée dans la *praxis* dont elle émane" (every work must be replaced in the praxis from which it emanates)[1] appears attractive when analyzing the sources that were written in relation to medical apprenticeship in pre-modern China. This is why while focusing on one particular text, devoted to teaching medicine at the beginning of the nineteenth century in China, I have found it important to reconstruct the social, material, cultural environment in which this text is thought to have been written and to replace the modalities of teaching and learning to which this particular text attests into the wider landscape of medical instruction in late imperial China. As it will appear in the discussion, this landscape was complex even though an imperial institution existed that was intended to set the rule.

[1] Hadot (2001, 116).

F. Bretelle-Establet (✉)
Laboratoire SPHERE, CNRS & Université Paris Diderot, Paris, France
e-mail: f.bretelle@wanadoo.fr

A. Bernard and C. Proust (eds.), *Scientific Sources and Teaching Contexts Throughout History: Problems and Perspectives*, Boston Studies in the Philosophy and History of Science 301, DOI 10.1007/978-94-007-5122-4_11,

In this chapter, by relying on historical studies and on a monograph devoted to this institution, I will first give a glimpse of how medical teaching and learning were carried out within the institution, in the capital. I will then differentiate the various environments of medical instruction in lay society as they appear from an analysis of a wide corpus of biographies of medical experts in the far south of the empire. Finally, by focusing on one particular medical text from the far south and for which many historical proofs provide evidence that it was written in a medical teaching and learning environment, I will try to see whether this part of evidence allows us to reconstruct the process of medical teaching and learning, and, in return, whether pedagogical purposes left their hallmark in medical writings.

11.1 Medical Education in the Central Institution Under the Qing Dynasty

Quite early in the history of imperial China, medical education was the subject of attention from central government. The first government-sponsored medical school was recorded in China in 443, but, as this school was abolished 10 years later, the Sui dynasty (589–618) is usually considered as the dynasty that first formalized government-endorsed medical education within a larger and comprehensive official system of education. From this time, and as it was not the case for mathematics education,[2] each ruling dynasty maintained a central medical school which belonged to a wider institution intended to deal with health matters within the empire. This central institution gave birth to numerous official sources which have allowed much research into this institution and its role in medical education in imperial China.[3] While this institution did not, over time, always follow the same rules, bore different names and came under different ministries, it did however always have the same goal: to regulate medical instruction by defining the branches of medical knowledge to be taught, the corpuses of required texts, the length of study, and the modalities of recruitment and of promotion within the institution. In fact, an important feature in the history of the medical instruction provided within this institution lies in the continuity in the modalities of the students' evaluation and promotion: examinations. Within the institution, what enabled a doctor to progress, what allowed him to obtain a post at the imperial court, within the central medical administration and, less prestigiously, in the army or, worse, with the general public, was passing examinations.

[2] Siu (2004).

[3] The most comprehensive study of this institution is (Gong 1983). See (Goldschmidt 2009; Hinrichs 2003; Shinno 2002) for a particular focus on the Song (960–1279) and Yuan (1277–1367) dynasties, (Chang 1998; Guan 2008) for the Qing (1644–1911).

11.1.1 Overview of the Pedagogical Methods Within the Imperial Medical Bureau Taiyi Yuan 太醫院 During the Qing Dynasty

Unfortunately, the official sources do not reflect with much accuracy how medicine was taught. While they give a glimpse of the list of books that had to be studied for each branch of medicine, "evidence of actual interactions between the instructors and the students is frustratingly thin" noted Reiko Shinno in her analysis of this institution under the Yuan dynasty (1277–1367).[4] Scant evidence of these interactions is also true for later periods even in sources which come from individual records and not from the institution itself. The *Monograph on the Imperial Medical Bureau* 太醫院誌 written in 1863 by the imperial physician Ren Xigeng 任錫庚 after 20 years spent working in the institution says little on the modalities of his medical instruction. In the sections devoted to "medical studies 醫學" and "examinations 考試" we only get some general elements on the rules surrounding examinations and on the trajectories of the pupils who entered the institution rather than a precise description of what this medical official had himself experienced in it.[5] Courses existed during the Qing dynasty to which two imperial doctors were appointed as teachers, and after 1866, taught for 15 h each month.[6] The lessons consisted in lectures and commentaries on medical texts (*zhusheng bing piyue* 諸生並批閱) and in reading the texts aloud (*songdu* 誦讀). The texts studied were those that had been canonized in previous times, the *Medical Classic of the Yellow Emperor* (*Huangdi Neijing* 黃帝內經) compiled between the final centuries B.C. and 2 A.D., for its major part, the *Treatise* on *Cold Damage Disorders (Shanghanlun* 傷寒論*)* and the *Synopsis of Prescriptions of the Golden Chamber* (*Jingui yaolüe* 金匱藥略*)* by Zhang Zhongjing (150–219*)*, the *Compendium of Materia Medica* (*Bencao gangmu* 本草綱目, 1578), by Li Shizhen and some specific books for each branch of medicine. At the beginning of the dynasty, there were 11 branches, but this number was successively reduced to nine and then to five, through reorganization or suppression. In 1744, a medical compendium was commissioned by emperor Qianglong in order to rectify errors in medical learning. The *Imperial Golden Mirror on the Orthodox Medical Lineage* (*Yuzuan yizong jinjian* 御纂醫宗金鑒*)*, which combined critical editions of classical sources from the Han dynasty (−206; +220) and short monographs on different diseases became, after 1749, the official textbook of the institution.[7] Unfortunately, we have no concrete detail on how the instruction was carried out between teachers and students. What emerges from Ren Xigeng's monograph and other official sources from the institution is that book

[4] Shinno (2002, 100–103).

[5] Ren ([1863]1923, 11–12 and 18–24).

[6] Gong (1983, 129).

[7] Hanson (2003).

learning and memorization occupied much of the medical curriculum.[8] Medical instruction within the institution thus borrowed from Confucian schools the method that consisted in emphasizing bookish culture and memorization of texts.

As recalled elsewhere in this volume, learning by heart is universal and satisfied, in different environments and times, different functions. In China, learning by heart had, since antiquity, been required of the youngest children, who were known to have fantastic memorization skills. The "Daily Schedule of Study in the Cheng Family School, Graded According to Age", written by Cheng Duanli (1271–1345) which became the standard educational framework for many elite Chinese schools from the fourteenth to the nineteenth centuries, provides evidence of this long process of memorization.[9] After having memorized two or three primers in childhood, the young elite continued memorizing texts until the age of 25. One had to learn by heart the Four Books (*The Great learning*, *the Doctrine of the Mean*, *the Analects* and *the Mencius*), and, as time went by, from one to maybe all five of the Classics (*Classic of Change*, *of Poetry*, *of History, of Rites, and the Spring and Autumn Annals*). A process that could take 6 years, at a rate of 200 characters memorized a day.[10] Even if memorization as a didactic tool has at times been condemned in history, massive memorization still played a crucial role in classical education until the end of the empire in 1911.[11]

Memorization which thus played an important part in the classical curriculum was also an important element in the medical curriculum within the central institution. It is true that one of the targets of these didactic methods was passing the examinations in both classical and medical studies, examinations that, moreover, shared some common types of tests. In fact, medical examinations paralleled many of the features of the civil service examinations. First, there was an examination to enter the Imperial Medical Bureau *Taiyi yuan* as a student just as there were school entrance examinations. Once admitted to the *Taiyi yuan*, successful candidates (*yiyesheng* 肄業生) had then to study for 3 years taking two or three examinations a year.[12] After 3 years of study, and success in the bi and tri-annual examinations, students (*yishi* 醫士) had to take the final examination in the Ministry of Rites. Successful candidates then obtained the title of *yisheng* 醫生, and after other numerous further examinations, they could finally reach the top of the hierarchy. Once admitted to the *Taiyi yuan*, it would take no less than 10 years to reach the highest titles and functions. Examinations consisted in both oral and written tests and were subject to precise rules. According to Ren Xigeng, by the end of

---

[8]Ren ([1863]1923, 11).

[9]Gernet (2003).

[10]Mizayaki (1976, 16), Elman (2000, 260–271), Elman and Wooside (1994, 111–149 and 525–560).

[11]According to Li (2003), reading aloud and recitation still played crucial role in children education in the 1970s.

[12]Gong (1983, 118–120 and 127–130).

the nineteenth century, the questions for examinations were mainly taken from the Classics and the *Golden Mirror on the Orthodox Medical Lineage* mentioned above.[13]

Medical examinations, under the Qing, were on the model of those carried out under the Song dynasty (960–1279) and consisted in six types of tests: a test called *moyi* 墨義 that in fact required the candidate to supply the phrase immediately following the one read out loud and taken from a medical classic; a question about the pulse *maiyi* 脉 義; a question about cosmology *dayi* 大義; a question about prescriptions *lunfang* 論方; a question about a diagnosis *jialing* 假令; and a last question to check the student's knowledge of the principles of Yin and Yang during the annual cycle and their impact on the human body.[14] Until 1185 candidates were authorized to enter their examination cell with their own notes or a copy of the Classics.[15] After this date, medical students had to learn the Classics by heart, be able to explain them, and for examinations, reproduce in writing what they had learned by heart.[16] As far as we can ascertain from the official sources on the institution and from Ren Xigeng's monograph, the classical curriculum and medical studies within the institution shared many common features, notably, the examinations for which great feats of memorization were required.

11.1.2 The Central Institution and Its Impact in the Provinces of the Empire

Despite several edicts that testify to the Qing court's intention to extend the modalities of medical instruction established at the capital to the entire empire, the central institution did not succeed in setting up an extensive network of medical schools in the provinces and districts.[17] Local gazetteers which provide a detailed account of what was done locally are crucial sources for evaluating the extent to which what was ordered at the center was actually put into practice at the local level.[18] Let us leave the capital for a while and look at the far south of the empire

[13]Ren ([1863]1923, 18).

[14]Zhang (1995).

[15]Zheng and Zheng (2005).

[16]Gong (1983, 113).

[17]On the court's policy for local official medical instruction, see (Gong 1983, 130; Leung 1987). The situation was probably different before this time. The medical manuscripts discovered in Dunhuang give evidence that local imperial schools of medicine existed as early as the Tang dynasty in the remotest areas of the empire (Despeux 2010). Shinno (2002, 109) found a biography of a local doctor who had learned in a medical school with an officially appointed instructor, during the Yuan dynasty. Gong Chun (1983, 116) reports the establishment of three medical schools in Guangxi, three others in Guangdong and six schools or six imperial physicians in Yunnan during the fifteenth century.

[18]On these sources (Will 1992).

and particularly at the three provinces of Yunnan, Guangxi, and Guangdong. Local gazetteers in these provinces do not mention any local medical schools for the Qing dynasty, confirming Ch'ü's thesis that under the late Ming and the Qing, the running of local institutions dealing with education or health matters no longer relied on initiatives from central government.[19] Rare also are the traces of those official medical teachers who, according to central edicts, should have been appointed in the provinces to promote official medical instruction. In the gazetteer of the Yunnan province, compiled at the end of the nineteenth century, we learn that Tang Jingzu 湯敬祖, a native from Yimen 易門 sub-prefecture, was chosen as a 醫學訓科, a title given to an official medical teacher at the level of the sub-prefecture, and that he received 25 *liang* of silver for this.[20] But we do not have more precise chronological clues other than that he was practicing during the Qing dynasty. Furthermore, if we now look at another section of the gazetteer, dealing more precisely with the administrative configuration of each unit of a province, we see that none of the sub-prefectures of the Yunnan province makes any mention of these official medical teachers. Finally if we now turn our attention to the men who practiced medicine in the provinces of Yunnan, Guangxi, and Guangdong, under the Qing dynasty, and whose biographies were recorded in local gazetteers, we can see that in these provinces at least, (but surely also in lay society as a whole), medical instruction did not have any relationship with the medical instruction delivered by the central institution or by locally appointed official medical teachers. Out of 422 biographies of people who were renowned in the far south of the empire for their involvement in medical activity, only four mention a link with the imperial institution of medicine, but do not always signal an imperial apprenticeship. Tan Zongyu 譚宗禹 (Qing, dates unknown), a native of Yongbei 永北, Yunnan, is reported to have studied at the *Taiyi yuan* while Li Zuolin 李作琳 (ca. 1723–1736), from Yongchang 永昌, Yunnan, is reported to have worked there. Yi Jingguo 易經國 (Qing, dates unknown), from Dianbai 電白, Guangdong, met Zhai Yuhua 翟玉華 (1587–1671), a physician working at the *Taiyi yuan*. Finally, Qu Zunde 屈遵德 (ca. 1786) had a post of imperial physician (*Taiyi* 太醫).[21] With the exception of these four men, the people involved in the art of healing in late imperial southern China had no contact with the central institution and none had ever learned medicine with official medical teachers appointed locally by the Court. Hence, the vast majority of people involved in medical assistance in lay society had learned medicine through private instruction.

The fact that medical instruction and practice escaped institutionalization probably explains why the very issue of how a medical apprenticeship should be carried out was raised so frequently in Chinese medical texts produced in late imperial times. Nearly all the prefaces in medical books addressed this question.[22] The issue

[19]Ch'ü (1962, 156–164).

[20]Wang (1901, 187, 25).

[21]Wang (1901,187, 31), Liu (1886, 50, 288), He (1991, 2, 32), and Guo (1987, 2027).

[22]Bretelle-Establet (2011).

of knowing how a doctor should learn medicine was raised much earlier, but the debate became more acute under the last two dynasties (Ming, 1368–1644 and Qing, 1644–1911), in relation to the increase in the number of people who claimed to have medical skills and the status of 'physician'.[23]

11.2 The Different Environments for Medical Instruction Outside the Central Institution: The Example of the Far South During the Qing Dynasty

Local historical sources do not mention local representatives of the *Taiyi yuan*. Until the very end of the nineteenth century, they do not mention local private schools of medicine (in the sense of a building nor as a group either). Only at the beginning of the twentieth century, do foreign as well as Chinese sources mention the foundation of local private schools of Chinese medicine. A Chinese medical school was created in Kunming 昆明, Yunnan's provincial capital, in 1904, and another in Guangzhou 廣州, in 1913. How these particular schools really functioned is unclear, and we cannot ascertain if they took the first modern-style school of Chinese medicine, the *Liji yiyuan xuetang* 利濟醫學堂, created in Wenzhou 温州, Zhejiang province, in 1885 as a model.[24] But before the opening of this type of modern-style medical school, which was likely to have been an answer to the proliferation of schools of Western medicine on Chinese soil during the last 30 years of the nineteenth century, medical instruction was mainly carried out in private learning environments.

11.2.1 Private Learning Environments

From the analysis of a broad corpus of biographies of men involved in medicine, three major ways for learning medicine can be identified. Experts in medicine usually learned medicine from a master, from their father, grand-father or uncle, or through the self-study of texts. A very small proportion of the biographies mention a fourth way: through religion, either Buddhist or Taoist. And an even smaller proportion of these mention magical revelation. In fact, these ways of learning were not mutually exclusive. We have many examples of medical experts coming from medical families who started to learn with a relative and later improved their medical knowledge by attending a master's lessons and reading medical texts.[25] While biographies allow us to differentiate different environments for medical instruction

[23] Chao (2000). For an analysis of the size of the medical community in late imperial times, see (Chao 2009; Bretelle-Establet 2002).

[24] For Kunming and Canton, see (Feray 1910; Liu 1986). For Zhejiang, see (Lin 1983). Scheid (2007, 193–198) reports the creation of such schools in Shanghai.

[25] Bretelle-Establet (2009).

in the late imperial Chinese society, they say little on the precise modalities of medical instruction. Written mainly to give exemplary models for posterity, they prefer to emphasize the talents, genius and moral qualities of the medical experts, and report edifying medical anecdotes.[26] We thus must content ourselves with rare and limited information that mainly deals with the preferred age for learning medicine, the length of instruction, and some of the features of the learning process.

11.2.2 Features of the Learning Process Gleaned from Biographies and Prefaces

When medicine was learned as a first career choice, the apprenticeship began quite early in childhood. The biography of Cheng Shichao 程士超 (1804–1888), a native of Guangxi province, recounts: "At more than 10 years old, he travelled into Guangdong province to look for famous masters." "Tan Zuoyan 譚祚延 (nineteenth century) learned medicine when he was very young", "Tang Zhenglian 唐征濂 (Qing, dates unknown) started learning medicine as a young child", "Huang Xixia 黃 錫遐 (ca. 1851–1874), as a young child, embarked on learning medicine".[27] In his detailed analysis of Menghe medicine, Volker Scheid also found many examples of boys starting to learn medicine at an early age.[28] Learning medicine at an advanced age, however, was frequent too. It was not uncommon for people who had attained various administrative functions in their life to decide to learn medicine, either in their vacations from their public life or when retired. And turning to medicine after successive failures in the imperial examinations, and thus at an advanced age, was even more frequent.[29]

Biographies also report that learning medicine was a rather long enterprise. Huang Zijian 黃子健 (post-1742) retreated to his home, where he studied the medical classics for 20 years.[30] Chen Biqin 陳必勤 (ca. 1729–1784) started to treat people after having studied for 10 years with the son of an imperial physician.[31] Yao Shi'an 姚時安 (ca. 1847) spent several years learning medicine with his father and reading the medical classics. Medicine was moreover considered as a discipline to be learned throughout one's life. This is what Qian Yinggen's 錢穎根 (ca. 1811–1890) and Huang Yinggui's 黃應桂 (ca. 1851) biographies report, stressing for the former that he studied until the very end of his life in order to

[26] Bretelle-Establet (2009).

[27] Guo (1987, 2025, 2019, 2031, 2032).

[28] Scheid (2007, 251). Menghe was a small town of the Yangzi delta which hosted a medical lineage that flourished in late imperial China.

[29] Hymes (1987). Medicine, from the Song dynasty onwards, was considered a good alternative to a career in officialdom.

[30] Guo (1987, 2245, 1934–1935).

[31] Guo (1987, 1931).

prescribe the appropriate treatments to ever-changing illnesses; and for the latter, that he constantly strove to learn good techniques from the physicians he met.[32]

Learning medical texts by heart is mentioned in several biographies: Ding Naiqian 丁乃潛 (1863–1928), learned medical texts during his vacations by rote chanting (*yin feng ziyu* 吟諷自娛).[33] In the late nineteenth century Yang Dingguang 楊鼎光, a doctor and native of Guangxi, could, as an adult, recite some texts of Chen Nianzu 陳念祖 (1753–1823) (*neng langsong qi ci* 能朗誦其詞).[34]

The indirect records on medical apprenticeship found in biographies thus mainly underline the difficulty, the long years of study required to learn medicine and indicate that memorization was part of the medical apprenticeship. If we now turn our attention to what medical experts wrote about their own apprenticeship, in direct records such as authorial prefaces to medical texts, we find quite similar information.

Learning medicine, in the words of the medical experts themselves, could take place at different ages, depending on whether medicine was chosen as a first career or as a secondary occupation. Huang Piaozhi 黃朴之, for instance, wrote in his preface (1933) that he started to learn medicine at 15.[35] But Liang Lianfu 梁廉夫 (1810–1894), who was a graduate scholar and had held several posts in the bureaucracy, started to learn medicine, at 36 years old. He continued studying medicine until he was 70, when he published his medical book.[36] While more and more people turned to learning medicine at an early age and for professional reasons, as Leung has stressed (2003), it was not uncommon to learn medicine in later life as a purely cultural and intellectual pursuit.

In their own words, learning medicine was a long and difficult enterprise. Yu Tingju 俞廷舉 (1780) reports that he had spent 8 years discussing medicine with a famous physician, and had gathered together a huge number of medical texts in his room that he then studied for several more years. Jin Jinghua 金菁華 (nineteenth century) decided to learn medicine as a result of his own sickness that physicians were unable to cure. After 3 years of reading medical texts, and study of medicine for a few more years, he finally read the 92 medical texts that had been compiled in the *Complete Books of the Four Libraries (Siku quanshu* 四庫全書), an imperially commissioned compilation, which contained texts dating from antiquity through to the eighteenth century. Lin Xianfu 林賢輔 (ca. 1888) for his part explains that after 20 years of learning medicine through book learning, he still could not consider himself an excellent doctor.[37]

We also find several traces of apprenticeship based on memorization. He Mengyao 何夢瑤 (1693–1764) wrote in 1751 that in order to compile his

[32]Guo (1987, 2032 and 1983).

[33]Guo (1987, 1991).

[34]Cheng (1920, 1684).

[35]Huang (1933, 10).

[36]Guo (1987, 2028) and Liang ([1881] 1936).

[37]See in this order (Guo 1987, 1960, 1939, and 2019).

book for students of medicine, he relied on the medical words he had learned by heart when he was young "*yuan qu xiaori suo song Qihuang jia yan* 爰取少日所岐黄家言".[38] At the end of the nineteenth century, Gong Pengshou 彭 (1862–1926) still underlined this aspect of his medical apprenticeship.[39] A controversy over memorization existed, but it was less the memorization process which was contested than the content of what was learned by heart.[40] Learning classical texts by heart was never questioned, but practicing medicine by relying on a handful of medical formulas learned by heart was highly condemned. In fact, this criticism served to reinforce the boundary between those, the scholar doctors *ruyi* 儒醫, who could recite classical medical texts and could claim to have followed a scholarly curriculum for an overall understanding of the Way (Dao), and those, often qualified as *shiyi* 市醫 (market doctors) or 世醫 (hereditary doctors) who could only recite the formulas that they had learned in town markets or within their family context, in a purely practical perspective.

While authorial prefaces are likely to give more direct access to the medical experts' voice than biographies, we must realize that unlike Elliot's image of prefaces in general, authorial prefaces in Chinese medical books are not the place where the writer comes to speak to his readers, makes confessions, speaks about himself and explains in great detail how he became a physician.[41] Indirect records (biographies) like direct discourse such as doctors' prefaces provide little and rather imprecise detail on how doctors learned medicine in late imperial China. Unlike anthropologists, who, working on contemporary China, can migrate from academic milieus to private medical circles in order to study the modalities of knowledge transmission in these different environments, historians lack precise accounts to help them decipher how medicine was taught and learned in these different environments, in late imperial times.[42] Would medical texts written with the purpose of teaching medicine give a better account of the didactic practices used or suggested by a teacher? Can we use textual clues to reconstruct an environment which is otherwise largely hidden from us? This is what we are now going to look at, while keeping in mind the anthropological work just mentioned and the oral tales recounted by two beginners in medicine at the beginning of the twentieth century, reported to historians Angela Leung (2003) and Volker Scheid (2007, 111). Indeed, as Charlotte Furth (1999, 4) stresses, "the ethnographic present can, if carefully used, illuminate possible meanings of historical evidence".

[38]He ([1751]1994, 1).

[39]Guo (1987, 2033).

[40]As (Li 2003, 49) stresses, the ability of memorizing and reciting great portions of texts is a constant praise in Chinese biographies.

[41]Bretelle-Establet (2011).

[42]See notably (Hsu 1999; Scheid 2007).

11.3 *The Essentials of Medicine, Yixue jingyao* 醫學精要 A Textbook for Learning Medicine, in 1800 Guangdong

11.3.1 The Question of the "Textbook": The Challenge of Identification

In late imperial China, the market in medical books blossomed. All over the empire the expansion of both number of medical experts and of printing houses favored the proliferation of medical books.[43] Identifying a particular genre of texts, here, medical textbooks, from among all these medical books is not easy and raises the same issues as the identification of textbooks in the field of chemistry or in any field of knowledge in eighteenth century France, for instance.[44] Firstly, as was the case in eighteenth century France, and recalled by (Leung 2003), the bibliographic category of "textbook" did not exist in China. Secondly, as in France before the revolution and the educational reforms that followed it, no institution in China defined how a book devoted to medical instruction should be written. There was no particular social group especially devoted to writing textbooks, either. It is, therefore, difficult in these conditions to distinguish books composed for pedagogical use from other medical writings. In fact, we can even make the assumption that all medical books, whatever their format, were pedagogical. Unlike novels, poetry, and literature at large, medical books were not written with the aim of diversion or entertainment, but rather, to teach the readers a set of doctrines and practices about the discipline. This does not mean that literature, behind a diversional objective, could not have educational targets as well. But the targets less or more hidden in the literature were not reflexive: the novels or poetry did not aim to educate readers on how they should go about writing novels or poetry. On the contrary, whether medical books were written with either purely intellectual or practical intent, in both cases, the objective was to pass medical knowledge on to the reader either for the understanding of bodily phenomena or for treating people or both. For these reasons, we can say that all medical books were pedagogical in nature.

Several clues can, however, help us to identify within this broad body of medical literature, some medical books that were clearly written in a teaching/learning environment. This is the case for the *Yixue jingyao* (*Essentials of Medicine*). A large body of evidence shows that this particular text was produced within an environment of medical instruction and was aimed at pupils who wanted to practice medicine. Firstly, the allographic prefaces explain that the author, Huang Yan 黄巖, (1751–1831) from Jiaying 嘉應 (present Meixian) in the province of Guangdong, taught

[43]Local gazetteers provide evidence that this phenomenon took place everywhere in the empire. See (Hanson 1997, 35 and 332, on the basis of Guo 1987).

[44]Bensaude-Vincent et al. (2003, 10, 26) and Choppin (2008).

pupils medicine. Li Guangzhao 李光昭, a local literatus and poet,[45] in his preface (pr.1831), writes that Huang Yan had some pupils who practiced medicine at his office (*you tu shu ren, ye yi yu men* 有徒人，業醫于門). In his post-face (p-f.1830), Wen Baochun 温葆淳, a Hanlin member, recounting his discovery of Huang Yan's manuscript clearly linked the book and its author to a pedagogical context:

> On a table, I saw the *Yixue jingyao* in four sections, very well copied and ready for engraving. I asked my cousin who told me: this is the book written by Master Huang Yan from our canton for teaching his pupils and among his disciples, the disciple Zhang from Heyu (in present Yongding county) was in charge of ensuring it was well copied and assembled for it to be engraved and made public 見案頭醫學精要四卷眷繕端好似欲付梓未果者因詢族兄云既故里黄明巖先生授徒著本也門下荷畲張子鈔装拟.[46]

Secondly, the author's own words provide evidence that he was teaching and that his book was written within a master-disciple environment. Huang Yan declares at the beginning of his book. "This book was originally prepared to teach disciples (*shi bian ben wei shou tu er she* 是編本為授徒而設)". Then, in the first and subsequent editions, there are marginal notes which, according to the author, Huang Yan himself, had been added to his own manuscript by his pupils and which he warmly recommended the reading of. "The commentaries added by my pupils have developed what I had not; they supplement what I had not found. One must read them to increase one's discernment (*zhu zi suo can zhu, duo wei you fa yu suo wei fa, bu yu suo wei ji zhe, yi yi jian yue zhi, yi guang shijian* 諸子所參注，多為有發與所未發，補余所未及者，亦宜兼閱之，以廣識見".[47] Moreover, the author frequently speaks directly to his readers through the text, by saying "You, my pupils", and, at several points in his foreword, the author makes reference to his readership by saying "he who studies *xuezhe* 學者". All this evidence gives hint that this book was written within a teaching environment and aimed at beginners in medicine. How a medical teacher envisaged the use of his didactic tool, what kinds of tasks and activities he expected beginners to carry out in order to learn medicine, is what we are now going to explore.

11.3.2 The Didactics of the Teaching: The Mentor's Goals, the Cognitive Processes Required, and the Method Advocated

Unfortunately, while the *Yixue jingyao* (*Essentials of Medicine*) was prepared by a teacher for the purpose of teaching, we cannot find a systematic pedagogical doctrine that would explain in detail how one should go about learning medicine,

[45]Miles (2006, 65, 92, 107, 136).

[46]Huang ([1800] 1918, Wen Baochun's post-face, 1).

[47]Huang ([1800] 1918, *dufa*, 7).

how to plan the work, such as Cheng Duanli's detailed schedule for education, mentioned earlier, or such as Yang Hui's schedule for mathematics education written in 1274.[48] There is nothing resembling "On the Method of Studying for Those who Pursue Medicine with Zeal" written by Girolamo Mercuriale, a professor, in late sixteenth century Padua, who provided a step-by-step method for those who aspired to devote themselves to medicine.[49] At first glance, thus, we do not find anything resembling this systematic and precise didactic discourse in Huang Yan's book. However, the reading guidelines *dufa* 讀法, a piece of paratext that the author wrote at the beginning of the book, contain a few pedagogical recommendations that are mixed with a series of important concepts about medicine that the reader had to be aware of. These reading guidelines give hints as to the methods advocated by the author, the cognitive processes required of beginners, and offer clues as to the overall goal of the apprenticeship.

In the first part of these reading guidelines, the author explains the global method and overall goal of the apprenticeship: "He who wants to devote himself to this discipline must read the book in its entirety and learn it by heart. That way, after a long time, his Heart/Spirit (*xin* 心) will merge with the Principle (*li* 理),[50] and he will be able to do marvelous things. (*You zhi si dao, yi jin du zhi, shu du zhi, jiu ze xin yu li rong, zi neng chu shen ru hua* 有志斯道者，宜讀之，熟讀之，久則心與理融，自能出神入化)". After this first statement that thus emphasizes reading and learning by heart, the author, in the following points, provides other methodological recommendations for the aspiring doctor. In order to succeed in making a good interpretation of the signs gleaned from a careful examination of the patient's body, "one should first of all learn by heart and practice extensively (*gu shoulie zhi xuezhe zui yi shu du shu lian* 故首列之學者最宜熟讀熟練)". Then, because Huang Yan, like many other medical authors, experienced particular difficulties when treating children and women, he states that he chose to present the parts of his book in the particular order that he recommends his pupils to follow: "He who wants to study, by (reading the book) from its beginning to its end will know these two difficulties (of medicine) (*yu xuezhe shi dong zhi qi nan* 欲學者始知其難)". Finally, Huang Yan stresses that one should read the commentaries added by his pupils to increase one's discernment.[51] For Huang Yan, the method for learning medicine thus consisted in reading and learning the book by heart, practicing extensively, and taking into account the commentaries added by his disciples. This latter point is noteworthy as it shows that for Huang Yan learning meant using a network of interconnected texts,

[48]Gernet (2003, 44–45) and Siu (2004).

[49]Durling (1990, 185, 191).

[50]The term *li* 理 has been the subject of much controversy in the history of Chinese philosophy. Neo-Confucians understand *li* as the "principle" of all things. Each thing has its own *li*, or principle, but all of these principles are in fact one Principle. *Li* was thus considered an omnipresent universal principle that maintained a sort of consanguineous link between all things.

[51]Huang ([1800] 1918, *dufa*, 1–2).

the main text, the marginal notes, and, as we will see later in more detail, excerpts from many other medical books. Huang Yan's recommended program for learning was contextual, in the sense developed in the introduction of this volume.

As noted above, the overall objective of this way of learning was that the reader's heart/mind *xin* merge with the principle of everything *li*, that is, being fully aware of the processes taking place in his patient's body, so that he could perform marvelous cures. Before reaching this ultimate goal, that is, that of the heart/spirit merging with the principle underlying the disease, and thus, literally, 'knowing' without recourse to any other form of aid, some intermediate cognitive processes must have taken place. We can try to understand the types of processes that were necessary by researching and isolating the expressions Huang Yan used in his reading guidelines *dufa* when he addressed his students. Huang Yan expected from his students that they would be able to: know *zhi* 知, differentiate very carefully *que bian* 确辨, identify carefully *xiang ren* 詳認, be clear about *neng ming* 能明, investigate thoroughly *zui yi shen cha* 最宜深察, distinguish *fen* 分, and increase their discernment *guang shi jian* 廣識見. What was required above all was that the student would be able to differentiate, to distinguish, to identify and to have a high level of discernment: to observe and merge. If we look at the words used by the author, actual understanding was not at the fore. This type of scientific inquiry which thus combined cognitive processes such as learning by heart and non-rational means such as intuition and contemplation, according to Sivin, characterized the type of scientific inquiry practiced in early China. It was explained by the belief that the deepest aspect of reality (the nameless Dao) was so subtle that it could not be reached by cognitive means only (Lloyd and Sivin 2002: 192). Strikingly, Huang Yan's explicit didactic goal — the reader's heart/mind eventually merging with the principles of everything in order to do marvelous things — as well as on the methods advocated — learning by rote and extensive practice — also find an echo in the discourse pronounced by the senior Chinese doctors E. Hsu met when she was studying the modalities of medical transmission in 1980s Yunnan. The pupil of a senior private doctor was told that "if he persisted in memorizing texts and applying them to medical practice there might be revelations. Without the need for explanation, he would just know".[52]

Book learning was crucial for Huang Yan, but practice was also capital. While practice, within the apprenticeship, was thus put on the same footing as learning by heart, we do not find a method or a discourse about the practice of medicine. There are no clues on what a student had to do when practicing what he had learned: no mention of the amount of time he should spend at the patients' bedside, which master he should follow, at which stage of his apprenticeship he should attend clinics, how he should behave with his patients, and so on and so forth. Whether the actual practice of medicine was considered as something that could be taken for granted and did not need to be discussed or whether it was seen as belonging to another epistemological order, leading medical instructors to feel it unnecessary

[52]Hsu (1999, 103).

to hold a discourse and suggest a method for something that was far from and did not fit the theories and doctrines they advocated, but rather pertained to individual experience and sensibility, is difficult to ascertain.[53] In any case, we must underline, together with Anne-Marie Chartier in this volume, and more widely with the various scholars who have worked on the relationship between orality and textuality, that we cannot reconstruct the whole learning process from the text alone.[54] That part of the apprenticeship related to practice involved instruction for activities and procedures that were probably given orally. The text can only shed light on the other part of the apprenticeship; that based on learning texts. Let us therefore go ahead in our reading of Yuang Yan's textbook, a prerequisite for one's "heart/spirit merging with the principle, and being able to do marvelous things," and try to understand what reading this book really meant.

11.3.3 Content: From the General to the Singular

When Hsu was doing research in a Chinese medical college in the 1980s, she reported how the idea of being systematic was at the fore in medical colleges and how this idea opposed the teachers belonging to the academic institution, and those who, outside the academic institution, in private office delivered medical instruction.[55] Being systematic, she argues, was introduced in the wake of the communist government's reforms to medical instruction in China. The teaching program, she observed, followed a hierarchical order and divided the complexities of knowledge into basic and advanced knowledge. Hsu argues that this teaching program relied on a certain idea of knowledge and followed the didactics of post-Enlightenment education: instruction was supposed to be based on explanation, according to a given method and a hierarchical order from the simple to the more complicated.[56] In fact, and even though a step-by-step program of learning was not presented by the teacher-author Huang Yan, we do find, in his reading guidelines and in the main text, the traces of a graduation of what had to be learned in a systematic manner and a certain logic linking what he thought important to know and the organization of his book. For the author, indeed, at a macro-level, there were two major difficulties in medicine: treating children and treating women. Accordingly, he chose to start his textbook with children and end it with women: "this is why this book starts with children and ends with women. He who wants to study medicine,

[53]On the question of the practice of medicine and medical experience often referred to by *jingyan* 經驗, see (Lei 2002; Sivin 1995; Farquhar 1994).

[54]See notably (Waquet 2003).

[55]In spite of the reforms undertaken under the Maoist regime to regulate and control the field of Chinese medicine, it was still possible, in the 1980s to learn in these different environments, namely, state college or private offices.

[56]Hsu (1999, 165, 228).

by reading from the beginning to the end, will know these two difficulties (*shi bian, shi xiao'er, zhong furen, yu xuezhe shi zhong zhi qi nan* 是編, 始小兒, 終婦人, 欲學者始終知其難)."[57] Then, at a micro-level, the author had a precise idea of what a student in medicine should learn first and to what extent he should learn it. The organization of the book in fact implies an order of priorities and its textuality implies the depth of learning required.

Let us first have a look at the organization of the book. As already mentioned, this textbook began with a piece of paratext written by the author, some reading guidelines 讀法,[58] that dealt with a series of important ideas in the field of medicine. As soon as he opens the book, the beginner in medicine had to be aware of these essential points of medicine in order to progress. The guidelines provided some methods for learning medicine, and they also indicated the most important notions used in medicine, such as yin/yang, deplete/replete, hot/cold, superficial/deep that a student was required to study. After this concise reminder of the most important notions to know, the first chapter opens with a series of basic elements that "ought to be known" *xu zhi* 須知. Once these 'elements to be known' have been presented, the author emphasizes the fact by repeating: "this is what you must know (*shi ze suo yi zhi ye* 是則所宜知也)".[59]

The first thing a student had to learn concerned the remedies, their properties and main functions. The student had to memorize the incompatible plants, the remedies that neutralize each other and the plants that could be dangerous either because of their toxicity or their hot nature. "Those who want to practice medicine, must examine [the plants] beforehand (*yu xi yi bu ke bu yu wei jiangqiu* 欲習醫不可不預為講求)". Students were then also required to learn the remedies that could serve as an antidote. After this first set of short sentences, entirely reproduced from earlier medical texts, the author describes the remedies for emergency care and urges his students to learn them before all the others: "He who practices medicine must know all these good remedies that cure acute diseases beforehand (*yi shang xie jiuji liangyao, xi yi zhe bu ke bu yu zhi* 已上皆救急良藥, 習醫者不可不預知)".[60]

The beginner in medicine thus had to learn about the medicinal uses of plants before anything else. Once he had mastered the practical knowledge related to plants, their use, their incompatibilities, and thus the practical side of any physician's mission, the beginner in medicine had to learn what in the body gave an insight into a patient's state of health. Here the Depots and Palaces *Zang/Fu* 臟腑[61]

---

[57]Huang ([1800] 1918, 2).

[58]Bretelle-Establet (2011).

[59]Huang ([1800] 1918, 1–5).

[60]Huang ([1800] 1918, 3).

[61]Depots and Palaces are used to avoid, in translation, viscera or organs, and to be closer to the ideas lying behind these terms: "depot" renders the idea of a storage unit in the body, (liver, heart, spleen, lung and kidneys) and "palace" for the idea of a processing unit in the body (gallbladder, small intestine, triple burner, stomach, large intestine, bladder). See (Unschuld 1985; [1988]2005, 400 and 467).

firstly required the students' attention: "He who studies the discipline of medicine, must first know (識) the Palaces and Depots, and then, learn how to distinguish (分) if they pertain to Yin or Yang, and to which of the Five Phases. He must be clear about (明) the principles of conquest and destruction and how what is operating inside (the body) takes an external form. Thus, one must know (知) how the Palaces and Depots are mirrored in the sense organs (*xi yi zhi dao, xian shi zangfu, ci fen yinyang, wuxing bian qi peishu, shengke ming qi liyao, you zhu zhong, bi you xing zhu wai, ze zangfu zhi miao qiao* 習醫之道, 先識臟腑, 次分陰陽, 五行辨其配, 生克明其理要, 有諸中, 必有形諸外, 則臟腑之苗須知".[62] A very clear and systematic presentation follows indicating what are Depots, and thus Yang, in the body and what are the Palaces and thus Yin. A similarly clear statement follows about the cycle of conquest and destruction between the Palaces and Depots and the different Phases. Then the author provides a clear description of the corresponding organ of sense, color, sound, etc., of each Depot and indicates which symptoms defines the different states of each Depot (hot, deplete, damp). He also gives a description of how the Depots resonate in the five parts of the face. By virtue of this system of correspondences between the internal and external parts of the body, "everything could be known (*wu bu ke zhi* 無不可知)".

The student then had to know the different states of the pulse, which, from the outside inform him about internal health. Huang Yan starts by giving the fundamentals of the six pulses and then provides descriptions of the pulses specific to several syndromes, under the form of a list of syndromes and below this the characteristics of the pulse. He also provides some generalities about the pulse of the four seasons, the pulse of women, children, of northerners and southerners, of young girls and nuns, and other pulse indications useful to establish whether the patient will survive or die. He provides some advice on making a good diagnosis, by recalling the sentence of Zhang Jiebin 張介賓 (1563–1640), a Ming physician whose work had widely circulated in the empire: "For a child's diagnosis, first listen, then look at him". After this first piece of advice, a few short chapters follow, dealing with how to make a diagnosis by listening, by looking, by distinguishing replete from deplete conditions. This first chapter ends on two short songs about a type of moxa applied to several acupuncture points and often used for childhood diseases.[63]

The second chapter begins with a discussion on pediatrics as a discipline. Huang Yan regretted that the knowledge and attendant practices of general medicine and those of pediatrics had often been perceived as two distinct fields, leading people to learn either medicine for adults or to learn how to treat children. For Huang Yan, this way of learning did not fit with the unicity of the principle *li*, evoked earlier. Moreover, and like many of his contemporaries, he considered that medicine for adults and for children was almost the same, there were just some peculiarities for each that one had to know. That way, the future doctor could treat anyone's illnesses. However, as a diagnosis was more difficult to establish in children, and as, in the

[62] Huang ([1800] 1918, 3).

[63] Huang ([1800] 1918, 3–10).

learning process, difficulties should be addressed first, learning medicine should start with learning pediatrics: "Students must first address what they find difficult until difficulties become crystal-clear. Then, easy things become clear. You who want to practice medicine, you first have to start with pediatrics (*ran xuezhe ze dang xian qi suo nan, shi nanzhe yi ru guanhuo, ze yizhe zike dongyuan yi. Gu erdeng yu xi yi, bi cong xiao'er xiashou* 然學者則當先期所難,使難者已如觀火, 則易者自可洞垣矣。故爾等欲習醫, 必從小兒下手".[64] In this chapter, he provides the beginner with some general principles on diagnosis. How to diagnose and treat cold disorders, tetanus, hot damage, jaundice, erysipelas, external syndrome, convulsions, spasms, chronic convulsions and deplete fever in babies. The chapter continues, dealing with how to diagnose and treat diseases that were not specific to children such as sudden disorders, intermittent fevers, and dysentery.[65]

We will not go further in describing the book's content which, after the second chapter and in all the six remaining chapters, was organized according to the same pattern: an enumeration of how to diagnose and treat particular diseases or pains, introduced by the title "how to diagnose and treat" a variety of syndromes, symptoms or pains. This enumeration does not seem to follow a specific nosological order and in the successive editions of the book while the content remained the same, the position in the book where each of the eight chapters started and ended were changed. This book, just like earlier medical texts composed for learning medicine in a professional perspective and well studied by Leung (2003), was very practical in character. It enabled the reader to know what to do when confronted with patients affected with diseases, that is, how to provide effective therapy to well identified syndromes, which was the exact literal meaning of the title to each entry. While this book relied on the cosmological framework of the systematic correspondences, the theoretical knowledge that lay behind it was not emphasized and was in fact only briefly described in the first section. Moreover while thermal properties and flavor of the remedies were explained at the start of the program, the author did not, however, feel it necessary to go deeper into this area of knowledge.[66] In fact, in his reading guidelines, the author explains that while he detailed extensively how one could make diagnosis and understand the principle lying behind any disease, he purposely provided only a short summary on the nature of plants and encouraged students to read other books of *materia medica* (*pian zhong mai yin zhengzhi zhi li, wu bu bao yu xiangming, er yaoxing du jian zhe, yu ren ling du bencao* 篇中因證

[64]Huang ([1800] 1918, 10).

[65]One must bear in mind that the anatomical and pathophysiological terms in Chinese medical literature cannot always be directly inferred from the English terms.

[66]The use of natural substances for therapeutic purposes goes back to the Zhou dynasty (1121–256). In the first *materia medica* written under the Han dynasty, natural substances, classified in three categories, were granted thermal properties and flavors. In the twelfth century, a science of the properties of drugs developed within the theoretical framework of *yin*, *yang*, and Fives Phases *wuxing*. However, because of its great complexity and numerous contradictions, it was often abandoned by physicians who continued to prescribe remedies by relying on ancient formulas and on empirism (Unschuld 1985, 114–115).

治之理, 無不包舉詳明, 而藥性獨簡者, 欲人另讀本草).[67] To summarize, the first two chapters addressed the basics of medicine that students had to master — the overall properties of remedies, the body's component parts and how they function and interrelate —, and the chapters that followed provided information allowing students to identify and treat some 65 diseases. Let us now turn to the textuality of the book which provides other clues to understanding what it really meant to read the book and to learn medicine through this reading process.

11.3.4 Verses and Citations: Vocalize, Memorize, and Believe

As already mentioned, for Huang Yan, learning medicine meant not only reading his book, but also learning it by heart. The author underlined that the textuality of his book was particularly appropriate for this purpose: "Characters and sentences are chosen after reflection. Rhymes are pleasant and tones arc harmonious. One can chant them out loud and recite them (*zi zhen ju zhuo, yun chang sheng xie, ke ge ke feng* 字斟句酌, 暢聲諧, 可歌可諷)".[68] Versification, as experienced everywhere in the world, facilitates memorization. Pan Mingxiong 潘明熊 (1807–1886), a medical author of the same place, writing a few decades later, explains it: "I think that it is easy to read books but difficult to memorize them ... this is why I wrote in four-character verses in order to facilitate memorization. (*yu nian kanshuyi er jishu nan ... zuo wei si yan gekuo shi zhi shu du* 余念看書易而記書難. ... 作為四言歌括使之熟讀)".[69] Huang Yan's text was thus submitted to rhyme and rhythmic constraints which were, in fact, very frequent in texts devoted to pedagogical uses, not only in the medical field as Leung (2003) has shown, but also in other fields of knowledge, such as mathematics (see Bréard, in this volume). What was reported as a part of the learning process in the biographies as well as in the physicians' own accounts of their apprenticeship, and also in the imperial curriculum – to chant aloud and to recite – is thus clearly attested in this book. All that one had to learn about the pulse was put into four-character verses. What should be learned about the pulse and conduits was first presented in prose and was then rewritten to be chanted aloud (歌日). What was to be learned about the remedies that should not be taken at the same time as others were recorded in seven-character verses. And what had to be learned about each corresponding phase, color, sound and so on and so forth of each *Zang* and *Fu*, were in three-character verses. However, it is noteworthy that not all the book was written in verse. These textual differences may indicate that the entire book did not have to be learned to the same extent, and that there was hence a graduation reflecting the depth to which a pupil should read and learn the

[67] Huang ([1800] 1918, *dufa*, 1).

[68] Huang ([1800] 1918, *dufa*, 1). See Bréard, in this volume for more details on how verses and rimes were vocalized.

[69] Pan ([1864] 1935, 3).

book. Not only was the pupil required to learn some aspects before others, as we have already seen, but also, some parts should be learned by rote, while others not. We thus see that by paying close attention to the organization of the book and to its textuality, a type of order emerges. This order does not seem to fit the analytic order—a progression from the most simple to the most complex, from the already known to the unknown—that was given preeminence by Abbé de Condillac in his *La logique ou l'art de raisonner* (1780) and that inspired the layout and organization used in chemistry textbooks in eighteenth century Europe, and more widely, in post-enlightenment pedagogy, including the medical textbooks used in modern China.[70] In the particular case of Huang Yan's medical textbook, we see other pedagogical orders: from the general to the singular (from basics to individual diseases), from the most complex to the most simple, from what had to be learned by heart to what had just to be read.

We may wonder why students in medicine were expected to learn by rote. At the beginning of the nineteenth century, is memorizing the trace of a time, before the Song dynasty for example, when access to books was difficult and when medical texts were transmitted by memorization and hand-copied? Is it the trace of a shift in the pedagogical methods from classical studies to medical studies? Or is it linked to a need specific to a discipline and activity such as medicine, where, treating a patient, perhaps far away from the doctor's office, one had to store the knowledge in oneself. Before providing elements of answer to this question, let us underline another aspect of this textbook's textuality. In his reading guidelines, the author wrote that his book was a compilation of the ancient sages' truths and of his own work. "This book has been composed for teaching my pupils. This is why all is gathered here comes from the best ancient sages and from what I have obtained through my own work (*shi bian ben wei shou tu er she, gu fan suo zuanji, leixie qianxian jingcui ji yu suo xin de* 是编 本為授徒而設, 故凡所纂輯, 類皆前賢精粹及余所心得)". This could be a rhetorical claim presented in the paratext to promote his book and make it sell better. It is well known that medical authors gained more authority and legitimacy when they framed themselves as heirs to and conveyors of past knowledge rather than as authors who so despise their heritage, that they brush it off and display only their own points of view.[71] But, when we look closer at the content, we find a quite systematic way of writing which actually and literally refers to the forbears. Under almost every heading, there are two parts: the first contains the author's words, introduced by "Huang Yan says ... *yue* 曰"; a second brings

[70] Bensaude-Vincent et al. (2003, 32–35).

[71] Bretelle-Establet (2011). In their comparative analysis of science and medicine in early China and Greece, Lloyd and Sivin noted: "Scientific pursuits in China [...] did not aim at stepwise approximations to an objective reality but at a recovery of what the archaic sages already knew" (Lloyd and Sivin 2002, 193).

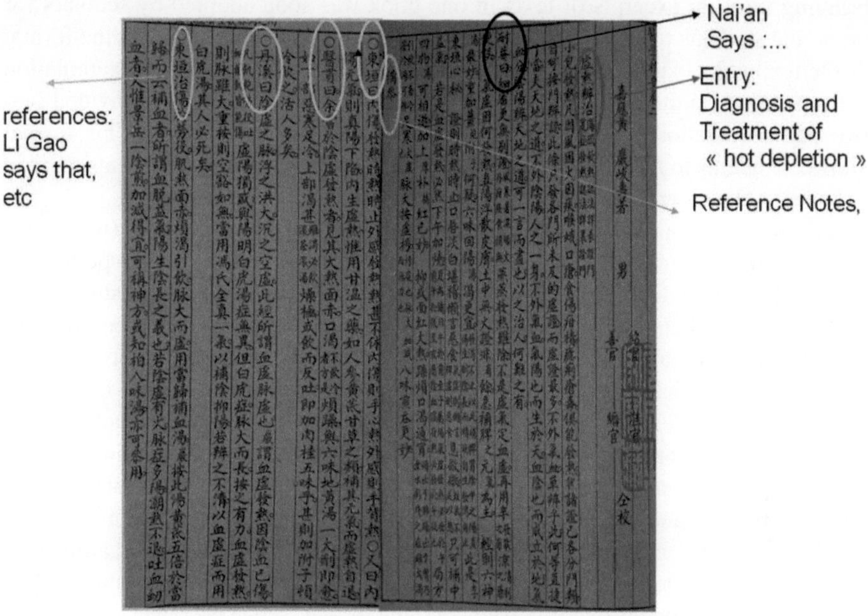

Fig. 11.1 Huang Yan's *Essentials of Medicine*, the *Yixue jingyao* 医学精要 (1800)

together, under a sub-heading "preparatory material (of reference) *beican* 備參", a wide or narrow range of citations, introduced by 'Mr. so and so says . . . ', or 'this or that book says . . . ' (Fig. 11.1).[72]

This way of writing, by making collections of related citations and excerpts, varying in both length and number, was far from uncommon in the Chinese literature. Compilation of citations was the main feature of a particular genre of texts that had appeared between the third and the sixth centuries and was given, under the Tang dynasty (618–907), a bibliographic category *leishu* that we usually translate as "encyclopedias" but which are best rendered as "classified books".[73] Compiling excerpts of texts must have had different functions. In China as in Europe, assembling excerpts from longer texts responded to the need to summarize and classify a literature that had grown very quickly and broadly.[74] In China as in Europe, compilations of excerpts may have also be linked to the need to provide access to parts of original texts that were not easily available.[75] But in China,

[72]Huang Yan cites 42 different texts, authors, local masters, from the most ancient times to his latest contemporaries.

[73]Drège (2007).

[74]Blair (2007) and Chartier (2007).

[75]On the issue of books' accessibility in late imperial China, see notably (Brokaw 2005; MacDermott 2006; Rowe 1994, 440; Bretelle-Establet forthcoming b).

bringing together excerpts of texts in one book was soon adopted by scholars for preparing students for the examinations. As Hilde De Weerdt has shown, the format of the textbooks that appeared in the Song dynasty, together with the examination culture, relied on the model of the "classified books": each entry was divided into two parts, quotations from primary documents and an explanation. This format allowed students to have all the relevant excerpts of the primary documents to hand to be learned and written out later, in their examinations.[76]

The way Huang Yan's medical textbook is written and organized (with, in a first part, the author's words, and in a second part, the citations) thus points out a parallel between the pedagogical material used in the field of classical studies and that used in medicine. The shift in teaching methods and tools from one field to the other was probably facilitated by the fact that many medical authors had first followed the classical curriculum. In fact, as I have shown elsewhere, 41 % of the medical authors identified in the far south of the empire were degree holders and had first-hand experience of the pedagogical methods and tools used in classical studies. Moreover, many of those who practiced medicine and who wrote medical texts often had functions either in officialdom or in the private training of the youth elite for the civil service examination.[77] Just as they used the examination encyclopedias for teaching and training students for the civil examinations, they wrote medical textbooks that followed the same format. When we read Huang Yan's textbook today, we are tempted to apply the suppositions formulated by Hilde de Weerdt about encyclopedias for examinations, that is, this type of textbook reflects what may have gone on in classes: "The section on primary sources relates to the secondary texts like a handout does to the lecture. In the explanatory part the author/lecturer explicitly refers to passages in the primary source texts."[78]

The large place given to citations in Huang Yan's textbook also reveals that teaching medicine could not be the transmission of one singular synthetic voice but more an interpretation justified by a series of proofs taken from the medical "archives". What Hilde De Weerdt has identified as an archival mentality in the classical curriculum and among Song literati was at the fore in medical education too: to speak about something or to do something needed a justification from the archives. Huang Yan's textbook suggests that the fostering of an "archival mentality" in the pedagogical practices and methods in the classical curriculum had penetrated into the pedagogical methods used in other fields of knowledge, such as medicine, and was still used in nineteenth century medical training. Just as he who reflected on the government needed to draw on past political history, he who learned and practiced medicine had to draw on past medical history. The lectures taught students how to diagnose and treat diseases on the basis of comprehensive archival research. The understanding of a disease and its treatment thus had to be based on referenced primary documents, on evidence to support the master's interpretation. In

[76]De Weerdt (2007, 82).

[77]Bretelle-Establet (1999, 562).

[78]De Weerdt (2007, 82).

the beginning of the nineteenth century, this evidence was mainly textual. It should be wrong, however, to say that only past medical texts were source of authority for a teacher. Huang Yan's textbook in fact also includes clinical cases *yi'an* 醫案 : his own, those of famous medical writers and some of his local masters.

Citation of forebears' texts and masters' experiences were, in Huang Yan's own words, a crucial point for being believed by his students: "Below each disease, I had to quote as proofs the sayings of the classics or the famous theories of ancient sages... If there was no quotation and that I speak as the main speaker, who would believe me? (*Qi wu xin na* 豈吾信那)".[79] Huang Yan refers to the word "xin 信", to have faith in, a notion which was of paramount concern in the transmission of knowledge in early China,[80] and which was still crucial in the private and non institutionalized context of apprenticeship of medicine in modern China. "Crucial was whether mentor and followers had faith in each other and could build a relationship of mutual trust", writes E. Hsu who consequently proposed to feature this style of teaching/learning as a "personal mode of (knowledge) transmission" (1999, 102). Interestingly and surprisingly for those familiar with modern textbooks, usually intended to present the state of a science, in its objectivity, Huang Yan's textbook keeps the trace of a highly personalized discourse.

11.3.5 A Personal Mode of Transmission Reflected in Huang Yan's Textbook

In her analysis of Chinese medical textbooks, Leung had noted that Li Chan 李梴 (late sixteenth to early seventeenth century)'s textbook *Introduction to medicine* (*Yixue rumen* 醫學入門), published in 1575, was written in such a lively manner that "one can almost hear Li's voice in a teaching session" (2003, 135). In his text, Huang Yan does the same and does not use neutral and impersonal language. Firstly, the author frequently addresses himself directly to his students by calling them, "you my pupils", "you who want to study". At many times in the text, he urges his disciples to take great care of the teaching in the work, which, as he emphasizes, is the sum of ancient sages' wisdom, his masters' secrets, his own research, and his own experience. "On this (particular point), the books of recipes do not contain what only a few people know. I will tell it to you, my disciples, you must know it (*fangshu bu zai shaoren zhi. Shuo yu tu xu zhi* 方書不載少人知。說與徒須知)". "For this, I obtained good results ... I cannot keep this secret for me (*yu zhi de xin yingshou ... jin bug an mi* 予之得心應手...今不敢秘)". "I entrust the totality of my work to you (*yu suo xin de zhe yi gao erdeng* 予所心得者以告爾等", "I teach to you, but you must take great care of it (*chuanshou erdeng, dan erdeng xu zhen zhi zhong zhi*

[79]Huang ([1800] 1918, 2).

[80]Sivin (1995).

傳授爾等。但爾等須珍之重之)",[81] Huang Yan declares, in a tone that evokes the rituals of the transmission of texts and knowledge in early China but that perhaps also bears the mark of a more universalistic trust in the strong cognitive value of oral teaching. The frequent addresses of Huang Yan to his students actually remind the famous address by Michelet to his students, in 1842, in which he emphasized the superiority of words on writings, and the specificity of oral teaching: "Je parle avec confiance avec vous, à vous seuls, et point aux gens du dehors. Je ne vous confie pas seulement ma science, mais ma pensée la plus intime sur le sujet le plus vital" (I speak to you with confidence, only to you, and not to people outside. I do not only entrust my science to you, but my most private thinking on the most essential subject).[82] Huang Yan's text is very often punctuated by such personalized advice and bring us very far from the impersonal and neutral discourse that is expected of textbooks as we understand them today, namely, as books devoted to presenting the state of a science objectively. This text bears the marks of a style of teaching and learning very close to that identified by Elisabeth Hsu in her anthropological experience of learning medicine with a senior doctor in the 1980s.

11.4 Conclusion

As Huang Yan's medical text shows, capacities required in a context of private medical instruction were not very far removed from those required in the *Taiyi yuan*, the imperial Bureau of medicine, nor from those required in the curriculum of classical studies. Pupils were expected to learn a large corpus of texts by heart and in this way use the Ancients' knowledge and experience in their diagnosis and treatment of diseases. Even though there was no physical link between the imperial institution evoked at the beginning of the chapter and the teacher Huang Yan, however, Huang Yan's textbook reveals that the teaching/learning praxis in the institution, largely influenced by the classical curriculum, has still an impact on the private pedagogical practices in the far south, during the nineteenth century.

To what extent this particular book of medical instruction was representative of the pedagogical methods and expectations in late imperial China is difficult to ascertain. Huang Yan's textbook shared many features of the medical textbooks that had emerged since the fourteenth century, studied by (Leung 2003): simplification of the language used, recourse to versification, emphasis on the practical side of this branch of knowledge.[83] However, not all the textbooks followed the pattern of Huang Yan's book. The ideal would be to gather together a large corpus of texts explicitly conceived to serve as textbooks before the institutionalization of the

[81] Huang ([1800] 1918, chap. 3, p. 13; chap. 2, p. 14; chap.1, p. 11).

[82] Sivin (1995) and Waquet (2003, 9).

[83] Such as the *Yijing xiaoxue* 医经小学 written by Liu Chunyu in 1388 or the *Yixue rumen* 医学入门 written by Li Chan in 1576 from which Huang Yan reproduces some parts. See (Gong 1983, 113; Leung 2003).

medical curriculum and the birth of national standardized textbooks. And highlight the differences and similarities between the pedagogical methods presented to pupils in the different parts of the empire, at different times and according to the authors' social and cultural backgrounds. But even without going so far, and taking, as an example, the *Stepping stone for medicine* (*Yibian* 醫碥), written in 1751, by He Mengyao 何梦瑶, a native of Nanhai in Guangdong, 50 years before the *Essentials of Medicine*, we can note great differences in format, content and language.[84] A first difference between the two texts that were nevertheless conceived as didactic books lies in their length. While Huang Yan's book contained a hundred pages, He Mengyao's book contained some four hundred pages, in a modern edition. Undoubtedly these two authors had different ideas as to the extent and the depth of the knowledge a pupil should learn and master before he practiced medicine. A second difference lies in the order in which the knowledge is presented. Both books started with generalities, Huang Yan, with a series of things that had to be known (*xuzhi*) and He Mengyao, with a series of sayings (*shuo* 說) and doctrines (*lun* 論). But while Huang Yan continued in the following chapters of his book by enumerating how to diagnose and cure 65 particular diseases, syndromes or pains, He Mengyao lays out his text in a different order. After a long first chapter that brings together all the various doctrines on the body and the cosmological patterns of its function, chapters 2, 3, and 4 discuss particular diseases, chapter 5 presents the principles of diagnosis and the last two chapters are a collection of the recipes referred to in the text. The systematic use of cross-references in the *Stepping-Stone for Medicine* indicates that He Mengyao had a very different idea of how a pupil would read his book compared to Huang Yan's text, which was conceived for a continuous reading. The pedagogical methods used and advocated by these two authors were also different. While Huang Yan used versification to facilitate rote learning, He Mengyao did not use it and did not even recommend it. While both claimed to be respectful of their forebears, the way they referred to them, however, was different. While Huang Yan based his discourse on a range of quotations or citations, introduced by the author's name, He Mengyao, in the perspective of simplifying the ancient masters' language, preferred to keep their ideas without reproducing the original text.[85]

What seems to be common to these two texts is the extent to which the presence of the author is emphasized. As linguists have shown, there are many ways for an author to be present in his text and thus to create degrees of personalization. In Huang Yan's text, this presence is manifested through the Teacher's frequent direct addresses to his Pupils, the quotation of himself ("Nai'an says"), the frequent use of the personal pronoun I " 我" which is not necessary in written Chinese, the possessive adjective "my family (*yujia* 予家)", and through typographical

[84]He Mengyao's biography is known through official biographies recorded in local gazetteers and several prefaces to his books. For He Mengyao's biographies and his disciple Chen Guodong, see (Guo 1987, 1933–4).

[85]He ([1751]1994: 54).

markers, either punctuation markers intended to emphasize some particular points, or small size characters for his own comments.[86] In He Mengyao's text, his presence is expressed through the frequent use of small size characters to add personal comments and by the high frequency use of discourse markers (the particles 也, 矣, 乎 "it's sure", "that's it", "but why then") which, in addition to helping the reader to segment this unpunctuated text, also provide various kinds of metalinguistic indications about the unfolding discourse, such as translating the speaker's emotion for what he is saying.[87] As I have shown elsewhere by comparing six excerpts of medical texts dealing with the same disease, but written for different purposes, not all medical texts bear the author's presence to the same extent. While some texts are totally impersonal and deliberately hide the author's emotions or feelings, his intellectual preferences for this or that master for the profit of a more general truth, the texts that are explicitly written in a teaching environment convey a highly personalized discourse, giving sometimes the impression that these texts were the written transcription of oral lessons, including the master's angers against common mistakes, questions and answers, anecdotes but also repetitions.[88]

But apart from this particular common aspect which seems to translate "a personal mode of transmission", we see important variations between these two didactic texts. These variations must be down to the same reasons as those highlighted recently by historians of chemistry in France: the social and cultural milieus of the authors were not unique, the public for which they wrote was not clearly segmented, and the ultimate goal of learning medicine was not univocal.[89] Moreover, until the 1960s, medicine in China was not constituted by a systematic and unified theory but a juxtaposition of individual voices. It became a unified and coherent theory, known in the West under the label "Traditional Chinese Medicine", only after Mao's reforms in the late 1950s, that the standardized textbooks, written and published soon afterwards, contributed to consolidate.[90]

References

Banks, David. 2005. *Les marqueurs linguistiques de la présence de l'auteur*. Paris: L'Harmattan.

Bensaude-Vincent, B., A. Garcia Bellimar, and J.R. Bertomeu Sanchez. 2003. *L'émergence d'une science des manuels. Les livres de chimie en France (1789–1852)*. Paris: Editions des Archives Contemporaines.

[86]According to Banks (2005, 40–43) typographical change constitutes a linguistic marker of the author's presence to the same extent than the use of personal pronouns.

[87]On discourse markers, see (Schiffrin 1988).

[88]Bretelle-Establet (forthcoming a).

[89]Bensaude-Vincent et al. (2003, 59, 64) who also reported that in spite of the government's efforts to impose an official book, the French chemistry textbooks written in the first years of the nineteenth century remained highly personalized.

[90]On the attempts made to regulate medicine under the republican and nationalist regimes, see (Croizier 1968; Xu 1997; Yip 1982, 1994, 1995); on the reforms undertaken under Mao's regime, see (Taylor 2005).

Blair, Ann. 2007. The Latin florilegium as a point of comparison. In *Qu'était-ce qu'écrire une encyclopédie en Chine ?* Extrême-Orient, Extrême-Occident, Hors-série, ed. F. Bretelle-Establet and K. Chemla, 185–204. Paris: Presses Universitaires de Vincennes.

Bretelle-Establet, F. 1999. La santé en Chine du Sud (Yunnan, Guangxi, Guangdong) à la fin de l'empire et au début de la république. Ph.D. dissertation, LCAO, Université Paris 7.

Bretelle-Establet, Florence. 2002. *La santé en Chine du Sud, 1898–1928*, Collection Asie Orientale. Paris: CNRS.

Bretelle-Establet, Florence. 2009. Chinese biographies of medical experts: What uses can we make of them? *East Asian Science, Technology, and Society* 3: 421–451.

Bretelle-Establet, Florence. 2011. The construction of the medical writer's authority and legitimacy in Late Imperial China through authorial and allographic prefaces. *NTM, Zeitschrift für Geschichte der Wissenschaften, Technik und Medizin* Heft 4 (2011), Seite 349–390. Band 19.

Bretelle-Establet, Florence. Forthcoming [a]. The issue of textual genres in the medical literature produced in late imperial China. In *Textology via scientific writings*, ed. Karine Chemla and Jacques Virbel.

Bretelle-Establet, Florence. Forthcoming [b]. Human mobility and books: Modes of circulation of medical ideas and doctrines, in the Far South (18th and 19th centuries). In *Itinéraires individuels et circulation des savoirs scientifiques et techniques en Chine 898 moderne (XVIe–XXe si'ecles)*, ed. Catherine Jami.

Brokaw, Cynthia J. 2005. *Printing and book culture in Late Imperia China*. Berkeley: University of California Press.

Chang, Che-chia. 1998. The therapeutic Tug of War: The imperial physician-patient relationship in the era of Empress Dowager Cixi (1874–1908). Ph.D. thesis, University of Pennsylvania.

Chao, Yuan-ling. 2000. The ideal physician in Late Imperial China: The question of *sanshi* 三世. *East Asian Science, Technology, and Medicine* 17: 66–93.

Chao, Yuan-ling. 2009. *Medicine and society in Late Imperial China: A study of physicians in Suzhou, 1600–1850*. Frankfurt: Peter Lang.

Chartier, Roger. 2007. La muraille et les livres. In *Qu'était-ce qu'écrire une encyclopédie en Chine ?* Extrême-Orient, Extrême-Occident, Hors-série, ed. F. Bretelle-Establet and K. Chemla, 205–216. Paris: Presses Universitaires de Vincennes.

Cheng, Daozhang. [1920] 1968. *Guiping xianzhi*, 桂平縣誌 [Gazetteer of Guiping]. Taibei: Chengwen chubanshe.

Choppin, Alain. 2008. Le manuel scolaire, une fausse évidence historique. *Histoire de l'éducation* 117: 7–56.

Chü, T'ung-Tsu. 1962. *Local government in China under the Ch'ing*. Stanford: Stanford University Press.

Croizier, Ralph. 1968. *Traditional medicine in modern China*. Cambridge, MA: Harvard University Press.

De Weerdt, Hilde. 2007. The encyclopedia as textbook: Selling private Chinese encyclopedias in the twelfth and thirteenth centuries. In *Qu'était-ce qu'écrire une encyclopédie en Chine ?* Extrême-Orient, Extrême-Occident, Hors-série, ed. F. Bretelle-Establet and K. Chemla, 77–102. Paris: Presses Universitaires de Vincennes.

Despeux, Catherine (ed.). 2010. *Médecine, religion et société dans la Chine médiévale*. Paris: Collège de France, Institut des Hautes Etudes Chinoises.

Drège, Jean-Pierre. 2007. Des ouvrages classés par catégories: les encyclopédies chinoises. In *Qu'était-ce qu'écrire une encyclopédie en Chine ?* Extrême-Orient, Extrême-Occident, Hors-série, ed. F. Bretelle-Establet and K. Chemla, 205–216. Paris: Presses Universitaires de Vincennes.

Durling, Richard T. 1990. Girolamo Mercuriale's De modo studendi. *Osiris* 6: 181–195.

Elman, Benjamin. 2000. *A cultural history of civil examinations in Late Imperial China*. Berkeley: University of California Press.

Elman, Benjamin, and Alexandre Woodside. 1994. *Education and society in Late Imperial China*. Berkeley: University of California Press.

Farquhar, Judith. 1994. *Knowing practice: The clinical encounter of Chinese medicine*. San Francisco: Westview Press.

Feray, Dr. 1910. Le poste medical de Yun-nan-fou. *Annales d'hygiène et de médecine coloniale*. Paris: Doin.

Furth, Charlotte. 1999. *A flourishing Yin. Gender in China medical history, 960–1665*. Berkeley: University of California Press.

Gernet, Jacques. 2003. L'éducation des premières années (du 11e au 17e siècles). In *Education et instruction en Chine,* L'éducation élémentaire, vol. 1, ed. Christine Nguyen Tri et Catherine Despeux, 7–60. Louvain: Peeters.

Goldschmidt, Asaf. 2009. *The evolution of Chinese medicine, Song Dynasty (960–1127)*. London/New York: Routledge.

Gong, Chun 龚春. 1983. *Zhongguo lidai weisheng zuzhi ji yixue jiaoyu* 中國歷代衛生組織及醫學教育 [Medical Teaching and Public Health Organization in Chinese History]. Beijing: Weisheng bu kejiao.

Guan, Xueling 关雪玲. 2008. *Qingdai gongting yixue yu yixue wenwu* 清代廷醫學與醫學文物 [Medicine and cultural objects of the Qing Dynasty Court]. Zijincheng chubanshe.

Guo, Aichun 郭靄春. 1987. *Zhongguo fensheng yiji kao* 中國分省醫籍考 [Catalogue of medical books by province]. Tianjin: kexue jishu chubanshe.

Hadot, Pierre. 2001. *La philosophie comme manière de vivre*. Paris: Albin Michel.

Hanson, Marta. 1997. Inventing a tradition in Chinese medicine. Ph.D. dissertation, University of Pennsylvania.

Hanson, Marta. 2003. The golden mirror in the imperial court. In *Early science and medicine*, ed. C. Jami, 111–147. Leiden: E.J. Brill.

He, Mengyao 何梦瑶. [1751] 1994. *Yibian* 醫碥 [Stepping-stone for medicine]. Beijing: renmin weisheng chubanshe.

He, Shixi 何时希. 1991. *Zhongguo lidai yijia zhuanlu* 中國歷代醫家傳錄 [Catalogue of biographies of physicians in Chinese history]. Beijing: Renmin weisheng chubanshe.

Hinrichs, T.J. 2003. The medical transforming of governance and southern customs in Song Dynasty China (960–1279 C.C.). Ph.D. dissertation, Harvard University.

Hsu, Elisabeth. 1999. *The transmission of Chinese medicine*. Cambridge: Cambridge University Press.

Huang, Piaozhi 黄朴之. 1933. *Xiaoyan liangfang* 效驗良方 [Excellent and experimented formulas]. Huang Piaozhi's printing house (zhuzuo qianyin).

Huang, Yan 黄巖. [1800] 1918. *Yixue jingyao* 醫學精要 [Essentials in medicine]. Shanghai cuiying shuju yinhang.

Hymes, Robert. 1987. Not quite gentlemen? Doctors in Sung and Yuan. *Chinese Science* 8: 9–76.

Lei, Hsiang-lin. 2002. How did Chinese medicine become experiential? The political epistemology of jingyan. *Positions: East Asia Cultures Critique* 10(2): 333–364.

Leung, Ki.-che Angela. 1987. Organized medicine in Ming-Qing China state and private medical institutions in the lower Yangzi region. *Late Imperial China* 8(1): 134–166.

Leung, Ki-che Angela. 2003. L'instruction médicale et sa vulgarisation dans la Chine des Ming et des Qing. In *Education et instruction en Chine*, L'éducation élémentaire, vol. 1, ed. Nguyen Tri et Catherine Despeux, 89–114. Louvain: Peeters.

Li, Yu. 2003. A history of reading in Late Imperial China, 1000–1800. Ph.D. dissertation, Ohio State University.

Liang, Lianfu 梁廉夫. [1881] 1936. *Bu zhi yi biyao* 不知醫必要 [What the ignorant person in medicine should know]. Zhenben yishu jicheng.

Lin, Ganliang. 1983. *Jindai Zhejiang de Zhongyi jiaoyu* 近代浙江的中醫教育 [The teaching of Chinese medicine in modern Zhejiang]. *Zhonghua yishi zazhi* 13(4): 224–226.

Liu, Yuke. [1886] 1967. *Yongchang fuzhi* 永昌府誌 [Gazetteer of Yongchang]. Taibei: Chengwen chubanshe.

Liu, Xiaobin. 1986. *Guangdong jindai de zhonyi jiaoyu* 廣東近代的中醫教育 [The teaching of Chinese medicine in modern Guangdong]. *Zhonghua yishi zazhi* 16(3) 148–151.

Lloyd, Geoffrey, and Nathan Sivin. 2002. *The way and the word. Science and medicine in early China and Greece*. New Haven/London: Yale University Press.

MacDermott, Joseph P. 2006. *A social history of the Chinese book. Books and literati culture in Late Imperial China*. Hong Kong: Hong Kong University Press.

Miles, Steven. 2006. *The sea of learning. Mobility and identity in nineteenth-century Guangzhou*. Cambridge: Harvard University Asia Center.

Mizayaki, Ichisada. 1976. *China's examination hell. The civil service examinations of Imperial China*. New Haven: Yale University Press.

Pan, Mingxiong 潘明熊. [1864] 1935. *Ye'an kuoyao* 叶案括要 [*Summary of the clinical cases of Ye*]. Guangzhou: Guangzhou linji shuzhuang.

Ren, Xigeng 任锡庚. [1863]1923. *Taiyiyuan zhi* 太醫院誌 [Monograph about the Imperial Medical Bureau]. np: np.

Rowe, William T. 1994. Education and empire in southwest China, Ch'eng Hung-mou in Yunnan. In *Education and society in Late Imperial China*, ed. B. Elman and A. Woodsite, 417–457. Berkeley: University of California Press.

Scheid, Volker. 2007. *Currents of tradition in Chinese medicine, 1626–2006*. Seattle: Eastland Press.

Schiffrin, Deborah. 1988. *Discourse markers*, Studies in interactional sociolinguistics. Cambridge: Cambridge University Press.

Shinno, Reiko. 2002. Promoting medicine in the Yuan Dynasty (1206–1368): An aspect of Mongol Rule in China. Ph.D. dissertation. Stanford: University of Stanford.

Siu, Man Keung. 2004. Official curriculum in mathematics in ancient China: How did candidates study for the examination? In *How Chinese learn mathematics: Perspective from insiders*, ed. N.Y. Wong et al., 157–185. Singapore/River Edge: World Scientific.

Sivin, Nathan. 1995. Text and experience in classical Chinese medicine. In *Knowledge and the scholarly medical traditions*, ed. Don Bates, 177–204. Cambridge: Cambridge University Press.

Taylor, Kim. 2005. *Chinese medicine in early Communist China, 1945–1963: A medicine of revolution*. London: Routledge Curzon.

Unschuld, Paul U. 1985. *Medicine in China: A history of ideas*. Berkeley: University of California Press.

Unschuld, Paul U. [1988] 2005. *Chinese life sciences. Introductory readings in classical Chinese medicine*. Taos: Paradigm Publications.

Wang Wenshao 王文韶. 1901. *Xu Yunnan tongzhi gao* 續雲南通誌稿 [Draft of the gazetteer of Yunnan].

Waquet, Françoise. 2003. *Parler comme un livre. L'oralité et le savoir, XVIe–XXe siècle*. Paris: Albin Michel.

Will, Pierre-Etienne. 1992. *Chinese local gazetteers. An historical and practical introduction*. Paris: Centre de Recherche et de Documentation sur la Chine Contemporaine.

Xu, Xiaoqun. 1997. 'National essence' vs 'Science': Chinese native physicians' fight for legitimacy, 1912–1937. *Modern Asian Studies* 31(4): 847–877.

Yip, Ka-Che. 1982. Science, medicine and public health in twentieth century China, health and society in China: Public health education for the community, 1912–1937. *Social Science of Medicine* 16: 1197–1205.

Yip, Ka-Che. 1994. The Kuomintang and China's medical modernization, 1928–1937. *Centennial symposium on Sun Yat-Sen's founding of the Kuomintang for revolution*, Taiwan, 19–23.

Yip, Ka-Che. 1995. *Health and national reconstruction in nationalist China: The development of modern health services, 1928–1937*. Ann Arbor: Association for Asian Studies.

Zhang, Hongsheng 張鴻聲. 1995. *Qingdai yiguan kaoshi ji tilie*. 清代醫官考試及題例 [Examination for medical officials and its question examples in Qing Dynasty]. *Zhonghua yishi zazhi* 25(2): 95–96.

Zheng, Xueyu 鄭學宝, and Zheng Hong 鄭洪. 2005. *Lüelun Songdai yixue kaoshi de tedian*. 略論宋代醫學考試的特點 [The particularities of medical examinations under the song]. *Zhongyi jiaoyu* 5: 74–77.

Part V
Postface

Chapter 12
On the Sources of the Historian of Science from the Perspective of a History of Education

Karine Chemla

The book that these few words conclude aims primarily, as its title makes explicit, at offering clarification on the issue of the relationship between documents that are sources for present-day historians of science and what is often called, for lack of a better expression, a "school context." This project is most welcome, since it addresses an area in which, undeniably, as I shall emphasize, a great deal of confusion reigns and that this confusion is detrimental to a genuinely historical treatment of our sources.

In the following lines, I will sketch reasons why I hold this view on the topic. This is the view of a historian dealing mainly with mathematics in the ancient world, that is to say, with a discipline plagued by anachronism and with an historical period for which sources are scarce. Critical reflection on the sources is thus vital, if we are to avoid gross misinterpretation. It cannot be incidental that most of the authors in this book, including primarily the two editors, carry out research in the same sub-field and have accordingly come to share the same concern. The general introduction that opens the volume makes this clear.

Fortunately, in the past few decades, research on the history of education and most importantly for us, research on the history of scientific education has taken

---

The book is the outcome of seminars and conferences organized by Alain Bernard and Christine Proust in the last few years. I wish to extend my thanks to the participants, since their questions, their suggestions and their contributions to this book helped me prepare this postface. I am particularly indebted to Bruno Belhoste, Alain Bernard and Christine Proust, with whom I continued conversation beyond these meetings. It is a pleasure to thank Richard Kennedy for his contribution in the preparation of this chapter.

K. Chemla (✉)
Laboratoire SPHERE, CNRS & Université Paris Diderot, Case 7093, 5 rue Thomas Mann, 75205 Paris Cedex 13, France
e-mail: chemla@univ-paris-diderot.fr

A. Bernard and C. Proust (eds.), *Scientific Sources and Teaching Contexts Throughout History: Problems and Perspectives*, Boston Studies in the Philosophy and History of Science 301, DOI 10.1007/978-94-007-5122-4_12,

off, providing historians of science with both safeguards against wild assumptions and a source of inspiration for new lines of research. This book can be read as part of this larger movement, and I would like to formulate research perspectives that it prompts us to consider and whose exploration could, in my view, contribute to shedding some light on the complex issue, for the history of science, of the sources and their relationships with contexts of teaching activities.

I have begun this postface with the expression "school context," and I now use the much wider expression of "teaching activities," which incidentally echoes the title chosen for the book. This derives from both an observation and a decision.

First, the observation. In fact, when anachronism can be perceived in the way documents are presumed to have been tied to educational practice, in the majority of cases the context assumed is that of school. What is projected onto the text is the type of institution, in which one had one's own personal experience of learning, as well as its mode of organization. This is exactly what Alain Bernard shows was Adolphe Rome's approach to Theon's commentaries on the *Almagest* and the *Handy Tables.* In addition, most often, the anachronism extends to assumptions made about the types of texts used and the practices in relation to which they were used. That is to say, even in cases when it is documented that teaching occurred in the context of schools, anachronism remains rampant. The opening chapter of the book, by Anne-Marie Chartier, provides a powerful antidote against such historical anachronisms. It brings to light that, in the contexts of schools for which there is documentary evidence that reading was taught at an elementary level, the progression proceeded in ways that are wholly different from those we have experienced individually. The types of text used and the ways of handling these texts are not the same. The aim of the teaching and the type of reading competence inculcated—that Anne-Marie Chartier calls, after Jack Goody, "intensive reading," as opposed to "extensive reading"—also differ. Furthermore, in the following chapter, Christine Proust exposes the ambiguities of the expression "school context," establishing a variety of situations arising from such a context and accordingly the distinction to be established between various types of documents, even though all of them can be attached to a school context. Echoing these ideas, the introduction by Alain Bernard and Christine Proust develops insightful remarks on the analysis of the historiographic notion of "context." We come back to these important insights below.

So much for the observation. The decision to deal with wider "teaching activities," and not only school contexts is unavoidable if we want to address, with the adequate breadth of scope, the issue of the impact of teaching activities on our sources and critically review the ways in which documents have been perceived to derive from such contexts. Indeed, schools represent only a small part of the educational contexts attested to in the past, be it distant or less remote. As Florence Bretelle-Establet emphasizes, for medicine in modern China, teaching or learning situations cover a continuum ranging from learning with books at any age, alone or with a master, or within the family, to following a course in a teaching institution with its more or less strict corpus of texts and teaching curriculum, its modalities of teaching and learning, and its criteria as well as its institutions for measuring success. In the same way, in his wide-ranging study on textbooks, Alain Choppin feels the need to broaden the scope of situations observed.[1] The remarks explain why I chose here to follow the editors of the volume and speak with them of "teaching contexts," or of "teaching situations," in an attempt to avoid making any a priori assumption on the contexts in question. In any event, it must be kept in mind that none of these expressions refer to a situation that can be taken for granted but, on the contrary, what exactly is meant requires, in each case, historical analysis.

12.1 Old and Lasting Prejudices

Of all the social contexts that have come to historians' attention, learning and teaching contexts were, until the 1980s, some of the most poorly treated. I shall concentrate my remarks here on the impact of this situation on the history of science and the way the available documents were dealt with. As the contributors to this book concur to conclude, despite the research conducted since the end of the twentieth century, confusion still often reigns in debates touching these issues, in particular with respect to the question of determining how our sources relate to such contexts and what this fact means for their interpretation. Indeed, the confusion started long ago.

School documents, or more generally documents used or produced in teaching environments, have long been under-studied. History of mathematics amply illustrates this statement. A huge amount of documents perceived as relating strictly to a school context have remained untouched by historical works until recently.[2] In contrast, when documents produced in the context of a teaching activity have been studied, they have most often been analyzed in and of themselves, without emphasizing, or taking into account, the context in which they were produced. A paradigmatic example is *Analyse Algébrique* (1821), produced by Cauchy in relation to his teaching at the *Ecole Polytechnique*, in fact, as volume 1 of his *Cours d'Analyse de l'Ecole Royale Polytechnique.*[3] We will come back to this.

One of the reasons that can be put forward to account for the lack of interest historians of science have shown for documents produced in educational contexts is the implicit assumption that masters simply repeat, or worse, transpose content created by others, while the students' clumsiness made their work seem unpromising. In the framework of a history of science dominated, since the nineteenth century,

[1]Choppin (2008).

[2]C. Proust, in this volume, pp. 69–70, comes to this conclusion with respect to the sources documenting mathematical activities in Mesopotamia.

[3]The benefits derived from putting the book in the context of the teaching activities from which it derived were made manifest recently, see (Gilain 1989; Belhoste 1991, 61–86). In his plea for a reassessment of the part played by education in the history of mathematics, (Belhoste 1998) analyzes the theoretical impact of the teaching context on the main features of Cauchy's contribution.

by a program of documenting novelty in mathematical concepts and results, and attributing "firsts," what has happened in what has been perceived to be the "forefront of research" has received more attention than what was, and has been practiced in teaching environments. Such a diagnosis would account for the fact that until the 1980s, work written by masters or students, be it Mesopotamian tablets or lecture notes from the *Ecole Polytechnique*, surfaced in historical writings mainly for the mathematical achievements to which they attested, with no enquiry into their relation to the teaching context.

Schools appear to have been perceived as places of second-rate knowledge, and the documents produced in this environment have been treated accordingly, that is, considered to be of secondary importance[4] with respect to texts written by those extending the boundaries of knowledge. They have also been associated with several types of conservatism: conservatism, or even dogmatism, with respect to the bodies of knowledge taught, as well as conservatism with respect to practices, when stress is laid on repetition, heavy memorization and learning by rote, that is, practices often contrasted with understanding and innovating.

Such perceptions had their effect on the conservation of documents.[5] Textbooks have, for instance, been kept to a lesser extent than other types of scholarly writings.[6] Similarly, when the participants in the Babylonian expedition (1898–1900) had to share out the tablets that had been excavated from Nippur, the students' drafts went into the part left to Ottoman authorities.[7] These drafts were not only repetitive exercises, but as objects they were considered to be less valuable than other types of tablets, especially those shaped more regularly and more elegantly written that were probably masters' texts. Neither the economic worth nor the symbolic importance of students' drafts gave them value in the eyes of the archeologists. As Issitt notes, one reason for textbooks to be "the first to be cast out in the interests of shelf space," can be that they were "fairly cheaply produced." The same logic eliminated ephemera in the first place, especially those produced by inexperienced hands, by comparison with more refined and attractive objects.

The lack of appreciation of documents produced in teaching situations has extended to writings assumed to have been composed in such an environment. The point can be illustrated clearly with the example of mathematics in ancient China, even though this is far from being the only example. As Alexei Volkov explains in his chapter, most of the Chinese mathematical treatises that have come down to us through the written tradition, and were composed between the first century BCE and the early seventh century CE, were collected into an anthology entitled *Ten Classics*

---

[4]This analysis echoes the one developed in (Issitt 2004, esp. p. 690).

[5]For a reflection on the interaction between actors' contrasted attitudes towards different types of documents and the divergent historical trajectories of writings, see (Bretelle-Establet 2010).

[6]Issitt (2004, 692) has interesting remarks on the material management of textbooks in libraries.

[7]On the history of the sharing, see (Proust 2007, 40–50).

of Mathematics and presented to the throne in 656.[8] Administrative regulations from the time show clearly that these books were used shortly thereafter as textbooks in the context of mathematical instruction given at the State University.

Note, however, that the fact that these books were used as textbooks does *not* necessarily mean that they had been written *for* educational purposes. In effect, we know next to nothing of the circumstances in which most of them were composed and we have hardly any information on their authors. These facts should serve as a warning when we consider which questions we can legitimately ask about them. As Anne-Marie Chartier brings to our attention, the fact that between the sixteenth and the eighteenth century in Western Europe, teaching materials for beginners included prayers does not imply that prayers had been written *in order to* teach reading and that they *were* textbooks. Nor does the fact that I read Molière or Balzac at school, as I actually did, imply that Molière and Balzac wrote *for* the classroom. However, the step is often taken with Chinese mathematical classics collected in the seventh century anthology. They are regularly considered as having been not only *used*, which is wholly evidenced, but also *composed* as textbooks.

This is what transpires from Jean Dhombres's foreword to Jean-Claude Martzloff's *History of Chinese Mathematics*, and it is manifest that this assumption about the books goes along with a note of disenchantment about them. Dhombres writes:

> (...) there is the question of a difference in status between the mathematics of practitioners and that of textbooks intended for teaching purposes. Broadly speaking, as far as China is concerned, it is mainly textbooks which have come down to us, worse still, these are textbooks which belong to an educational framework which placed great value on the oral tradition and on the memorising of parallel, rhyming formulae. How could we describe 18th-century French mathematics if we only had access to the manuals due to Bezout, Clairaut or Bougainville? Moreover, should not textbooks be written in such a way that they adhere to the practice of mathematical research of a period, as Monge, Lagrange and Laplace deigned to believe during the French Revolution? Should greater importance be placed on metonymy, the passage from the particular to the general, based on the a priori idea that local success should reveal a hidden structure, even during the training procedure? Is a vague sense of analogy a sufficient basis on which to found an education at several successive theoretical levels? Thus, the history of the mathematics developed in Hangzhou, or any other capital, gives the teacher something to think about.[9]

Parallel to the perception that the earliest Chinese mathematical books that have come down to us were textbooks, comes the idea that they are second-rate witnesses of mathematical "achievements" from China's past, just as French textbooks written by practitioners working at less notable heights would be with respect to the whole range of innovation seen in eighteenth-century mathematics in France. This assessment certainly derives from a generous assumption, a regret that we, historians, are not in a position to document the genuinely brilliant innovations

[8]On the nuance I make here, see (Chemla 2010), where I analyze the process of transmission of this anthology.

[9]Dhombres (1997, [rev. ed. 2006], XI).

that must have been achieved in China and are now lost.[10] However, the assessment also reveals the perception of a clear hierarchy, projected onto historical documents, between "the mathematics of the practitioners" and that of "textbooks", when they were not produced by the greatest minds and in an appropriate educational framework. In that assessment, what relates to teaching activities is downplayed and offered to the reflection of other teachers rather than to that of historians. What matters to us here, for the sake of the argument, is the fact that, in this view, the earliest Chinese mathematical books to have been handed down are depreciated due to the fact that they were seen as textbooks.

Here two completely different sets of issues get entangled, as incidentally they often have been, and this is one of the causes of the confusion to which I have alluded above. It is important to disentangle them in order to introduce some clarity into the debate. I will differentiate them by formulating the two separate objections that one could use to oppose the previous assessment.

The first objection runs as follows: there is no a priori reason why textbooks should be downplayed as such, and especially in regard to their value to the historian. At the very least, even though not only, knowledge and documents that have been produced in teaching environments are an invaluable source of historical research. I argue further about these two points separately below. The refutation bears here on the historical judgment placed on school material.

The second objection addresses another dimension of the previous assessment. There is no a priori reason to assume that the earliest Chinese books that have come down to us were composed *for* educational purposes, distinguishing again between the fact that they were used as such and whether, in fact, they were composed to that end. The value attached to school documents does not license historians to freely associate a teaching context and didactic purposes to any piece of writing they like.

The reason the latter objection is important relates to questions of interpretation. Many authors in this volume adopt the requirement, inspired by Wittgenstein, which Hadot formulates as follows: "*toute oeuvre doit être placée dans la praxis dont elle émane*." If one concurs with them in this respect, as I do, then the determination of both the context and the use is essential to *interpretation*. Here, by context and use, I mean those in relation to which a given piece of writing was produced *as well as*, and as distinguished from, those in relation to which it was reproduced and re-appropriated.

The difficulty lies in the fact that, first, frequently documents are casually assumed to have been produced in a school environment and, second, the historians' present-day school experience is brought into play to determine the documents'

[10]In other cases, this kind of generosity is not always overriding. Mathematics in China is equated with what has been handed down. Furthermore, what has been handed down is interpreted and assessed in the same way as schoolbooks. At the end of the day, "civilizations" are "ranked" on that basis. By extension of the disparagement of documents related to educational situations, "civilizations" which have bequeathed to us only texts assessed as being school texts were judged accordingly, as if the extant material allowed a judgment to be passed and as if such judgments were appropriate.

use and finally interpret them. The various chapters of this book illustrate these widespread practices amply. I will only pick a single example from the many, chosen from the writings of some of the best historians. In her chapter, Christine Proust recalls how Neugebauer and Sachs interpreted series texts as teaching aids.[11]

More broadly, the fact that a text contained a problem or a series of problems has often been sufficient, without any further justification, for the problem to become an exercise, and the series of problems a pedagogical progression. In other words, a use has been projected onto them, without adducing historical evidence. In the latter case, that of sequences of problems, we thus see how the assumption that the writing is produced in a school context is essentially put into play in accounting for the logic of its composition in terms of didactic intention. One such case is illustrated in the way Jean-Claude Martzloff reads the chapter devoted to the rules of false double position in the classic from the first century CE, *The Nine Chapters on Mathematical Procedures.*[12]

In these examples, in my view, problems are read in terms of what problems were for us when we studied mathematics, and they are believed to reflect the environment in which we personally experienced problems: the classroom. Likewise, another type of text has also regularly been spontaneously associated with teaching, namely, commentaries. And, being freely associated to a school context, commentaries have thus been read anachronistically, as reflecting lecture notes and students' reactions. Alain Bernard illustrates the fact, in this volume, with the example of Rome's interpretation of Theon's commentaries.

More generally, documents produced in educational contexts have often been perceived in broad and somewhat dogmatic categories. They have mainly been categorized as either masters' textbooks, written in order to expound a theory, or teaching aids for masters, or exercises written down by students. The book that the reader has in his or her hand amply illustrates why these documents call for a finer historical analysis based on evidence.

To recapitulate, historians have perceived documents produced in school contexts as second-rate sources. Further, they have rarely felt the need to provide evidence when they have associated educational contexts with their documents. Finally, even when evidence has proved unquestionably that sources were produced or used for, or in school contexts, historians have seldom attempted to find historical evidence to restore how in fact these documents were *used*. They have often, instead, projected their own present-day experience of learning and their perception of pedagogy onto the sources. And yet, identifying a school context does not exempt us from the requirement of determining how documents were produced and used. I have differentiated these three arguments for the sake of the analysis. In debate, they are often mixed up, which contributes to increase the confusion reigning on these issues.

[11] See Chapter by Christine Proust, in this book, p. 88.

[12] Martzloff (1997 [rev. ed. 2006], 134).

12.2 Exposing Prejudices in the Context of a Turn in the History of Education

In the last few decades, new questions have been raised, and accordingly new values attached to the historical study of teaching contexts, and new attitudes developed towards it. This book illustrates this wider trend.

The research work carried out has exposed the limits of the historiographic attitudes sketched above. First, history of education has become institutionalized. One may think of the *Service d'histoire de l'éducation* at the *Institut National de la Recherche Pédagogique*. One may also think of the various journals and monographic series that have emerged and are devoted to these questions, for instance, *Histoire de l'Education* or *History of Education,* or else the History of Universities series. Second, history of education has addressed many new issues. This development has had an influence on history of science and the steady growth of the history of scientific education.

Historians have started to inquire into the institutional[13] and material[14] settings of education, into the curricula,[15] curricula reforms and their impact.[16] They have developed an interest in the format and impact of examinations.[17] Historians have emphasized the variety of teaching institutions and their publics.[18] They have studied the distinct processes of the institutionalization of primary, secondary, technical and higher education, and the various social structuring that went along with the establishment of these levels.

Closer to the concerns addressed in this book, historians have inquired into the various types of documents produced in educational contexts and at different levels of education, that is, not only the treatises composed in relation to lecture courses, or students' notes, but also, for instance, the sheets designed to be handed out to students, researched by Anne-Marie Compère, or the variety of visual supports used while teaching.[19] Furthermore they have offered means of distinguishing between similar kinds of texts related, for some, to school contexts, and for others, not directly so. For instance, in this volume, Philippe Clancier suggests means of discerning between commentaries related to a teaching environment and other

[13]See the pioneering collective work in (Taton 1986).

[14]See the articles collected in (Bruter 2011). See also the publications mentioned in Christine Proust's chapter, footnote 3.

[15]For the modern time period, see, for instance, (Belhoste 1995). For ancient time periods, see (Siu and Volkov 1999) as well as Volkov, in this volume. This trend had a significant impact on Assyriology, a field in which research on school curricula has brought about quite exciting results, as explained in C. Proust's chapter.

[16]See, for instance, (Gispert et al. 2007).

[17]For ancient time periods, see (Volkov 2012), and for modern times (Belhoste 2002).

[18]d'Enfert (2003) is an example for that trend of research, which is also evoked in the chapters by F. Bretelle-Establet and S. Lamassé in this volume.

[19]Tomic (2011, 65–68).

types of commentaries composed in first millennium BCE Mesopotamian societies. Assyriologists have been more generally quite resourceful in devising methods of distinguishing between documents attached to distinct aspects of teaching activities, as is explained in the chapters by Philippe Clancier and Christine Proust here.

Finally, the history of methods of both teaching and learning has been the focus of research, with due attention paid to the aims of teaching. These aims include the competences intended to be acquired, the bodies of knowledge to be studied, as well the ideals to be inculcated. In this book, Anne-Marie Chartier gives a clear example of the various goals that have been attached to the teaching of literacy, whether these included studying reading and writing, with the aim of allowing a child to carry out extensive reading, or simply reading, and not writing, in relation to the aim of inculcating intensive reading and religious teachings.

In the wake of the multiplication of historical studies devoted to educational activities, several prejudices towards texts produced in these contexts, to which I have alluded above, have been exposed.

Several case studies have shown how many scientific innovations took shape in the specific context of the preparation of lectures or the composition of textbooks. A very well-known and even paradigmatic example of such a fact is the breakthrough in mathematical analysis made by Cauchy in the context of his lectures at the *Ecole Polytechnique*, evoked above. As historians have argued, the educational context was essential for Cauchy to promote rigor as a chief value and address the problem of providing a systematic and reliable approach to analysis. What is important here is that the teaching context is not conceived as a mere environment, in which new ideas emerge by mere contingency. The resources this context provides, the constraints it establishes, the values it promotes, and the aims that it dictates go hand in hand with the content produced. This is the thesis defended by Bruno Belhoste, in his reflexive article devoted to how historians of mathematics could benefit from the study of educational contexts. Consequently, teachers' lectures, far from being the pale repetition of what had been previously found out before by pioneers, appear to have been, as such, an important site of mathematical innovation. The fact that, as Belhoste and Gilain both emphasize, due to their complexity, Cauchy's lectures were barely understood, and his texts hardly used, by students at the *Ecole Polytechnique*, and that finally his *Analyse algébrique* became a treatise widely used by his fellow mathematicians, does not affect the conclusion. It rather illustrates an interesting point, which we will come back to in a moment.

The type of work and the resources mobilized in the production of such textbooks can be examined in a wider perspective and become the topic of specific historical inquiry. This is precisely one of the aims of Caroline Ehrhardt's study devoted to Sylvestre-François Lacroix, in which she analyzes the specificities of the scholarly work engaged in by Lacroix in the composition of his numerous treatises.[20]

[20]Ehrhardt (2009). The description of Lacroix's work can be contrasted with the one given by Jean Itard, in his biography of François-Sylvstre Lacroix, published in the *Dictionary of Scientific Biography*, vol. 7 (1973), 549–551 (see http://www.encyclopedia.com/doc/1G2-2830902420.html,

However, the type of relationship that texts produced by teachers may have had with their teaching activities and practices remains a complex historical problem. In his chapter, Alain Bernard addresses the issue in the case of Theon's commentaries, paying due attention to the scholarly environment in which Theon was active. Christine Proust, for her part, examines the question from another point of view, when she establishes that, in the context of schools in which they were active, teachers may have written compositions whose aim was disconnected from their teachings even though they drew on resources elaborated for teaching, and whose virtuosity was mainly meant for their peers. In other words, teachers do not only write for their students!

Specific historical inquiry has also been devoted to the writings and activities of students in the context of the learning environment.

In the monograph she has devoted to students' drafts and other school writings of Old Babylonian Nippur, Christine Proust has shown the insights that history of mathematics could gain from concentrating on exercises carried out in the context of elementary mathematical education. In particular, she has illuminated, through these documents, mathematical practices that helped her shed interpretative light on the more advanced texts that had until now largely monopolized historiographic attention. In the same vein, in the article she has devoted to candidates' papers for the mathematics section of the 1829 entrance examination for the Paris *Ecole Préparatoire*, and in particular to Evariste Galois's paper, Caroline Ehrhardt aptly discusses the question of the historiographic import of such documents. In her view, they allow the examination of the training received by Galois and the type of practices he was taught, thereby casting light on the research work he would later carry out. The papers further, and perhaps more importantly, also allow us to perceive how much Galois shared with his contemporaries, in view of understanding the reception of Galois's work in a historically situated way.[21]

The two examples mentioned above emphasize, each in its own way, the historiographic import of different types of documents produced by students. By contrast, the innovations that could be brought about through students' schoolwork is the topic of the study Bruno Belhoste has devoted to Charles Dupin, at the time when the latter was a student of the *Ecole Polytechnique*.[22] Here again, the prejudice that what happens in an educational context, and more precisely in school exercises, is mere repetition is exposed. The context in which Dupin obtained important insight on cyclides, today named after him, is quite specific. It relates to the type of teaching Monge devised in geometry for this educational institution. Outside the lecture room, students were given geometrical problems, which had been chosen from among the classic problems of geometry explored since Antiquity or were extensions of them. These problems had been revived in the previous centuries and

(consulted on May 12, 2013)). The same project animates the various articles gathered in Bruter (2008). See also (Ehrhardt 2011).

[21] See, respectively, (Proust 2007) and (Ehrhardt 2008).

[22] See (Belhoste 2009, esp. 88–96).

reused, generalized, to prove the power of analysis. Monge resumed working on them to prove the power of new geometrical methods. Students had to explore these problems through the means taught to them, working together and confronting their solutions. On the case studied by B. Belhoste, Dupin was not the only one to obtain interesting results in the context of this activity, and Monge's teaching assistant for these matters, Hachette, selected the best among the students' highly innovative solutions, to present in 1804 in an article published in an internal journal devoted to such purposes, the *Correspondance sur l'École impériale polytechnique: à l'usage des élèves de cette école.* This article is our first published evidence of Dupin's work on cyclides as a student. In the same way, several students at the *Ecole Polytechnique* published, in the same journal, works that they had developed while they were at school. In Dupin's case, he was still a student when he introduced the geometrical surface to which he himself came back later in his career, and to which numerous mathematicians would devote attention and research throughout the nineteenth century. Students' schoolwork has had a clear impact on mathematical research.

The studies evoked above have cast light on the historiographic import of writings produced by both masters and students in educational environments. I have devoted some attention to other assumptions that appeared to me as problematic, that is, to those guiding the interpretation of mathematical problems. Are we allowed, when we read a problem in an ancient document, to assume directly that the problem was an exercise and that it was attached to a teaching activity? To address this question, I have taken the example of the Chinese mathematical classic completed in the first century CE, *The Nine Chapters on Mathematical Procedures.* Here, the interpretation of the book is determined by the answer to the question of what a problem is, since the book, like many other ancient Chinese mathematical texts, is mainly composed of problems and algorithms solving them.

As I have recalled above, we do know that the book was used as a textbook in the seventh century, in the context of the State University. This does not mean, however, that we have evidence on *how* it was used. To determine what the problems making up *The Nine Chapters* in fact were in the eyes of the practitioners who wrote these texts or used them, I have thus looked for evidence on practices with problems in their writings. I have found pieces of evidence in a third century commentary, selected in the seventh century, to be handed down with the text of *The Nine Chapters,* as well as in the sub-commentary written in the seventh century, in the context of the preparation of the anthology of mathematical classics. Note that we do not know the context of the composition of the commentary, and we know very little on the circumstances in which the sub-commentary was composed. However, the commentators use the framework of problems in their mathematical activity, as can be observed through their commentaries. They thus provide evidence to address the issue of what a problem meant for them. The most important thing here is that, on the basis of this evidence, a problem appears *not* to be an exercise.[23] The interpretation consisting in reading a modern school environment and a present-day

[23]See the argument in Chemla (1997a, 2009).

type of exercise behind each mathematical problem has thus its limits. Let me stress the fact that this conclusion tells us a priori nothing about what happened in the State University in terms of teaching. The reason for this is that we do not know the uses to which problems were put in that framework. This conclusion only provides information that could be useful in addressing this second question.

Closer to us, we have similar historical evidence that the choice of composing a book on a given topic with problems as its basic elements cannot be interpreted as the production of a mere collection of exercises for the classroom. Or, to put it differently, that the use made of problems and the way of writing them down are subject to historical variation that are meaningful and reflect the various authors' projects. To illustrate this caution, let me quote a few words of the magnificent preface of *A Hilbert Space Problem Book*, in which Paul Halmos explains *how* he composed the text of his problems and how he expected the reader to use them.[24]

> The *only way to learn* mathematics[25] is *to do* mathematics. That tenet is the foundation of (...) the method in which the teacher plays the role of an omniscient but largely uncommunicative referee between the learner and the facts. Although that method is usually and perhaps necessarily *oral*, this book tries to use the same method to give a *written exposition* of certain topics in Hilbert space theory. (...)
>
> This book was written for the *active* reader. The first part consists of problems, frequently preceded by definitions and motivation, and sometimes followed by corollaries and historical remarks. Most of the problems are statements to be proved, but some are questions (is it?, what is?), and some are challenges (construct, determine). The second part, a very short one, consists of *hints*. A hint is a word, or a paragraph, usually intended to help the reader find a solution. The hint itself is *not necessarily* a condensed *solution* of the problem; it may just point to what I regard as the *heart of the matter*. (...) The third part, the longest, consists of solutions: proofs, answers, or constructions, depending on the nature of the problem.
>
> The problems are *intended* to be challenges to thought, not legal technicalities. A reader who offers solutions in the strict sense only (this is what was asked, and here is how it goes) will miss a lot of the point, and he will miss a lot of fun. Do not just answer the question, but try to think of related questions, of generalizations (...), and of special cases (...). What makes the assertion true? What would make it false?[26]
>
> Problems in life, in mathematics, and even in this book, do *not necessarily* arise in increasing *order* of depth and difficulty. (...) Do not be discouraged if the solution of an early problem borrows from the future and uses the results of a later discussion. The logical error of circular reasoning must be avoided, of course. An insistently *linear view of the intricate architecture of mathematics* is, however, *almost as bad*: it tends to conceal the beauty of the subject and to delay or even to make impossible an *understanding* of the full truth. If you cannot solve a problem, and the hint did not help, the best thing to do at first is to go on to another problem. (...) If, on the other hand, you solved a problem, look at the hint, and then the solution, anyway. You may find modifications, generalizations, and specializations that you did not think of. (...)

---

[24]Halmos (1967 [rev. ed. 1974, 1982]), vii–x. I quote from the 1982 edition. The emphasis is mine. I can only quote a few words from the preface, but I invite the reader to read it entirely.

[25]Let me emphasize that the type of learning meant here is not merely that which occurs in the classroom. It was not intended to be so, and the multiple citations of the book in research publications in the subsequent years establish the fact (Note KC).

[26]Interestingly enough, the following paragraph was added in the 1982 edition (Note KC).

Such evidence invites caution, showing why the historian should not rush to interpret similar texts without looking for evidence of what problems actually meant in the eyes of their authors. The problems in Halmos' book are as far removed as is possible from those assembled in books to be used in the competitive context of tripos examinations.[27] Similarly, as Stéphane Lamassé shows in his chapter, Chuquet considered problems as a means to expound knowledge.[28]

Halmos also, most interestingly, describes practices of using mathematical problems in a given context, and they too are quite different from those evidenced in competitive examination contexts. Observing the reactions to the book in subsequent years would highlight how various types of practitioners have used and read the book, but this would take us too far. We shall only stress a single point: the previous remarks do *not* mean that these problems have *not* been used in a school environment.

To summarize our conclusions on the matter, problems cannot a priori be limited to school exercises, even when they were used in teaching environments. One can document various types of use of mathematical problems in different contexts: their meaning vary accordingly. In this book, C. Proust convincingly shows how the same problem has a completely different status in different texts (her part 2). She also argues why the problems contained in different kinds of Mesopotamian sources are not the same type of textual objects and were not all intended to be solved. S. Lamassé further remarks how the same problem could have reflected an actual situation in some places, and have completely lost its practical value in another.[29] Mathematical problems cannot be interpreted at face value. Since they undergo transmission, circulation and recycling, they have had different types of publics and accordingly different purposes.

The issue of how problems were handled in the context of different activities appears to be a meaningful one. What is at stake is to determine, in each context, what a mathematical problem was, how it was used and read, in which environment, and for which purpose. A first essential outcome of such research is to establish when two problems were the same, or were variations of each other, *in actors' perception*, not in ours. The key point is that there can be historical variation on this issue.[30]

Establishing the conditions of identity in each context is indeed an essential prerequisite before one formulates any assumption regarding the intention governing the way sets of problems were composed. To assume that behind a sequence of

[27] I rely here on the talk given on March 12, 2007, by June Barrow-Green with the title "The problem-experience in Cambridge: the Mathematical Tripos examination in the 19th century." This talk was given in the context of workshops REHSEIS had organized on the issue of the variety of the types of problems in mathematical activity.

[28] See the chapter by Lamassé, p. 147.

[29] See Chapter by S. Lamassé, where it is explained how for Chuquet the mercantile problems he used had lost their practical value, pp. 143–144.

[30] This is a conclusion that I have been led to in Chemla (1997a, 2009).

problems there is a didactic progression, one has to establish that *in the actors' view* the problems belonged to the same sequence. I have shown elsewhere that the problems through which rules of false double position are expounded in *The Nine Chapters* belong to two categories irreducible to each other, and that this feature is correlated with the structure and properties of the single algorithm solving them both.[31] This conclusion seems to me to cause difficulty to an interpretation that reads a didactic progression in the sequence of problems, as Jean-Claude Martzloff has suggested.

Naturally, this is not to say that there are no sets of problems arranged with the intention of displaying a didactic progression. However, if didactic progression is to be read in documents, the fact has to be established, as several authors in this book do. In the paragraph he added to his preface in 1982, which is quoted above, Halmos particularly stresses the fact that there is no linear progression in the composition of his problems.

Prejudice has proved persistent on another issue often attached to school practices. It is the issue of memorization, regularly envisaged from the perspective of repetition, dogmatism and learning by rote. In the last few years, various interesting attempts have been made to address the topic in a historical way. Let me evoke some of them insofar as they relate to the topic of this book. In research work devoted to sutras of Sanskrit mathematical texts, that is, precisely the type of text whose purpose was to be handled through memorization, Agathe Keller has raised the question of how knowledge was compressed in these writings and how the commentators, who are the earliest readers of the sutras we have evidence for, made sense of these texts. What she has been able to show is that ancient readers seem to have expected sutras to provide the gist of understanding. In other words, far from being contrary to understanding in this case, simply in order to work, memorization relies on understanding in an essential way.[32] This conclusion echoes the description of intensive reading, practiced by early modern readers, as described by Anne-Marie Chartier.

In this volume, Andrea Bréard shows, through the study of the use of versification in several Chinese mathematical texts, that memorization was used in different ways, with different types of verses, and for different purposes. Memorization was accordingly expected to play different roles in the acquisition of knowledge and the use of knowledge acquired in this way. Similarly, in her study of the evidence left by the physician Huang Yan, who wrote for his students in the context of private instruction, Florence Bretelle-Establet focuses on the issue of memorization. In his book, Huang Yan puts, in various types of verses, only passages that had to be memorized.[33] This piece of evidence backs up Andrea Bréard's argument, that versification and learning by rote have been perceived as being connected to each other. In some sense, learning texts by rote evokes the type of learning practiced in

[31] Chemla (1997b).

[32] Keller (forthcoming).

[33] See Bretelle-Establet's chapter in this book, pp. 292–293.

official institutions teaching medicine at the center of the empire. However, Florence Bretelle-Establet stresses the actors' perception that there were different types of learning by rote, the difference in practice being a basis on which social distinction is established.[34] This remark prompts us to pay attention, beyond the fact that learning involves memorizing, to *how* one memorizes, *what* one memorizes and for *which* type of later *use*. Huang Yan's explicit explanation that he has versified some parts of his text so that they could be more easily memorized draws attention to another important fact: the book written for students, like those described by Andrea Bréard, does not have a uniform composition, but it has different types of passages, referring to different types of activities and these various passages are expected to be handled in different ways. We will come back to this remark below.

In this volume too, Anne-Marie Chartier does not only approach memorization in a historical way, but she also sheds light on the historiography of memorization, through suggesting a hypothesis on the historical contexts in which some lingering ideas about memorization took shape. She argues how in the shift from intensive to extensive reading, the types of competences that readers had to achieve changed radically. The shift in the competences acquired brought about a shift in the methods used, the old method being blindly judged for its performance in relation to the new reading required. More generally, such historiographic research on the origin of persistent ideas about educational activities is likely to shed important light on the issues addressed in the book.

12.3 Promising Issues

The previous remarks on the types of text produced and used in teaching contexts indicate how research carried out in the last decades has undermined prejudice on these issues. It is all the more crucial to elucidate in a historical way what happened in educational contexts, that the activities taking place in these environments have produced many documents, and that these activities have also left their imprint on many of the texts that historians of science can use in their research. It is thus essential that clarification be achieved on such issues as which of these sources were once attached to education and how they were actually used, which type of marks a teaching context has left on these documents, and which types of documents have been produced in such contexts —the ephemera as well as the writings meant for a longer life span. As Christine Proust and Philippe Clancier both emphasize, clarification here is essential for interpretation. This holds more generally true especially for Antiquity, for which documentation is the sparsest!

Let me now, to conclude, emphasize some of the promising directions of research that this book indicates with respect to its main topic: the historian of science's sources.

[34] See Bretelle-Establet's chapter in this book, p. 283.

12.3.1 From Text to Context

Several chapters in the book raise the same issue: how can we rely on a document to approach features of the context in which it was produced or used, in particular regarding educational contexts? All the authors concur in admitting that this is an issue which is as promising as it is tricky. A.-M. Chartier provides the best of all warnings as to the difficulty of the matter. In the practices she describes, the texts used to teach literacy are not of the type the present-day historian would expect. The ways in which actors are shown to relate to these texts are quite specific. In other words, she shows respects in which a text cannot directly reflect a teaching situation, or even the way in which it was to be used in such an environment. The text, and even the object, does not count for much without any evidence about its use.

And yet, some clues allow the historian to approach features of the context. For instance, Lamassé offers interesting insight into how manuscripts as objects can reflect their use in a teaching context. Some clues, among which the size of the object, the notes added to it, and the layout, allow, for example, a countertop book to be distinguished from a school book, and then allow enquiry into the type of "class situation" or teaching in which they would have been used, whereas other manuscripts appear to have mainly been objects of prestige.

Florence Bretelle-Establet addresses the issue of how the book reflects the teaching situation in another way, when she evokes the various additions students made to Huang Yan's book, while learning. As Huang Yan himself explains, the text is an object through which the master and disciples interact, in written forms that leave their mark in the successive editions of the book. We thus have here the example of a joint production of a text, which again reflects how teaching and learning are practiced. This example introduces the next issue worth pondering.

12.3.2 Features of Textual Production in Educational Contexts

The case just examined is not the only one in which the production of a treatise is multi-handed. The nineteenth century has seen the publication of numerous treatises whose composition owed as much to the students as to their professor.[35] In addition, it suffices to think of anthologies and numerical tables to highlight the fact that teachers have not merely produced expository treatises and that educational contexts recycle other types of written production. More generally, the type of written material produced for and used in the context of education deserves further study. As Florence Bretelle-Establet indicates, compilations—another all too quickly depreciated form of writing—have occupied pride of place among these texts, and they still await historical research allowing us to appreciate the type of work required to produce them.

[35] Some examples are outlined in Ehrhardt (2011).

Compilations are probably only one form of a much more general phenomenon. In the description Stéphane Lamassé offers of various mathematical manuscripts used in medieval France, he highlights an interesting fact in this respect. He shows how the same clusters of problems sometimes appear in various manuscripts, in which they are placed in different textual environments and bespeak different intentions from the authors. This remark highlights how this type of writing recycles pieces taken from other similar writings as elements for a new composition.

At another level, we can observe a similar phenomenon. Christine Proust shows how, in the various "school contexts" that can be distinguished in Old-Babylonian scribal schools, clusters of texts were formed, that is, coherent sets of documents used in parallel at a given time. Some of these clusters were used each at a specific level of education and allow historians to define finer types of contexts. This is a major contribution assyriologists have brought to the debate. Other clusters were, in the past, more specifically linked to a given task, which they allow historians to understand better. How educational activities have led practitioners to shape bits of writings and clusters of books appears to me as an issue that could enable us to approach the production and use of texts in these specific environments. This remark leads to two other interesting ranges of questions.

12.3.3 Non-uniformity of Documents and How They Were Handled

The examples outlined above bring to light that most of these writings are made up of sections of text and that these sections can be of different types. The book by Huang Yan, as described by Florence Bretelle-Establet, brings together entries structured into two parts: quotations and texts written by the master. To this, commentaries by disciples are added throughout the editions, illustrating the general features of the multi-handedness and fluidity[36] of most of these books. Huang Yan, we have seen, versifies some of the passages that are to be memorized. Likewise, the texts described by Andrea Bréard show, in the first instance, a non-uniform composition in relation to the different types of versification that one finds in them.

In my view, this feature invites research on how texts shaped in relation to teaching are composed in pieces, and also on how these pieces were to be handled in different ways. The reading of such documents was not continuous, even when the texts were, as is demonstrated in the case analyzed by Anne-Marie Chartier. Readers pick-up, consult, focus on one part then learn another by heart. Their ways of relating to the text, when evidence can be found, could be an excellent entry into the issue of how writings are produced and used in educational contexts.

---

[36]I borrow this latter term Irène Rosier-Catach's and her colleagues' study of the *Glosulae in Priscianum* (project Histoire des savoirs, 2003–2007, http://www.cnrs.fr/prg/PIR/programmes-termines/histsavoirs/synth2003-2007Histoiredessavoirs.pdf, 14).

12.3.4 Recycling

Writing in pieces and producing non-uniform documents; both features echo a general phenomenon. As C. Proust's chapter shows for the case of Old-Babylonian scribal schools, mathematical problems there are first evidenced in students' work, then in masters' compilations, where perhaps they contribute to shape curricula, and finally, later, in series bringing together statements of problems on the basis of logic, and with concrete intentions, different from those inspiring catalogues of exercises to be solved.

This is but one example of the fact that the history of textual recycling could be promising for the project of better understanding the production and use of texts in the context of educational activities! We have seen several other examples above. How fragments of texts and inscriptions are recycled, how their meanings and use are thereby reconfigured, what effects this operation has on the production of new meanings and new bodies of knowledge are issues awaiting reflection.

Recycling is an even more general phenomenon, one that is closely connected with the issue of interpretation. When assyriologists like Clancier ask the question of whether a text is an administrative text or a school exercise, recycling has taken place and makes the interpretation a delicate issue for the historian.

Pieces are recycled, and whole books as well. This is what happens when Cauchy's *Analyse algébrique* becomes a treatise for his peers, or when theatrical plays by Molière, novels by Balzac or the Qur'an are used in the classroom. Recycling in these cases can be materialized by the production of specific editions, at least partly enlarged by introductions, endnotes and other marks of educational use. It can also be materialized by inscriptions produced on the board, whose use A.M. Chartier describes. Texts are recycled for the class, unless writings produced for the class are recycled elsewhere.

All these remarks appear to converge towards a same conclusion. They show in fact why it is difficult, if not impossible, to provide straightforward answers to the question of the relationship of sources to a teaching context. In many cases, a document is probably neither entirely attached to such a context, nor completely detached from it. Only, perhaps, its uses can be.

References

Belhoste, Bruno. 1991. *Augustin-Louis Cauchy. A biography*. New York: Springer.

Belhoste, Bruno. 1995. *Les Sciences dans l'enseignement secondaire français. Textes officiels. Tome 1: 1789–1914*. Paris: INRP & Editions Economica.

Belhoste, Bruno. 1998. Pour une réévaluation du rôle de l'enseignement dans l'histoire des mathématiques. *Revue d'histoire des mathématiques* 4: 289–304.

Belhoste, Bruno. 2002. Anatomie d'un concours: l'organisation de l'examen d'admission à l'Ecole polytechnique de la Révolution à nos jours. In *L'examen. Evaluer, sélectionner, certifier. XVIe—XXe siècles*, ed. Bruno Belhoste. Paris: INRP.

Belhoste, Bruno. 2009. Charles Dupin et l'héritage de Monge en géométrie. In *Charles Dupin (1784–1873). Ingénieur, savant, économiste, pédagogue et parlementaire du Premier au Second Empire*, ed. Carole Christen and François Vatin, 81–97. Rennes: Presses Universitaires de Rennes.

Bretelle-Establet, Florence (ed.). 2010. *Looking at it from Asia: The processes that shaped the sources of history of science*. Dordrecht: Springer.

Bruter, Annie (ed.). 2008. *Le cours magistral. XVe–XXe siècles. I. Publics et savoirs, vol. 120, Histoire de l'éducation*. Paris: Institut National de Recherche Pédagogique.

Bruter, Annie (ed.). 2011. *Le cours magistral. XVe–XXe siècles. II. Le cadre institutionnel et matériel, vol. 130, Histoire de l'éducation*. Paris: Institut National de Recherche Pédagogique.

Chemla, Karine. 1997a. Qu'est-ce qu'un problème dans la tradition mathématique de la Chine ancienne? *Extrême-Orient-Extrême-Occident* 19: 91–126.

Chemla, Karine. 1997b. Reflections on the world-wide history of the rule of false double position, or: How a loop was closed. *Centaurus. International Journal of the History of Mathematics Science and Technology* 39(2): 97–120.

Chemla, Karine. 2009. On mathematical problems as historically determined artifacts. Reflections inspired by sources from ancient China. *Historia Mathematica* 36(3): 213–246.

Chemla, Karine. 2010. A Chinese Canon in Mathematics and its two Layers of Commentaries: Reading a collection of texts as shaped by actors. In *Looking at it from Asia: The processes that shaped the sources of history of science*, ed. Florence Bretelle-Establet, 169–210. Dordrecht: Springer.

Choppin, Alain. 2008. Le manuel scolaire, une fausse évidence historique. *Histoire de l'éducation* 117: 7–56.

d'Enfert, Renaud. 2003. *L'enseignement du dessin en France. Figure humaine et dessin géométrique (1750–1850)*. Histoire de l'éducation. Paris: Belin.

Dhombres, Jean. 1997. Foreword. In *A history of Chinese mathematics*, Jean Claude Martzloff, IX–XII. Heidelberg: Springer (Rev. Ed. 2006).

Ehrhardt, Caroline. 2008. Evariste Galois, un candidat à l'école préparatoire en 1829. *Revue d'histoire des mathématiques* 14: 289–328.

Ehrhardt, Caroline. 2009. L'identité sociale d'un mathématicien et enseignant. Sylvestre-François Lacroix (1765–1843). *Histoire de l'éducation* 123: 5–43.

Ehrhardt, Caroline. 2011. Du cours magistral à l'entreprise éditoriale. La "Collection Borel", publiée par Gauthier-Villars au début du XXe siècle. In *Le cours magistral. XVe—XXe siècles. II. Le cadre institutionnel et matériel*, ed. Annie Bruter, 111–139. Paris: Institut National de Recherche Pédagogique.

Gilain, Christian. 1989. Cauchy et le cours d'analyse à l'Ecole Polytechnique. *Bulletin de la SABIX (Société des Amis de la Bibliothèque de l'Ecole Polytechnique* 5: 3–145.

Gispert, Hélène, Nicole Hulin, and Marie-Claire Robic (eds.). 2007. *Science et enseignement. L'exemple de la grande réforme des programmes du lycée au début du XXe siècle*. Paris: Vuibert & INRP.

Halmos, Paul R. 1967. *A Hilbert space problem book*. Graduate texts in mathematics, vol. 19. New York: American Book Company & Springer International Student Edition (Rev. ed. 1974, 1982).

Issitt, John. 2004. Reflections on the study of textbooks. *History of Education* 33(6): 683–696.

Keller, Agathe. Forthcoming. Ordering operations in square root extractions. Analyzing some early medieval Sanskrit mathematical texts with the help of Speech Act Theory. In *Texts, textual acts and the history of science*, ed. Karine Chemla and Jacques Virbel.

Martzloff, Jean Claude. 1997. *A history of Chinese mathematics. With forewords by Jacques Gernet and Jean Dhombres*. Trans. Stephen S. Wilson. Heidelberg: Springer (Rev. ed. 2002).

Proust, Christine. 2007. *Tablettes mathématiques de Nippur: reconstitution du cursus scolaire. Edition des tablettes conservées à Istanbul. Translittération des textes lexicaux et littéraires par Antoine Cavigneaux. Préface de Christian Houzel*, Varia Anatolica. Istanbul: IFEA, De Boccard.

Siu, Man-Keung, and Alexei Volkov. 1999. Official curriculum in traditional Chinese mathematics: How did candidates pass the examinations? *Historia Scientiarum* 9: 85–99.

Taton, René (ed.). 1986. *Enseignement et diffusion des sciences en France au XVIIIe siècle*. Paris: Hermann éditeur des sciences et des arts.

Tomic, Sacha. 2011. Le cadre matériel des cours de chimie en France au milieu du XVIIIe siècle. In *Le cours magistral. XVe—XXe siècles. II. Le cadre institutionnel et matériel*, ed. Annie Bruter, 57–83. Paris: Institut National de Recherche Pédagogique.

Volkov, Alexei. 2012. Argumentation for state examinations: Demonstration in traditional Chinese and Vietnamese mathematics. In *The history of mathematical proof in ancient traditions*, ed. Karine Chemla, 509–551. Cambridge: Cambridge University Press.

Author Index

A. Bernard and C. Proust (eds.), *Scientific Sources and Teaching Contexts Throughout History: Problems and Perspectives*, Boston Studies in the Philosophy and History of Science 301, DOI 10.1007/978-94-007-5122-4,

Subject Index

A. Bernard and C. Proust (eds.), *Scientific Sources and Teaching Contexts Throughout History: Problems and Perspectives*, Boston Studies in the Philosophy and History of Science 301, DOI 10.1007/978-94-007-5122-4,

Q

R

S

T

MIX
Papier aus verantwortungsvollen Quellen
Paper from responsible sources
FSC® C105338
www.fsc.org

If you have any concerns about our products,
you can contact us on
ProductSafety@springernature.com

In case Publisher is established outside the EU,
the EU authorized representative is:
Springer Nature Customer Service Center GmbH
Europaplatz 3, 69115 Heidelberg, Germany

Printed by Libri Plureos GmbH
in Hamburg, Germany